ACTION, MIND, AND BRAIN

ACTION, MIND, AND BRAIN

An Introduction

David A. Rosenbaum

The MIT Press
Cambridge, Massachusetts
London, England

The MIT Press would like to thank the anonymous peer reviewers who provided comments on drafts of this book. The generous work of academic experts is essential for establishing the authority and quality of our publications. We acknowledge with gratitude the contributions of these otherwise uncredited readers.

This book was set in Stone Serif and Stone Sans by Westchester Publishing Services. Printed and bound in the United States of America.

Library of Congress Cataloging-in-Publication Data

Names: Rosenbaum, David A., author.
Title: Action, mind, and brain : an introduction / David A. Rosenbaum.
Description: Cambridge, Massachusetts : The MIT Press, 2022. | Includes bibliographical references and index.
Identifiers: LCCN 2021018188 | ISBN 9780262543392 (paperback)
Subjects: LCSH: Movement, Psychology of. | Motor learning. | Expectation (Psychology) | Brain.
Classification: LCC BF295 .R667 2022 | DDC 152.3/34--dc23

10 9 8 7 6 5 4 3 2 1

Contents

Preface

When I was a kid in college, I read a book that made such an impression on me I decided to go into the field the book described. The book, whose citation is given in the References section, was *Eye and Brain.* It was by a British vision researcher named Richard Gregory. The book was full of fascinating conclusions and conjectures, and the book's thesis was a revelation for me as a college kid. According to Gregory (1966), what we see is largely constructed. We don't passively receive inputs through our retinas. Instead, we *construct* reality.

I was enthralled by this heady claim and by the demonstrations supporting it. What also intrigued me was the possibility of learning how the mind works—the deepest question of all, one might say—by doing simple behavioral experiments. I loved the juxtaposition of simple experiments with profound inferences. I was also attracted to the notion that the field of experimental psychology lets one build apparatus, interact with participants, analyze and graph data, present results in public forums, and write about the results.

Why do I tell you about Gregory's book and the thoughts it triggered? Gregory's book was, in many ways, a model for this one. Gregory's book was for students and was filled with pictures. I came to see that an analogous text would be useful for the field I ended up in: *Action, Mind, and Brain.*

This is the first book on the interplay of action, mind, and brain written mainly for students, especially students in psychology and kindred fields: biology, cognitive science, computer science, ergonomics, kinesiology, human factors, industrial engineering, neuroscience, linguistics, and philosophy. I hope this book will be as inspiring for you as Gregory's book was for me.

If you are scientifically inclined, you might be skeptical that this is the first book on this topic pitched for such a wide range of students. Aren't there other books like this? There are some, but they're either for a different audience than the one I have in mind or are long technical treatises. I've written a couple of those myself, as you can see in the References section. No volume covers the interplay of action, mind, and brain as this one does, with an eye toward capturing the interests and concerns of students, especially undergraduates and beginning-level graduates. In addition, no other book that I know of covers the present range of topics in as fun and personal a way as that one does (or tries to do). In my many years of teaching—I started teaching in 1981—I have found that fun and personal presentations count as much as, if not more than, punctiliously accurate, up-to-date summaries. Until the fun-and-personal threshold has been crossed, few students care about the ins and outs of hems and haws by scholars in the field at the time.[1]

Is a fun-filled text on action, mind, and brain actually possible? Isn't this topic horribly complicated, understandable only by the nerdiest of nerds and the geekiest of geeks? The answer is no. Students have told me through their anonymous ratings (which the students know I only get to see *after* grading) that they were surprised by how entertaining and fascinating this subject turned out to be. They also exclaimed how astonished they were by how much of modern culture and technology stems from work related to this field.

Here's an example: movies were not initially made for entertainment purposes. The motion picture camera was invented to answer a question about the running of horses: when horses run, is there a time when all the hooves leave the ground? People hotly debated this topic for years. Only after the motion picture camera was invented was the question finally answered. Not for several years did people come to see that motion pictures could be used for entertainment. One of those people was Thomas Alva Edison, inventor of the lightbulb, phonograph machine, and other devices.

Students who have taken my class on Action, Mind, and Brain have not only said that they found the material fun and fascinating but also appreciated how many deep and important problems arise in this area. The students told me they found it sobering to learn about people who suffer from a host of neuromuscular disorders like autism, stuttering, polio, and Parkinson's disease. And they said they came to see how critical this field is for the design of safer, more efficient technology. Action research, they understood, can save lives.

How is it possible that a book like this hasn't appeared before? To approach this question, it helps to observe that if you take a tour of a library or bookstore (brick or electronic), you'll find lots of books on other related topics in scientific psychology: memory, learning, attention, and visual perception (the topic of Gregory's *Eye and Brain*). You'll see many books on these topics because they're widely taught in colleges and universities. The reason for that comes from a long tradition of research on these subjects. Ever since the ancients wondered whether we can know the world as it truly is, scholars have tried to understand how we perceive and learn. They reasoned that by understanding perception and learning, we may better understand understanding itself. Meanwhile, the sibling of perception and learning—*action*—got less attention. This is the main reason why only now, at this time in history, is the world getting its first general introduction to Action, Mind, and Brain, written at an often elementary level with the aim of inviting in as many people as possible.

But why has action per se gotten such short shrift? One reason is that perception seems easier to study than action. To study perception, you can present people with stimuli and ask them what they see, hear, taste, smell, or feel. You can vary the nature of the stimuli and track changes in subjects' reports. And you can trace how their reports change as you wait various amounts of time to check their memories. While conducting such experiments, you can enjoy feeling control as a scientist.

Studying action seems harder. You can watch people or animals move around and marvel at what they do, or you can shake your head in disbelief or disappointment as they do less well than you think they could or should. If you want to, you can record what they do—for example, by taking movies. But wonderment and moviewatching only go so far. If you're a scientist, you need to get measurements to test scientific hypotheses in controlled situations. Experimental control seems harder in the study of action than in the study of perception. If you study action, your participants could run amok.

There's a solution to this problem. You can put limits on what your subjects do, and you can restrict what you look at in the behaviors they perform. You can also narrow the behaviors you record, even in the everyday environment, where lots of wonderful behaviors can be observed. By focusing on particular features of performance, you can learn what affects them.

There's a deeper reason why action research has gotten less play than perception, learning, and memory. A major tenet of psychology has been functionalism—the idea that the proper topics of scientific study in psychology are functional relations between stimuli and responses. Want to make progress in psychology, some ask? Present stimuli and record responses so you can create a big table of input–output pairings. With a big enough table, you should be good to go.

That approach, which has been pursued in a more sophisticated way than just described, has been pursued in thousands of studies. Researchers taking this approach have viewed behavior as a set of reactions to stimuli arriving at sensory receptors. Meanwhile, a small but growing number of investigators have objected to this approach, saying that behaviors aren't just *reactions*; instead, they are *actions* leading to perceptions. For those researchers (of whom I am one), it makes more sense to endorse a continual perception–action *cycle* than a one-way street from stimuli to responses.

According to the perception–action cycle way of thinking, there's no clear start and end to behavioral events, nor is there a real division between acting and perceiving. Action encompasses perception and perception encompasses action. Even the most basic reflex is best understood as embodying perception and action in a cyclic manner.

Because action encompasses movement as well as perception and learning, I will use the term "action" here in a sweeping way, applying it to movements of the body as well as long-term goal-directed behavior. Some authors have drawn a distinction between actions and movements, reserving the word "movement" for what the body does and "action" for what individuals (actors) intend to do. I won't subscribe to that distinction here because I find it hard to sustain the distinction in practice. Even when lowly movements are studied in detail, they turn out to embody a vast amount of skill and planning. And actions of a higher-order kind turn out to be more body based (embodied) than was previously appreciated. Consequently, I will use the word *action* to refer to physically realized behavior, be it the downward sweep of a conductor's baton or the concerted effort by a large political bloc to change social policy.

Actions reflect plans, and because you are now embarking on the act of reading this book, you may want to see the plan for what's to come. Chapter 1 will be about the intellectual background to this field. Chapter 2 will be about applications and methods. Chapter 3 will be about the physical substrates of action: bones, tendons, ligaments, muscles, and nerves. Chapter 4 will address a question that may never have occurred to you, at least if you're a neurotypical individual: "How do you control your actions in space and time?" How, in other words, do you manage to move your hand to the left and then to the right rather than the reverse? This just happens without much thought, you might say, and I would agree. But how exactly does this work? How is sequencing and positioning of action achieved by the brain? How can you sing songs like "Happy Birthday" or map out a route for going shopping? These actions occur over time, yet they are represented in the brain all at once.

Chapter 5 will be about learning. The overarching question will be, How do you acquire skills? What is the role of inheritance (Nature), and what is the role of experience (Nurture)? How can skills be taught effectively? If skills are compromised by stroke, disease, or accident, how can they be relearned more effectively? Exciting advances have been made here.

Chapter 6 will be about feedback. When you take an exam, you get feedback that can help you gauge your study habits. Similarly, when you throw a baseball, you get feedback that can help you modify your subsequent change-ups and fastballs. In both cases, the feedback comes after the fact, yet you can benefit from it afterward. For example, if you walk down a path, the sounds of your footsteps provide you with information about your gait. It turns out that people feel heavier or lighter if the sounds they're fed from their footsteps are transformed through software. Via clever experimental tricks, a young woman has shown that high pitches associated with footfalls cause people to feel lighter, whereas low pitches cause people to feel heavier. She has also found that where you think your own hand is can be influenced by where you hear your hand to be. In work by others, it has been shown that where your hand seems to be can be dramatically affected if you feel your hand being stroked while the hand you see being stroked isn't actually your hand but instead is a rubber replica. The fact that you can be fooled into thinking your hand is several inches away from where it actually is shows that you are more strongly tuned to feedback than you might have imagined.

Chapter 7 will be about feedforward rather than feedback. Feedforward is anticipated feedback. It's what you expect to get in return for the actions you're about to perform. Before you speak, you expect to hear your voice. Before you write, you expect to see your writing. The speaking or writing you expect may not be as fully formed as the speaking or writing you finally produce. Therefore, it's reasonable to suppose that the actions you produce are designed to fill those incomplete feedforward representations. These representations are particularly important in this book because they are mental representations—the stuff, so to speak, of mind.

Chapter 8 will be about the degrees-of-freedom problem. The issue here centers on the fact that there are multiple ways of getting things done. Your body has more degrees of freedom (ways of varying) than task descriptions you're typically given. If you're told to touch your nose, you can do so without a moment's thought. What you do will be sensible and efficient. Yet there are endless ways of touching your nose. If you're in a clownish mood, you can generate some very silly nose-touching patterns. Ironically, though, and in a way that rains on this parade, such patterns may be exactly like ones that patients with cerebellar damage can't help but produce. Their movement patterns are decidedly unfunny. We want to help these patients get better.

The fact that neurotypical individuals can decide to move in funny ways means that they (you probably) have more means available for moving than you might realize. Otherwise, you would be unable to compensate for changes in the way you normally move, as when a pebble enters your shoe and you switch from an amble to a limp. What are the factors that help you narrow the alternatives? Chapter 8 will focus on three. The first is biomechanical. Here the idea is that the physical makeup of your body, interacting with the external environment, helps reduce the options. The second method is coupling. Here the idea is that your limbs interact in ways that limit what you can do. For example, if you try to draw a circle with one hand while drawing a square with the other, you'll produce a squared-off circle and a rounded-off square. Only if you are one of a tiny number of people who have had their two cerebral hemispheres surgically disconnected—so-called split-brain patients—will you be able to avoid this sort of bimanual interference. The third approach uses soft constraints. These are optional rather than obligatory limits. An example of a soft constraint is your capacity either to amble down the street, feeling happy and bubbly, or to traipse along, head bowed, feet barely lifting off the ground, because you feel so terribly depressed. "I'm sorry you're having a bad day," a friend says to you as you pass them this way. If you bop along, bubbly and ebullient, your friend may delight in your joyfulness. The point is that your behavior reflects your mood. From the point of view of the control of physical action, this means that the way you act isn't just hardwired or governed completely by physical mechanics—what may be called hard constraints. How you behave also depends on soft constraints, psychological factors *par excellence*.

Chapter 9, the last chapter, is called "Onward." As its name implies, the aim of this chapter is to draw your attention to new issues, hopefully launching you in one or more exciting directions for your own life going forward.

Some other matters remain for this Preface. First, I want to explain the tone of the writing. The tone will sometimes be informal and chatty, and I will include free use of side material such as jokes and occasional definitions of terms that might be unfamiliar. The longer I have taught, the more I have seen that students savor humor and a personal touch as much, if not more, than deadly-serious recounting of facts. I have also come to see that it's important to acknowledge that a course on a topic like this be core to students' general education. Vocabulary tips and other general information can be provided. Explaining the meanings of words and phrases can be especially important for students whose first language isn't English or whose English, regardless of whether it's their first language or not, isn't stylistically identical to the style of academics and authors of books like this one. The name-of-the-game for educators is to *invite* students in, to make them feel comfortable and respected, and to help them use material that will help them develop general thinking skills as well as specific facts. With this perspective, I spend a fair amount of time on hypotheses needing careful scrutiny, inviting students to think through the evidence with me. Sometimes I go off on tangents, however, such as one that appears in Chapter 1 on the development of cinema and one in Chapter 2 on Benjamin Franklin. Tangents like these are interesting, I think, and countless students have told me they find them so. If you are a super-serious, never-crack-a-smile person who knows every word of English and only teaches students who already know everything there is to know in this book, then this book isn't for you. In addition, if you feel that what's most important for a textbook is the very latest research on every topic covered, then again I'm afraid this book isn't for you either. In that connection, I have become concerned that students seldom learn about older work, as if research began in the twenty-first century or, roughly speaking, when the internet became central to our lives. It's important, I think, to know where current research comes from and to understand how ongoing work fits into long-term cultural and intellectual streams. Just as you are part of a family tree, the work you will read about here emerged from a deeply rooted genealogy of ideas. Knowing about the past as well as the present will help you put the work into perspective and help you see how your own interests and concerns fit into a larger cultural river.

Another matter is that this book is replete with figures, including lots of photographs of people, some of whom have contributed to the field and others of whom did not but are included for other reasons. Seeing the people behind the work, especially when students can identify with them, is very important. Other pictures are designed to intrigue and attract attention, hopefully arousing curiosity about the surrounding text. In my teaching, students have told me repeatedly that they appreciate the human touch, especially in connection with a topic that can seem quite technical and therefore scary for some. People like looking at other people. Seeing them here may help them see more about the deeper reasons for doing so.

A third comment is a disclaimer about topics that won't be covered here in as much detail as might be expected. Attention, task switching, executive processes, and stimulus–response compatibility won't be discussed very much, and neither will agency or the vicissitudes of embodiment. These are important areas to be sure, but they are vast and I prefer to leave these topics to others to delve into them (see, e.g., Clark, 1997; Hommel, Brown, & Nattkemper, 2016; Prinz, 2012; Proctor & Vu, 2006; Radman, 2013). Several domains will be mentioned in passing: sports, biomechanics, robotics, physical therapy, human factors, ergonomics. One can only know so much. I know too little about these topics to cover them as much as might be optimal.

Fourth, none of the chapters in this book could have materialized without help and support from colleagues, friends, and relatives. I am indebted to Phil Laughlin of MIT Press who signed me to write this work. After I sent Phil the proposal, he checked with experts who suggested improvements to the plan. I appreciated their comments and, for the most part, took them to heart. One of the reviewers identified himself: Robert W. Proctor. I thank Robert as well as the anonymous referees who helped make the work better. Amy Brand, the president of MIT Press, kindly sent me a note when I signed with MIT Press, wishing me well with this project. I hope her brand will be well served by this text. To the extent it will be, that was due in no small measure to the person who led the team responsible for production of this book: Alex Hoopes, who helped me jump through all the hoops and hurdles needed to bring this work to completion. Deborah Grahame-Smith, Gillian Dickens, and Roger Wood helped with copy-editing. I appreciate their contributions as well as all those whose names I don't know who played some role in the final production of this work.

Ann M. Tenglund, director of Library and Faculty Development at St. Bonaventure University in Allegany, New York, helped me early one Sunday morning when I tried, online, to get a monograph by an early twentieth-century researcher whose work on perceptual adaptation was unknown to me until I happened to come across her name in a book I was reading. A passing remark in the book tossed off the fact that similar experiments to classic work of a famous man in the field had been done earlier by someone named Margaret Wooster. Thanks to Ann Tenglund, I learned that Margaret Wooster actually did the groundbreaking work for which the man took credit. In recognition of Wooster's contribution, I have shown Wooster's picture here and spent some time on her work. I don't know whether the man was aware of her work, so I don't mean to cast aspersions on him.

Another woman whose research is highlighted here, although her contributions were unknown to me before, is Lucia Galvani. You may already know that muscles can be activated through electrical stimulation, which is why you stay away from downed electric lines during thunderstorms and don't stick your fingers into electrical sockets. The electrical nature of muscles was said to have been discovered by Luigi Galvani, but historians of science now suspect that his wife, Lucia Galvani, may have been the brains of the operation. As a result, when I talk about electricity and muscles, the person whose picture is shown is Lucia, not Luigi. In general, the people who have advanced the fields of action, mind, and brain shown here reflect a wide range of demographics. I hope this will galvanize (a word derived from the name Galvani) the belief that everyone, no matter what their personal characteristics, can see themselves as welcome participants in science, technology, engineering, and mathematics (STEM).

A fair amount of my own research is presented here. I have been pleased to add to what's known about action, mind, and brain and have tried to present my work in proportion to its relevance to the themes unfolding here. To colleagues whose work I may have missed, I beg forgiveness.

My research has been supported by grants from the National Science Foundation, the National Institutes of Health, the Dutch National Science Foundation, the German

National Science Foundation, and the John Simon Guggenheim Foundation. The support of these agencies is gratefully acknowledged. So too is support I received as a graduate student in the form of an NSF Graduate Fellowship, which gave me the time and confidence to go off in a new direction—the one that is the focus of this book. I pursued the work with the blessing of my thesis adviser, Gordon Bower, at Stanford University. Gordon died when this book was being written. I will miss his big booming voice. He might have expressed satisfaction with the upshot of what his mentoring wrought in my case.

Any mistakes that have slipped into this work are my fault and no one else's. If you spot any bloops and would like to tell me about them, please do. My email address is david.rosenbaum@ucr.edu. I'll welcome your critical input.

The final matter for this Preface is the dedication. I dedicate this book to three people who, at the time of this writing, are too young to read it: my three grandchildren—Benny Karpman, Greta Wohl, and Eli Wohl, aged ten, five, and two at this time. If other grandkids come along, this book will be for them as well. I feel blessed to have a wonderful family—especially my wife and best friend, Judith Kroll—and to be able, after forty-three years since getting my PhD, to write this book along with all else I'm doing. How lucky I am to plunge into this work every day, finding the material endlessly fascinating. I hope you will find it so as well.

Culver City, California
December 25, 2020

1 Intellectual Background

How do you do what you do? Whether the acts are mundane like combing your hair or large-scale like completing your college education, the small acts and large acts are interconnected. The small acts serve the larger ones, and the large acts shape the smaller ones. Acts at both levels are interwoven. Their goals and behaviors support each other. Understanding how these goals and behaviors are created, how they are pursued, and how they are evaluated can help you understand how behavior is controlled. Ultimately, it may give you a better understanding of yourself and others.

How can you get started on this topic? A good way is to consider the intellectual background for the material at hand. By "intellectual background," I mean the ideas forming the foundation of this field. The word "intellectual" may be off-putting; it may make you feel ill at ease. You may not consider yourself "an intellectual"—someone who walks along, peering at the ground, lost in thought, stroking your chin, speaking of lofty ideas that only a few people can understand. Instead of thinking of yourself as an intellectual, you might consider yourself as someone with practical concerns. You might be preparing for a career in a profession like physical therapy, say, or occupational therapy, or clinical psychology. Those pursuits aren't typically associated with deep intellectual thought. Nevertheless, they *are* intellectual in two important senses. First, they require a great deal of intelligence and training, and second, they're rooted in years of thought by others. By learning about that thought, you will be able to better understand this field and the field you enter.

Plato

Begin with an ancient Greek philosopher named Plato (Figure 1.1). Plato lived about 2,500 years ago, from 428 or 427 b.c.e. to 348 or 347 b.c.e. The letters b.c.e. stand for "Before the Common Era," a secular way of saying b.c. or "Before Christ." Plato was interested in many questions: among them, How do you know? How do you go from perceiving to knowing? How can you truly know the essence of the world?

Why start this discussion with a philosopher? Plato is not typically thought of when people consider physical actions like making jump shots, playing the trombone, or scrubbing floors. Plato was a philosopher concerned with *thinking*. When people make jump shots, play the trombone, or scrub floors, they're not typically viewed as engaging in deep thinking.

Are acts like these really so thoughtless? If you suppose that you do what you know, then even mundane acts reflect a huge amount of hidden knowledge. By this way of thinking, studying action becomes a way of studying knowledge itself. That being the case, it makes sense to consider what people have thought about knowing.

To learn more about Plato's claims, consider a play that Plato wrote. I carry out the play in the course on which this book is based. I ask student volunteers to help me

Figure 1.1
Plato.

re-create the drama, which stars Plato's teacher, Socrates, and two others: Meno (a colleague of Socrates) and a slave whose name never appears.

In the play, Socrates invites Meno to listen to a conversation that he (Socrates) plans to have with the slave. Socrates wants to show Meno that, even though the slave has had no formal schooling, the slave will manage to tap into deep knowledge using reason alone. Socrates plans to show that if the slave can tap into that knowledge without having been taught the material explicitly, the knowledge must be inborn. That view is called *nativism*.

When the play gets going, Socrates asks the slave a series of questions on a topic he expects the slave to know about, albeit intuitively. The topic is mathematical. It concerns the relation between the area of a rectangle and the rectangle's height and width.

You know through your schooling that the area of a rectangle is the product of the rectangle's height and width. You were taught that fact explicitly: "Area equals height times width." However, Socrates (or Plato, the author of the play featuring Socrates) believed that one needn't be taught the formula. Instead, it or the truth it conveys can be arrived at through pure reason. To the extent that's so, Plato argued, the knowledge must be inborn.

In the play, Socrates questions the slave and draws him out, showing through questioning that the slave understands the relation between width, height, and area. Because no teacher ever stood at a blackboard and wrote out the equation for the slave or inscribed the formula in sand, the slave must have known it from birth, or so Socrates (Plato) argued. The slave must have been born knowing it.

Extracting knowledge through questioning is called the *Socratic method.* It is so named because Socrates relied on questioning to evoke knowledge that he thought would exist in latent (not yet expressed) form.

When a professor lectures, recounting facts for students to learn, he or she is not using the Socratic method. Instead, he or she is simply telling students what they should learn, making the students more passive than might be preferred. By contrast, when a professor *asks* students questions, gradually prodding them toward an answer, the instructor is using the Socratic method. The professor hopes that by the end of the question-and-answer session, students will see that they knew the material all along. In addition, if they get to that point without having read the textbook or consulted another source, then, Plato would argue, they must have known it innately.

How does this argument sound to you? Are you convinced? Maybe not. You might not be sure the slave really understood the concept just because he told Socrates he did, which is more or less what happens in the play. By modern standards, it would never pass muster just to ask people if they understand something and take their word that they do. Plato wrote thousands of years ago, however, long before psychological research methods or educational psychology came along, so he can be forgiven for not knowing about modern research standards. That aside, Plato must have been idealistic to say the things he did. For him to believe that there are inborn abstract ideas that can be arrived at through pure reason, he must have subscribed to the notion that ideas exist in heritable form.

Plato was, in fact, the first, or at least the most famous, idealist. He is so strongly identified with idealism that his name is associated with interpersonal relationships that are entirely idea based. If you tell someone that you want your relationship to be Platonic, you convey to them that henceforth, you just want to be friends.

A final point is that when Plato wrote the *Meno,* he was not contemplating the passage of knowledge from one generation to the next via genes, for he knew nothing of genes, as they were not discovered until centuries later. Plato arrived at his claim through a seeming paradox, later called *Meno's paradox*. The paradox can be expressed this way (paraphrasing from a Wikipedia article on the subject):

> If you know what you're looking for, inquiry is unnecessary. If you don't know what you're looking for, inquiry is impossible. Therefore, inquiry is either unnecessary or impossible.

To escape this paradox, Plato argued, knowledge must be inborn.

René Descartes

Plato's nativism survived him. His claim about inborn knowledge was revived later, in the 1600s, by a French philosopher named René Descartes (pronounced "run Nay day Cart," where capital letters show which syllables are accented). Descartes (Figure 1.2) lived from 1596 to 1650. He agreed with Plato that reasoning alone can expose ultimate truths. In this respect, he was not just a nativist. He was also a *rationalist,* someone who subscribes to using reason to arrive at truth.

Descartes thought that the capacity for reasoning is unique to human beings and that the capacity to reason is linked to free will, which, he thought, was something that only human beings have. Animals, Descartes argued, are just "automata" (unthinking

Figure 1.2
René Descartes.

machines). Free will, Descartes speculated, is transmitted by God. Because free will comes from the Divine, free will is spirit alone. By believing in disembodied spirits as well as physical bodies, Descartes was a *dualist*, someone who believed in two distinct forms, material and mental.[1]

It is worth considering two further ideas that Descartes advanced. One concerned certainty. The other concerned mathematics. Regarding certainty, Descartes asked, "What can I be sure of?" He posed this question near the start of a full-scale investigation of reality itself. His answer (and I paraphrase it) was, "I'm thinking, so I can be sure I exist." In the book where Descartes described this insight, he wrote "*Cogito ergo sum*," which is Latin for "*I think, therefore I am.*" In French, the sentence translates to "*Je pense, donc je suis*." No matter what language is used to communicate the sentence, it is one of the most famous sentences in all of thought. It is so famous that there is a joke about it.[2]

The second idea from Descartes was that it is possible to represent objects abstractly with symbols. To locate a point in a geometric plane, for example, you can refer to its coordinates—where the point is along the horizontal and vertical axes. These numerical values are *Cartesian coordinates*, named in honor of Descartes himself. The word *Cartesian* is the adjective meaning "from Descartes." With Cartesian coordinates, it is possible to plot points in a graph, as in Figure 1.3, which shows points forming a circle.

The most important feature of Cartesian graphing is that it allows for the representation of geometric relations in mathematical terms. This approach is called *analytic geometry*. In analytic geometry, a shape such as a circle can be represented by an equation, $x^2+y^2=c$, where c is a positive constant whose magnitude determines the size of the circle. With this equation, it is possible to express the basis for recognizing or generating circles as such. Most important for the story to be developed here, circles are circles by virtue of the fact that they are described by the abstract relation expressed in the equation $x^2+y^2=c$. Descartes proclaimed that if patterns can be described with abstract equations that use symbols, then the mind uses symbols, too. By extension, manipulating and representing symbols is what the mind does.

This way of thinking led to the idea that the mind is a symbol-manipulating device. How and whether that view leaves room for physical action is a topic we'll return to.

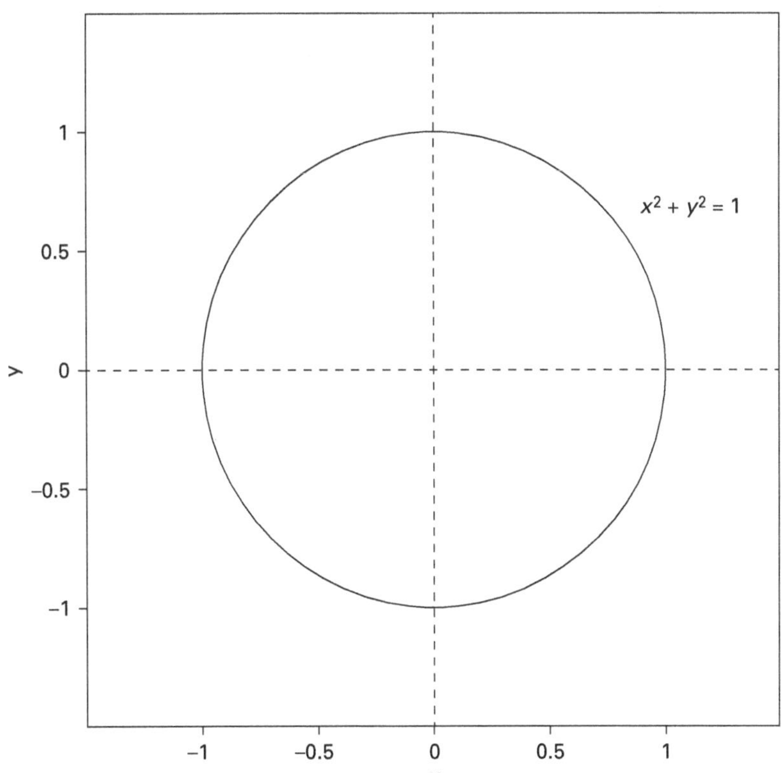

Figure 1.3
Cartesian coordinates, a circle, and the equation for that circle.

Figure 1.4
Omar Khayyam.

Suffice it to say that this perspective didn't do much to propel the study of action to the forefront of cognitive science. Descartes said, "I *think*, therefore I am." He didn't say, "I *act*, therefore I am." It took more steps to bring the study of action to the study of cognitive psychology and related fields.

Omar Khayyam

Descartes lived in the 1600s, and thanks to him, we have concepts that are core to science, technology, engineering, and mathematics today. The most prominent of these concepts is the use of graphs and their underlying mathematics to characterize real events and objects. When you look at a graph of stock market prices over time, say, or the relation between salary and body height, you're looking at a form of representation that René Descartes invented. It's a bit of a mind-bender to think that before Descartes, no one (at least in Western culture) had ever thought of representing points this way. Graphs locating points with pairs of numbers in a plane didn't exist before Descartes, at least in the West, but it turns out that in ancient Persia (now Iran), someone invented analytic geometry 500 years or so before Descartes did.

That individual (Figure 1.4) was Omar Khayyam (1048–1131). You may have heard of Khayyam in connection with his collection of poems, *The Rubaiyat of Omar Khayyam*. It's important to be aware of contributions to thought and technology from non-Western cultures. For that reason, I introduce Khayyam here, although he is seldom thought of in connection with cognitive science today.[3]

John Locke

Return now to René Descartes, noting that no matter how much one may revere him, one can also question what he said. Descartes, like Plato, was a nativist. He said that

Figure 1.5
John Locke.

people are born knowing things. That seems like a strange idea. Probably it's not the first idea you would come up with when you first contemplate the nature of knowing.

Was there some other influential thinker who had a different take on the matter, a more down-to-earth idea, so to speak, that granted a central role of experience?

There was indeed. John Locke (Figure 1.5), a British philosopher who lived from 1632 to 1704, asserted that everything we know is learned. There is no inborn knowledge, he insisted. The mind at birth, Locke said, is a blank slate:[4]

> Let us then suppose the mind to be, as we say, white paper void of all characters, without any ideas. How comes it to be furnished? Whence comes it by that vast store which the busy and boundless fancy of man has painted on it with an almost endless variety? Whence has it all the materials of reason and knowledge? To this I answer, in one word, from experience.

Locke's experience-based view of knowledge is called *empiricism*. Empiricism is the tenet (the principle) that knowledge is gained through experience. Those who collect data are called empiricists, and research done in lab and field studies exemplify the empiricist approach. If you serve as a participant in a lab or field study, you contribute data to empirical research. Virtually all research described in scientific journals is empirical. In fact, scientists who do experiments to demonstrate that nativism is correct (e.g., Spelke, 1994) take the empirical approach.

Immanuel Kant

How can it be that scientists who try to demonstrate nativism are empiricists? Isn't that a contradiction? Aren't the two views—nativism and empiricism—mutually exclusively? If one view is correct, mustn't the other view be wrong?

You may already know about a debate that hinges on this supposed contradiction: the Nature–Nurture debate. According to the Nature view, behaviors and other tendencies are inherited, bestowed by Nature (genes). According to the Nurture view, behaviors and other tendencies are learned, shaped by experience or "nurture."

Figure 1.6
Immanuel Kant.

Given the strong polarity of these two positions, you might wonder whether anyone ever blended them. Someone must have, you probably think. There was such a person, Immanuel Kant (Figure 1.6), a German philosopher who lived from 1724 to 1804. Kant (whose name rhymes with "font") wrote a book called *A Critique of Pure Reason* in which he rejected Plato's and Descartes's argument that pure reason can be used to discern ultimate truths. In the book, he also rejected Locke's assertion that our minds are blank slates. Instead, Kant argued, there are aspects of reality that we cannot reach by reason alone because we have inborn ways of perceiving. For example, if everything you see is rose-colored, then you cannot see the world as anything *but* rose-colored. No amount of pure thought will let you see the world another way. Paradoxically, then, the fact that the world looks rose-colored makes it impossible for you to know that it appears that way. Reason alone will never get you to that truth. You can gather data (engage in empiricism) until you are blue in the face, but your perception will always be rose-tinged. How you collect and appreciate data (empiricism) depends unavoidably on your innate tendencies (nativism).[5]

This integrative view, often called the Kantian view, is what most scientists accept today as they consider learning, perceiving, and knowing. These scientists grant the empiricist position that knowledge is acquired through data gathering, and they accept the nativist position that aspects of data gathering are innately influenced. They say that we come to the world with inborn *core knowledge*, with innate concepts concerning objects, actions, number, space, gravity, and causality (Spelke & Kinzler, 2007). Yet another kind of core knowledge, which has been advocated by the linguist Noam Chomsky, is the disposition to learn language (Chomsky, 1968).

Karl Marx

The Western thinkers covered above—Plato, Socrates, Descartes, and Kant—were philosophers. Before the 1900s, all branches of research were carried out by people who called themselves philosophers. There were few if any biologists as such, or physicists as such, or psychologists as such. Those fields and others came along after academic specializations developed. Today, there are scores of specialized professions, but there is a fact about most of them that shows that all or most of them were once called philosophy: the most advanced research degree you can get is called the "doctor of philosophy" (the PhD).

The main method used by traditional philosophers is thinking. Plato, Socrates, Descartes, and Kant thought about thinking per se, and even now, what philosophers mainly do is *think* (as well as speak and write, of course). Philosophers who get paid for their philosophizing do *intellectual* work. What they *don't* get paid for, at least if they earn their salaries by philosophizing, is busing tables, changing bed sheets, or picking lettuce. Doing such tasks is not at the top of their personal agendas. Perhaps for this reason, physical action tends, for better or worse, to be lower in their agendas in their contemplations about knowledge. A consequence is that philosophers may be biased against giving much thought to the vast amount of knowledge that is brought to bear in blue-collar occupations.

Now that was a political remark. I made it because of whom I'll discuss next, a German philosopher named Karl Marx (1818–1883). Marx (Figure 1.7) upped the ante for action. He wrote about the role of wealth in human welfare. As he studied history and witnessed the growth of the Industrial Revolution, he came to realize that as some people got richer, others got poorer. The way things were arranged, he observed, the distribution of wealth was what would later be called a zero-sum game.

Objecting to this state of affairs, Marx proclaimed that wealth should be shared by everyone. He offered a maxim to suggest how wealth should be shared: "*From* each according to his (or her) ability; *to* each according to his (or her) need." You may have

Figure 1.7
Karl Marx.

encountered this slogan before. It's the mantra of *socialism*, the economic and political system proposed by Marx. The basis for the term is that wealth should be shared by all in society. Marx argued that no one person or group should decide how wealth is distributed.

A key idea for Marx was putting the idea of socialism into practice. He used the word *praxis* to refer to the enactment of socially useful ideas. You needn't be a socialist to appreciate the importance of praxis (action), political or otherwise. Taking action is always important, whether it's to share wealth with others or to keep it for yourself if that's your preference. (There have been some capitalists who have done great good in the world, just as there have been some not-so-generous socialists.) In general, whether in the political/economic sphere or in daily life, a principle that seems indisputable is that to know is not enough.[6] Cognitive psychology, neuroscience, and related fields should, accordingly, focus on acting as well as knowing.[7]

John Dewey

Taking action isn't always political or economic, but no matter where action is taken, it's useful to take into account the good or bad that can come from it. Some have urged the view that actions should be as useful as possible; actions should maximize utility. This is the core claim of a school of thought called *pragmatism*.

An influential pragmatist was the American philosopher John Dewey (Figure 1.8). Dewey lived from 1859 to 1952 and was a political progressive who believed strongly in the rights of working people, including, and especially, the right of all people to get an education. Perhaps more than anyone, Dewey changed the way schools are run in America. The belief that all children in the United States should go to school for free and benefit from professional teachers came from Dewey. So impactful was Dewey's thinking that the U.S. Postal Service issued a stamp in his honor. The stamp was issued in 1968 and is shown in Figure 1.8.

Dewey is relevant to this book because he was one of the founders of scientific psychology. In that role, he rebelled against reflex-based psychology—the idea that each element of behavior is simply a response to a triggering stimulus (Figure 1.9, top panel). (The reflex concept will be reviewed in some detail in Chapter 4.) In a prescient article published in 1896, Dewey argued that "reflexology" is misguided (Dewey, 1896). It is

Figure 1.8
U.S. postage stamp showing John Dewey, issued October 21, 1968.

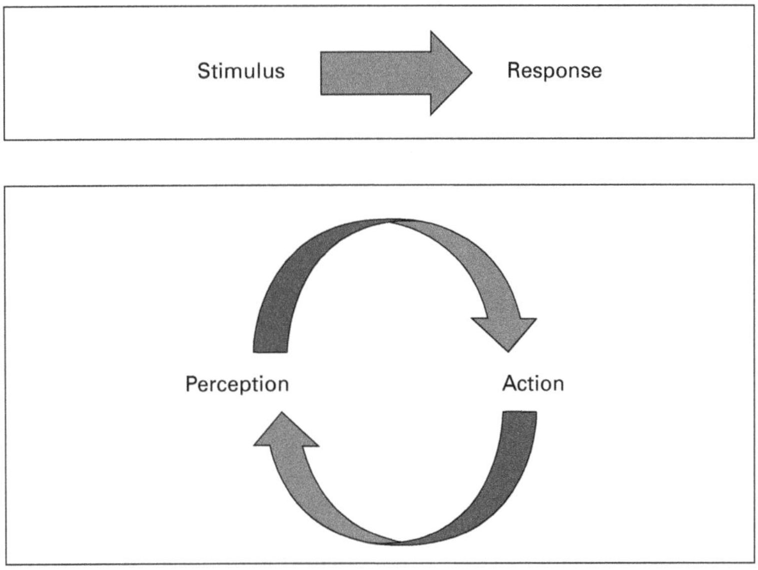

Figure 1.9
Two views of behavior. Top panel: the reflex view. Bottom panel: the perception–action cycle view.

wrong, he said, to think of behavior as a series of reactions to stimuli. Rather, it makes more sense to think of perception and action as occurring in a *cycle* with no predefined start and end (Figure 1.9, bottom panel).

The term *response*, Dewey argued, suggests passivity—waiting for a stimulus and then reacting to it. In everyday life, Dewey emphasized, we actively explore and shape the environment. Based on this idea, he lobbied for an analogous model of education: Students shouldn't be passive recipients of information. Rather, they should learn by doing.

Knowing How and Knowing That

The kinds of actions that will be discussed in this book—mostly small-scale everyday actions—have an intriguing property. They are right there in front of you, but you probably don't know how we do them. If someone asks you, "How exactly do you comb your hair?" you might give a general answer like "I move the comb through my hair." If pressed further, there's a good chance your answer could devolve to "Uh, I don't know. I just do it. Leave me alone."

The mismatch between doing things and not knowing how you do them is one of the greatest sources of fascination here. If you can comb your hair but can't say how you do, that reveals a distinction between forms of knowing worth exploring. You can show that you know how to comb your hair by combing it, and someone can confirm that you know how to perform the act by watching you. If that someone is your little brother or sister or your young son or daughter, they can learn how to comb their hair by watching you. That, indeed, is how children learn much of what they do—by watching others.

Virtually all physical acts are like this. They can be done but can't be described in detail. You may able to ride a bicycle, but you almost certainly can't say how you do.

You may be able to change your facial expressions to convey subtly different emotions, yet it's doubtful you can say what you do to communicate slight surprise, hints of doubt, subtle encouragement, and so forth. If you're asked how you put on a sweater, hit a baseball, speak, or swallow, your answer, perfectly reasonably, will probably be, "I don't know. I just do it."

It wasn't until the twentieth century that people caught on to the distinction between the two kinds of knowing discussed above—the kind that can be discussed and the kind that can't. The philosopher most responsible for this distinction was a British academic named Gilbert Ryle (1900–1976). Ryle, whose picture appears in Figure 1.10, drew a distinction between *procedural* knowledge and *declarative* knowledge, or what he also called *knowing how* and *knowing that* (Ryle 1949).

Knowing *how* (procedural knowledge) is the main kind of knowledge underlying fluent physical action. More often than not, procedural knowledge is *implicit*. It can't be put into words except in the broadest sense. General aspects of procedures can be talked about, as in recipes for cooking, where general guidelines suggest ingredients, the time those ingredients should be combined, and so on. But the exact means of preparing the ingredients—how the carrots should be diced, how the potatoes should be mashed—are left to you.

Could it be otherwise? Could more details be provided? Probably not. The reason is that your body, your kitchen, your carrots and potatoes are all unique. No one can tell you precisely what you must do to cut your carrots or mash your potatoes with the body you inhabit or the kitchen you occupy. Ironically, neither can *you* articulate all you

Figure 1.10
Gilbert Ryle.

know to slice and mash. The knowledge isn't missing. It's there for sure, just reflected in your performance if not the words you can provide to describe that performance.

Knowing *how* (procedural knowledge) is different from knowing *that* (declarative knowledge). Declarative knowledge is knowledge that can be put into sentences. For example, you can say, "George Washington was the first president of the United States," or "Yesterday I had a fresh tomato and mozzarella salad for lunch." Both sentences convey facts or beliefs summarizing things you learned, experienced, or imagined. They provide explanations to others, expressed as propositions concerning facts or putative (supposed) facts. The facts may be *semantic*, as in the example about George Washington, or *episodic*, as in the example about yesterday's lunch, which concerned an event or episode in your life.

Plato, Descartes, Locke, and Kant didn't concern themselves with the distinction between knowing how and knowing that. It wasn't until 1949, when Ryle published his book *The Concept of Mind* that this distinction came to be widely known. Soon after Ryle's book, a philosopher named Michael Polanyi promoted the idea of *personal knowledge*. His idea was that only you can know what pertains to you specifically. Only you can know what it feels like to bend your elbow as you do when you peel carrots, only you can know what it feels like to hold your wrist as you do when you mash potatoes, and so on. These sorts of knowledge are private (known just to you) and so they are *tacit*, the word Polanyi (1958) used to refer to knowledge that is typically available only in unconscious form. One of the great delights about the study of action, mind, and brain is that the goal of the study is to articulate what is known but can't be said.

The Paradox at Hand

That last remark suggests a paradox. The aim of this book is to lay out the knowledge that is brought to bear when actions are performed. The knowledge is tacit, not available for public discussion, yet the intent is to present facts that have been amassed about that knowledge. If tacit knowledge is tacit, shouldn't it be impossible to articulate it?

The answer is no because you can strive for informed descriptive accounts. Virtually everything that happens in the world—all the baseball pitches that are thrown, all the pizza pies that are tossed, all the javelins that are hurled—illustrate principles that can be described. The principles don't prescribe what *should* happen. Rather, they capture what *does*. If an apple falls from a tree, the apple follows Newton's law of gravitation. That law describes the apple's motion but isn't a fiat. It's not a law in the sense of a government-issued regulation. The apple doesn't go to the books to find out what it should do. It just falls and does so in accordance with the law of gravitation.

The principles governing behavior are no different. They describe what you *do* but don't prescribe what you *should* do unless they happen to be explicit rules for games or etiquette, and even then, what is done rarely complies perfectly with the verbal prescriptions.

Describing principles that characterize actions means identifying regularities of action. The job of action scientists is to try to approximate, ever more closely, the descriptions they provide. There is no more of a paradox in describing human or animal action than there is in describing the behavior of inanimate objects. The fact that an apple cannot tell you how or why it drops to the ground doesn't mean its motion is indescribable. The motion is describable, as Newton showed. The same sorts of explanation can be sought, and will be sought here, for the actions of living beings.

Embodied Cognition

A last concept needs to be introduced as you round out the coverage of the intellectual background for this field. This is the concept of *embodied cognition*. Think back to René Descartes and recall that he was a *dualist*, someone who believed that there are *two* sorts of elements when it comes to psychology: mind or spirit, on one hand, and brain or body, on the other. Descartes was, in fact, the most famous dualist ever.

Whereas *dualists* think there are two kinds of psychologically relevant elements, *monists* think there is only one—either mind/spirit or brain/body. You may recognize the prefix *mon-* as related to one thing, as in *monotheism*, the belief that there this is just one god, or *monologue*, a talk (often a comedy routine) given by just one person. Descartes was a monist when it came to characterizing animal behavior, for he believed that animals lacked minds and just had bodies, making them machines or automata. But Descartes was a dualist when it came to characterizing human behavior, for he believed that people have minds as well as bodies.

By considering dualism, you can begin to anticipate what embodied cognition might be. If you allow that knowledge includes knowing how to do things, including knowing how to do things with your *body*, then you need a way to connect mind to body. This is called the mind–body problem.

The mind–body problem is a knotty one if you believe that knowledge is *disembodied*, that it exists in abstract form as symbols, the way Descartes did. The problem is knotty (difficult) because if knowledge is expressed symbolically, you need to translate those symbols into some form of action-appropriate knowledge. You need to jump from the mind to the body to get from mentally represented intentions to physically realized actions. Then you need to jump back from the body to the mind to get information about what you did for the mind to comprehend your action. That's a lot of jumping! There's a way to avoid it, however. That way is to express knowledge in bodily terms. Express knowing in terms of what your body does and can do.

How can you make sense of this? One way is to consider a simple example. Think of a violinist watching another violinist play the violin. The observing violinist might watch the performer by directly translating the associated sights and sounds into impulses corresponding to what the observing player would do while playing themself.

There are telltale signs that this happens. If you observe one violinist watching another, there's a good chance you'll see the watching fiddler make subtle motions.

Figure 1.11
Jessica Witt (left panel) and Sian Beilock (right panel).

Their left-hand fingers may twitch. Their right hand (the hand used for bowing) may jostle. Their head may bob. There's no reason why extraneous movements like these would occur if the perception of violin playing simply led to the activation of abstract, symbolic representations. Therefore, if you see a violinist making sympathetic movements while watching another violinist, you can reasonably infer that something else is going on. Advocates of embodiment, such as two of the leading scientists in this area of research—Jessica Witt and Sian Beilock (Figure 1.11)—say that that something else is embodied perception/cognition. The research of these two distinguished women and others will be discussed later in this book.[8]

2 Applications and Methods

Because actions are carried out in all walks of life, they have useful things to say about virtually all that you do. Start with sports. Each year, millions of dollars are spent on methods for helping runners run faster, broad jumpers jump farther, pole vaulters vault higher, and so on. Sports have seen huge advances in the past century, with new records being set at an astonishing pace. Records have been set because new approaches have been taken to sports training as well as sports equipment, sports clothing, and the venues where sporting events take place. If you look at clothing worn by athletes in past times (Figure 2.1), it's no wonder that players today can perform so much more easily and effectively. Add to the new ways of dressing, changes in eating, and the emergence of sports psychology (an entire discipline devoted to helping people prepare mentally as well as physically for competitions), you can see how greater enlightenment has led to better athletic performance.

Medicine is another area where action research has played an important role. Many action-related disabilities beset people. Problems arise from genetic misfortune, disease, accidents, maladaptive lifestyles, and violence. Preventive measures must be sought, and effective treatments must be developed to meet these tests. Someone who has lost a limb needs a prosthesis, and someone whose spinal cord has been severed needs a new way to move the paralyzed limbs. Similarly, someone who has suffered brain damage from a stroke, accident, or infection needs to regain lost function.

Another domain of application is human factors. This area of work is concerned with safety and efficiency of human performance. Typical domains where human factors findings are applied are the workplace, schools, transportation, and the home. Human factors research has led to the specification of the height of kitchen stoves, the placement of kitchen stove knobs, the mapping of stove knobs to burners, and many other functions not just related to kitchen appliances. The shape of your car seat has also been based on work by human factors engineers, as has the size of your steering wheel, how much you have to turn the steering wheel to get your car to turn some number of degrees, and so on. The layout of your computer keyboard also reflects human factors research, or ergonomics, as the field is also called. The layout of your computer screen, how much your computer cursor moves depending on how far or fast you move your finger, the fonts that are used, and the sizes of the icons shown all depend on years of work by human factors engineers, many of whom were trained in psychology, industrial engineering, and related disciplines.

The arts and entertainment also benefit from studies of action. Considering that the arts depend on performance, this is hardly surprising. A guitarist must place her fingers on the fingerboard to get the chords she wants. A violinist must hold his bow in a way that lets him produce legato notes or staccato notes depending on what the

Figure 2.1
Charlotte Cooper, the first female Olympic tennis champion and first individual female Olympic champion, in 1900.

music calls for. To this list you can add your own examples. Whether the activities involve dancing, singing, telling jokes, acting, drawing, sculpting, or juggling, by understanding how actions are carried out, you can improve your own performance or think of ways of improving the performance of others. Teaching and coaching techniques can be enhanced by exploiting human factors principles, and practice and rest schedules can be designed to foster better performance to avoid needless tolls on the health and well-being of those engaging in whatever acts are of interest.

Naturalistic Observation

Consider the methods used in action studies. The most fundamental way to learn about action is to observe it in everyday life. Watching what others do is how most of us learn what we can do. As infants and toddlers, we watch others carry out actions and learn to act accordingly. At these early ages and beyond, we pick up on ways that individuals act in various circumstances—how they speak with one accent or another, how they gesture one way or the other, and so on. Learning through observation is one of the most important methods of learning.[1]

Observing actions is also used by researchers to learn about the nature of action control. Via *naturalistic observation*, it is possible to discover properties of action control that might be harder to find through lab studies. I will discuss an example from my own work that illustrates the point because it happens to point to some core principles that will come up repeatedly in this book.

One day I was in the main concourse of Grand Central Station in New York City, waiting for my wife to arrive by train. I was to meet her at the main information desk

Figure 2.2
Main staircase in Grand Central Station, New York City.

and got there a few minutes early. While milling around, I watched people descending the staircase leading down to the main concourse (Figure 2.2). I noticed that people looked down at the staircase as they first stepped onto the stairs. That wasn't surprising because one has to look down onto stairs when one first steps onto them. What caught my eye was something more interesting. As people went down the stairs, they tended to look down at seemingly random times during the descent, but when they were three or four steps from the bottom, they were extremely likely to glance down and then look up again. On the last two steps from the bottom, they almost always looked out ahead of them.

Why did the people do this? What was special about the third of fourth step from the bottom of the staircase? I had a hypothesis, but before trying to present it to the public in the form of a scientific report, I felt it was important to make sure this was a real phenomenon. I had no objective record of what I had seen—just casual observations. Nowadays, no respectable scientific journal will publish casual observations alone. I decided to get more naturalistic observations of people walking down stairs, but this time I would gather video data of their stair descents.

I worked with a student who, as part of a research practicum, agreed to sit with a camcorder at the base of a staircase in a public space—the Hetzel Student Union Building at Penn State University (University Park campus)—to record people walking down the stairs of that building. I got permission from the Penn State Institutional Review Board (IRB) for the study. Penn State's IRB, like any IRB, is responsible for ensuring that data from living subjects (human and animal) are obtained and processed ethically.

To get permission from the IRB, I explained that the video camera would be out of view (most likely under a newspaper on a table near the base of the staircase), and I promised that no one except the student recording the videos, a few other select individuals in our lab, and I would ever watch the videos. We would never reveal any personal information about any of the people we taped, nor would we categorize the people by gender, weight, height, or other personal characteristics since we had no hypotheses about those variables in this context. We would only analyze and report how many people looked down depending on which step they occupied. We would ask independent judges (different research assistants in the lab) to determine when the stair descenders looked down, and we would ask the research assistants to watch the videos frame by frame to complete their analyses.

The coded video data confirmed the initial observation. As shown in Figure 2.3 (left panel), there were quite a few downward looks at the top of the stairs (eight steps from the bottom), but the number of downward looks peaked at four steps from the bottom. At just one step from the bottom, there were virtually no downward looks. These results confirmed the initial observations made at one place (Grand Central Station in New York City) as checked with recording equipment at another (the Student Union Building at Penn State University, University Park, Pennsylvania).

Why was this observation interesting? Why, in fact, do I think it is so interesting that I bring it front and center here? The reason is that the hypothesis developed to account for the data illustrates an idea that will come up over and over again in this book. The idea is that action control depends on the interplay of low-level factors and high-level factors. According to this hypothesis, in the case of walking down stairs, there is a simple trigger for the last downward look. That trigger, as shown in the right panel of Figure 2.3, is the disappearance of the stairs from the bottom of the field of view. While looking straight ahead, at a certain point, the image of the staircase in the lower visual field suddenly vanishes; what appears instead is the floor below. The disappearance of the stairs from the lower visual field triggers the downward glance. This is the low-level aspect of control hypothesized to explain the data.

Meanwhile, there is also a high-level aspect of control. This is updating of the stair walkers' position within a memory representation of the staircase. The idea is that after

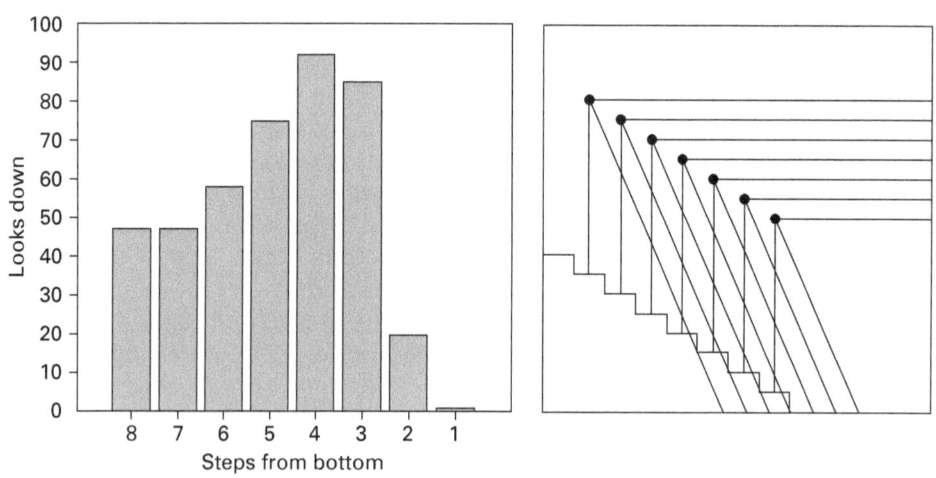

Figure 2.3
Number of looks down by people descending a staircase plotted as a function of the number of steps from the bottom of the staircase (left panel) and an idealized person standing on the steps of the staircase looking straight ahead (right panel). In the right panel, the horizontal dashed lines show the straight-ahead line of sight, while the solid downward lines show the bottom edge of the lower visual field.

the final look down, walkers complete their descent by keeping track implicitly (without conscious awareness) of where they are on the stairs. They take a kind of "walk down memory lane," a phrase that became the basis for the title of the article where this work was described (Rosenbaum, 2009).

These two processes—one based on visual input and the other based on memory—illustrate how one act can draw on two quite different resources, one low level (optical in this case) and the other high level (memorial in this case).[2]

Ecological Validity

The last section was concerned with naturalistic observation. It will be just one of many such observations presented in this book. You saw how an observation made in a casual way and more or less by accident (while hanging out in Grand Central Station) was documented via videotaping and subsequent analysis. I will say more about videotaping and moviemaking later in this chapter, but first I want to say more about naturalistic observation and ecological validity, a term explained momentarily.

Recall that I made the observation of stair walkers in the natural environment, albeit in a human-crafted one (Grand Central Station). The term "naturalistic observation" doesn't mean observation out in the middle of a forest or meadow. Instead, it means observations made where people (or animals) behave as they ordinarily would. Such observations are said to have *ecological validity*, a term that means true-to-the-environment. Studies with ecological validity have a good chance of reflecting features of behavior in the real world.

Evaluating ecological validity becomes important in laboratory experiments. If people or animals perform laboratory tasks that seem highly atypical of what they do day by day, some may complain that the lab results don't apply elsewhere. These complaints must be carefully considered. Criticizing a study on the grounds that it lacks ecological validity just because the outcome doesn't accord with one's prejudices isn't good, of course. Such a criticism is said to be ad hoc (Latin for "not essential"). If the charges are directed at the *person* who conducted the study, the criticism is said to be ad hominem (Latin for "directed at the person"). Ad hoc and ad hominem criticisms should be avoided. However, if laboratory studies seem really remote from anything that happens in everyday life, then more credence can be given to the complaint that the studies lack ecological validity.

In general, lab work should be as ecologically valid as possible. Naturalistic observation should be done in the hope of revealing interesting phenomena. Action science is particularly attractive from this point of view because it affords the opportunity to see interesting behavior in the everyday world. I encourage students in my Action, Mind, and Brain class to be on the lookout for aspects of behavior that they think might be important, and I encourage you to do the same. Lab studies are fine, but if deep principles can be derived from everyday observations, those observations may lay the foundation for future work, including detailed, equipment-intensive work in the lab.[3]

Lights, Cameras, Action!

As you saw in the description of the stair-walking observation, video recording was used to go beyond a general impression of where people looked. It's possible to gather data in naturalistic settings and have confidence in the data immediately without the need for further evaluation via video or other techniques if the aspects of interest are easy to spot and if independent observers can score the data in a way that shows high interobserver agreement. For example, if the data are whether people walk around the left or right side of a table, video recording isn't required. Observers with clipboards can watch the walkers and put a checkmark in a left box or right box. Presumably, their judgments will match up nearly perfectly and no one will question the measure unless the findings are weird or suspicious, in which case those doubting Thomases can attempt to replicate the study.[4]

The reason I used videos in the stair-walking case was that my eye wasn't quick enough to let me be sure where people looked when they were on each step, nor was my writing quick enough to let me write down everything in real time. Consequently, for the stair descent study, my student assistant and I used video freeze-frames to determine what stairs people stood on when they looked down.[5]

Having spoken of videos and their benefit, I now want to discuss a second method for studying action. The method uses movies, videos, and related technology for analyzing action. This topic brings up one of the most surprising and most entertaining facts about action research—which is that Hollywood movies, the Oscars, and all the rest came from an interest in the detailed features of everyday action patterns.

Imagine what it would be like if there were no motion pictures. You wouldn't be able to go to the cinema and watch your favorite action heroes fly through the air. You wouldn't be able to look at your computer screen and watch gunfights, car chases, partings of waves by Moses at the Red Sea, or kids in the Alps singing "Doe, a deer, a female deer."[6] There would be no blockbusters, no Oscars, no movie stars. Even more basically, there would be no way for you to watch people and animals moving in the same situation repeatedly, in slow motion, with stop action, or in reverse. Prior to the invention of movies, the only way you could watch action was to be where the action was; at the event itself. Even then, you very well may have been frustrated by the limits of what you could see. Replays were impossible. Overly quick actions would be a blur.

The invention of the motion picture can be traced to people's desire to see actions in greater detail. A particularly strong desire concerned the observation of horse running. If you had lived in the 1800s, you might have been curious to know whether horses ever have all their hooves in the air when they run. Without motion pictures, you'd have no way of checking. You could look long and hard as horses run past you, hoping to see the state of their hooves at some critical instant. No matter how hard you'd look, though, getting down onto the ground if necessary, letting yourself get all dusty just to satisfy your curiosity, you'd be hard pressed to say for sure whether the equines (horses) racing past you ever became fully airborne. Their motion would be too quick for you to tell.

In the 1800s, the question of whether horses ever "fly" was hotly debated. In saloons and kitchens, on farms and at county fairs, people argued about the matter. A wealthy man named Leland Stanford was curious about the matter, too, especially as it pertained to his prize horse, whom he adored. He hoped his horse would be like Pegasus, the mythical horse who could soar through the sky. Stanford hoped that a recently invented device, the still camera, could be adapted to provide pictures of his horse that might show its flightworthiness.

The name Leland Stanford might sound familiar to you. You might say, "Isn't a university named after him?" You'd be close but not exactly correct. Leland Stanford founded a university, but he named the school after his son, Leland Stanford Jr., who died at the age of sixteen of typhoid (in 1884). On Stanford University's seal and on all the diplomas it issues, Stanford University is actually called Leland Stanford Junior University. Leland

Stanford Senior, the founder of the university, was the governor of California. Governor Stanford and his wife were so saddened by the death of their only child that they decided to bankroll a university so "the children of California shall be our children."[7]

The person who took the critical photograph that Stanford coveted was named Eadward Muybridge. (Yes, there is an "a" between the "E" and the "d" in his name, which was pronounced "Edward My-bridge.") Muybridge's picture, along with Governor Stanford's, appears in Figure 2.4. The series of pictures that Muybridge took is shown in Figure 2.5. It shows that when a horse runs, there are, indeed, times when it is aloft.

The appearance of photographs like those in Figure 2.5 did more than resolve a question about horse running. Movies revolutionized people's capacity for studying motions they had only seen fleetingly before: a horse galloping, a person sneezing, a workman swinging a sledgehammer, a woman walking down stairs, and so on. Collections of photographs showing stop-action images became popular in the late 1800s. Many examples from this period are available on the internet; use as your search term "Muybridge." People marveled at such images. Even more, they marveled at movies, which they could see, thanks to a device called the kinescope, which let them look through a slit at a series of still images that passed by quickly enough that the images melded together, seeming to move.[8]

The motion picture camera was not invented to make Hollywood movies. There was no such thing as entertainment cinema, and there was certainly no town devoted to film production (Hollywood). It took a while for people to realize that they could put actors in front of motion picture cameras and have those actors play characters in dramas, comedies, musical numbers, and the like. The idea of making movies, showing them to others, and charging them to see the cinemas came from Thomas Edison, the inventor of the lightbulb, phonograph, and other devices, including the motion picture projector. Edison's picture is shown in Figure 2.6 to draw attention to the fact

Figure 2.4
Leland Stanford (left) and Eadward Muybridge (right).

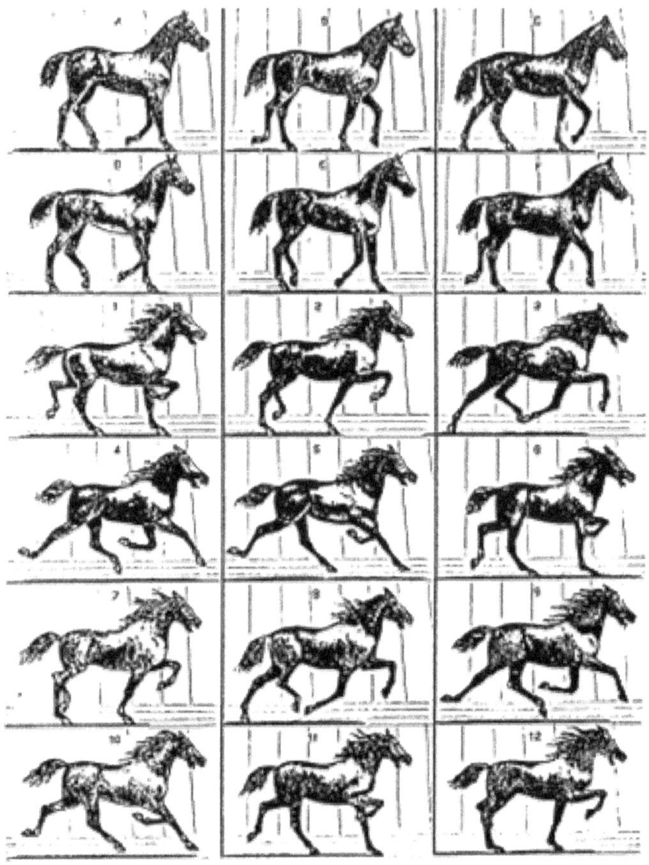

Figure 2.5
Muybridge's photographs of a horse in motion.

that famous people are associated with the triptych of action, mind, and brain one way or another. The main point is that interest in biological motion yielded results that are not just scientifically interesting but also central to our culture and history.[9]

Cinematic Analysis

When photographs of people and animals in motion first became available, there was widespread belief that scientists and others might gain a deep understanding of the control of action by studying the photos. That hope was realized. The first example, shown in Figure 2.7, is a bird diving into the sea to go after a tasty-looking fish. The bird is a gannet. These birds can be seen diving off the Scottish coast, which is where a Scottish researcher named David Lee observed them. Watching the birds dive into the sea, Lee wondered how the birds controlled the retraction of their wings as they approached the water surface. What was the trigger for the wing retraction?

To answer this question, Lee and Reddish (1981) relied on single-frame cinema analysis to study the birds' dives. The researchers focused on the moment that the gannets fully retracted their wings prior to plunging into the deep. Full wing retraction was required for streamlined water entry.

Figure 2.6
Thomas Edison.

Think, as David Lee did, about when the birds might retract their wings. One possibility is that they would do so when they reached a critical distance from the water. That would be a logical possibility but not necessarily an adaptive one. The birds might drop at too low or too high a speed to prepare for the dive when they happened to be the critical distance from the surface.

Another possibility is that the wing retraction would occur when the birds were some number of seconds or fractions of a second from the moment of contact with the water. That would be a time-based option rather than a distance-based option.

Through cinematic analysis, Lee and a colleague showed that the gannets pulled in their wings at a roughly constant time before contact with the water, consistent with the time hypothesis. Lee and Reddish called this time "tau" or τ, the Greek letter for the letter "t," the letter commonly used to denote time or moments of time in physics and other disciplines.

How could the birds tell the time to contact? Some math will help. Time equals distance divided by rate, so time to contact could be specified by the ratio of distance to rate. The distance to the water could be optically registered by the image of the water surface, with the rate being registered as the rate of optical expansion of the water. The closer the birds got to the water, the more quickly the optically registered texture of the water surface would expand. Time to contact could then be based on the rate of optical expansion. Once the rate exceeded some critical value, that would be the signal for the birds to close their wings and literally take the plunge.[10]

You may recognize a similarity between my idea about stair walkers and Lee's ideas about diving gannets. Like Lee, I appealed to optics to explain an aspect of action. For me, it was the disappearance of the image of the stairs in the lower visual field that

Figure 2.7
Drawings of a diving gannet based on movie frames.

(by hypothesis) triggered walkers' last downward looks. For Lee, it was the rate of optical expansion exceeding some value triggering closure of the gannets' wings (again, by hypothesis).

Is it good or bad that a similar principle—reliance on changing visual information—is said to govern action in species as seemingly disparate as gannets and people? Most scientists would say it's a good thing because, from the standpoint of scientific explanation, it is desirable to have the same principles apply in different contexts. One of the great attractions of Darwin's theory of evolution is that it applies across all species and assumes or leads to the expectation that Nature retains solutions. If sensitivity to optical flow was useful for birds when they evolved (long before people did), it might have also been useful for people or our forebears to retain that sensitivity. If you jump off a cliff to dive into a deep pool of water, it will help you to have some fairly precise information of when you will land. If you lacked such sensitivity, it is unlikely you would be able to ride a bicycle, drive a car, or fly an airplane. The idea of optic flow came, in fact, from studies of airplane pilots (Gibson, 1950; https://en.wikipedia.org/wiki/Optical_flow).

Movies with Markers

The invention of the movie camera was a natural development in the refinement of the still camera, for a movie camera is just a still camera that takes pictures quickly enough that when the images are replayed quickly, one sees movement. Over the years, the number of pictures that can be taken per second has increased dramatically, making it possible to see things that are incredibly brief: a water balloon at the moment of bursting, a corona of tiny milk droplets forming after a drop of milk lands in the liquid, or droplets of a sneeze being spread, often at peril to those downwind.[11]

Such high-speed photography has enabled new insights into action control. For example, an article in *Science* magazine described a copious analysis of the flight maneuvers of dragonflies darting about in various directions, including—amazingly—*backward*. The dragonflies were filmed with ultra-high-speed photography, making it possible to reveal aerodynamic tricks used by these insects that had been unknown before. Such aerodynamic maneuverability is interesting for flight enthusiasts and for developers of artificial flying machines.[12]

Besides taking pictures more and more rapidly, researchers interested in motion have also widened the kinds of recordings they take, hoping to get numerical data from the recordings. The motivating idea is that scientific analysis requires quantitative analysis. Watching movies is entertaining and suggestive of underlying principles, but getting quantitative data is often required as well.

With traditional movies, it's generally hard to get quantitative data about body movements immediately or automatically. More laborious methods are often needed, such as looking at frozen images on a screen and measuring points of interest. This approach was pioneered by a French neurologist named Marc Jeannerod, who was the first to characterize the opening and closing of the fingers (the index finger and thumb) during the simple act of approaching an object with the hand and closing the fingers around it. Figure 2.8 shows an image from one of Jeannerod's studies. Jeannerod (pronounced "jon-a-row" with a soft "j") showed that before grasping the object, the fingers spread apart farther

Figure 2.8
Hand approaching an object.

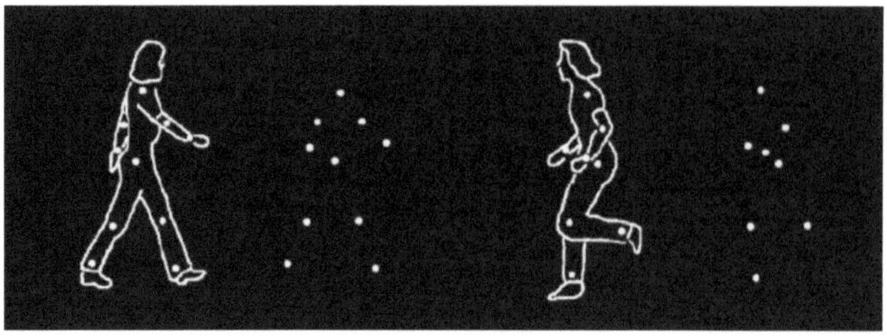

Figure 2.9
Biological motion perception based on moving dots alone.

than the object and then closed in on it. The timing of the finger spread and closure was well coordinated with the hand's approach to the object and depended on the size of the object as well as its distance from the hand at the start of the prehension act, that is, in the act of reaching and grasping (Jeannerod, 1981).

Anticipating the need for precision with such recordings, researchers often put markers on actors' bodies. Such markers can reduce the variability of the measurements. The markers can be informative in their own right, as shown in a study of visual motion perception. The researcher who did this study, a Swedish vision scientist named Gunnar Johansson, attached reflecting dots to people's bodies and had the people walk, run, hop, or dance (Figure 2.9). Johansson (1973) filmed the actors with special lighting, which made it possible to show the movies in such a way that only the reflective dots were visible. Even with such sparse information, observers could see full human motion. In fact, the observers were so adept in their observations that they could immediately identify the gender of the actors and, if the actors were people the observers knew, who the walkers were. Seeing two people dance together was also easy. Again, all of these visual percepts were gleaned from moving dots and nothing more.[13]

High-Tech Motion Recording

In the preceding sections, I spent some time unpacking the issues related to cinematic analysis to sensitize you to the issues that arise in gathering and analyzing motion data. The remainder of the chapter will go a bit more quickly. I'll cover more high-tech motion recording techniques and then turn to other methods used in the field of action, mind, and brain.

As you might imagine, a valuable capability for movement scientists is getting numerical data from actors in motion as soon as the actors move. That's much more convenient than measuring points of interest in frozen movie frames and getting measurements of points of interest in the static frames. Instant data gathering of body positions became a reality several decades ago. A method that proved useful was and continues to be picking up locations of individual flashes of light emitted by lights attached to actors' bodies.

One such system was developed in the 1980s at the University of Waterloo in Waterloo, Ontario, Canada—the so-called WATSMART system.[14] Here, three cameras peered into spaces occupied by people wearing wires leading to tiny lightbulbs on points of interest on their bodies. The bulbs were infrared light-emitting diodes (IREDs), which meant that when the IREDs were activated, their illumination was invisible to the human eye because wavelengths below the wavelength of red light (hence "infrared") are not detectable by the human visual system. Critically, just one IRED lit up a time, and three cameras, linked to a computer, triangulated the position to give the Cartesian coordinate (*x*, *y*, *z* coordinates) of the IRED that turned on. Because the IREDs were activated cyclically at a high rate relative to the motion of the body, it was possible, for all practical purposes, to immediately record where all the IREDs were at each functionally relevant moment. For example, if each functionally relevant moment was one-tenth of a second (a sampling rate of ten samples per second, or 10 Hz), then with ten markers on the body, it was possible to find out where each marker was in one second. With a sampling rate of 100 Hz, it was possible to find out where each of 10 markers was every one-tenth of a second.[15]

A disadvantage of such a system is that IREDs can get occluded. If the subject's hand passes in front of an IRED, blocking the cameras' view, there are missing data. Precautions can be taken to avoid this problem, but it is an inconvenience. So too are the wires that must be worn to connect the IREDs to the hub. Wireless systems have since been developed, such as ones that use reflecting markers, but occlusion remains a problem for them, too.

Another method is to use electromagnetic signals broadcast by emitters attached to the body. With electromagnetic systems, challenges arise in connection with electromagnetic interference. Moreover, wearing electromagnetic emitters can be inconvenient as they add weight and bulk.

Motion-capture systems are used widely in human factors research and in the entertainment field. When you watch a movie with computer animation, it is typically based on actors wearing tight-fitting suits with emitters or reflectors attached to the suits at key places on the actors' bodies (Figure 2.10). Once the data have been recorded, the motions are artistically enhanced.

As of this date, it remains a pipe dream (still a distant hope) to get accurate position data from people moving about in the three-dimensional environment with no special apparatus such as markers attached to them. Progress is being made on this front, yet ironically, one of the principal reasons why movies were developed in the first place—to enable the analysis of biological motion—remains a challenge to this day.[16]

Reaction Times

It is not always necessary to capture data about full body motion. Often all that's needed is information about some emergent aspect of performance such as reaction times (RTs). RTs are useful for basic science and for practical concerns. In practical contexts, RTs can determine who leaves the starting blocks first in a sprint race, whether one can stop a vehicle quickly enough to avoid plowing into the car ahead, and whether one can leap away from a falling rock. In the laboratory, RTs can shed light on fundamental mechanisms of action control. At the most basic level, RTs can be used to measure the speed of nerve conduction.

In the mid-1800s, a German scientist named Hermann von Helmholtz (Figure 2.11) did just that. He measured the speed of nerve conduction with RTs. Helmholtz, who lived from 1821 to 1894, made many contributions to medicine and science. You may not have heard of him before, but you have encountered an invention of his when you had your eyes checked. Helmholtz invented the ophthalmoscope (Figure 2.12), a contribution so noteworthy that, much as the U.S. government issued a stamp honoring

Figure 2.10
Actor Jason Gatt in a motion-capture outfit for the character Kratos from *God of War II* and *III* on the mocap stage for SCEA in San Diego (left panel) and Kratos himself (right panel).

Figure 2.11
Hermann von Helmholtz.

Figure 2.12
Eye exam with an ophthalmoscope.

John Dewey (Figure 1.8), the German government issued not one but *two* stamps honoring Helmholtz. One of them is shown here.

Why do I bring up Helmholtz in connection with reaction time? Besides inventing the ophthalmoscope, he determined the speed of nerve conduction by doing a simple experiment. He touched a frog on two different places and measured the time for reactions in the brain. He touched the *hindlimb* of the frog in one condition and the *forelimb* of the frog in another. The time for the brain to register the forelimb touch was shorter than the time for the brain to register the hindlimb touch. Helmholtz did some math with the two times. He took the difference between the times and the difference between the distances. Dividing the distance difference by the time difference, he estimated the speed of nerve conduction to be 50–60 meters/second, or about 112–134 miles per hour.

Simple and Choice Reaction Times

Helmholtz was a nineteenth-century German scientist-physician who studied the eyes and measured RTs. It happens that another nineteenth-century scientist—this one *Dutch*—also had a strong interest in the eyes and did research with RTs. The Dutch investigator (Figure 2.13) used RTs to estimate the durations of mental processes. His name was Franciscus Donders (1818–1889).

Before talking about Donders's RT work, I'll tell you about an important discovery he made about the eyes. He noticed, as shown in Figure 2.14, that for every gaze angle, the eyeball has a unique position. When the gaze angle is straight ahead, the eye has a unique combination of three rotation angles (degrees of turns about the up–down axis, the left–right axis, and the twist axis around the line of gaze). When the gaze angle is different—upward and to the left, say—the eye has a different combination of three rotation angles around these same three axes. Donders noticed that every gaze angle has a unique set of eyeball angles. This principle is called Donders's law (Donders, 1847; Tweed & Vilis, 1990).

Figure 2.13
Franciscus Donders.

What about RTs? Donders distinguished between two main types of RTs: *simple* RTs and *choice* RTs. He defined simple RTs as reaction times from tasks in which subjects have just one response to make (typically, a button press) when one possible stimulus is presented. The defining feature of simple RT tasks is that there is no uncertainty about which stimulus is presented or which response is needed; all that is unknown is *when* the response must be made. For *choice* RT tasks, by contrast, there *is* uncertainty about which stimulus will be presented and which response will be required. In choice RT tasks, there are several possible stimuli and several associated responses. Those stimuli and responses are typically announced before each set of experimental trials. Critically, the participant does not know which response will be needed per trial until one of the stimuli is presented.

Donders discovered that choice reaction times are longer than simple reaction times. In general, he found that simple reaction times are about 250 ms (250 milliseconds, or a quarter of a second), whereas choice reaction times are about 500 ms (half a second). The difference between those times provided an estimate of the time for a mental event, choosing a stimulus–response alternative.

The great elegance of Donders's approach lies in the fact that the difference between choice and simple RTs is all in the subject's head. The mere *possibility* of a stimulus or

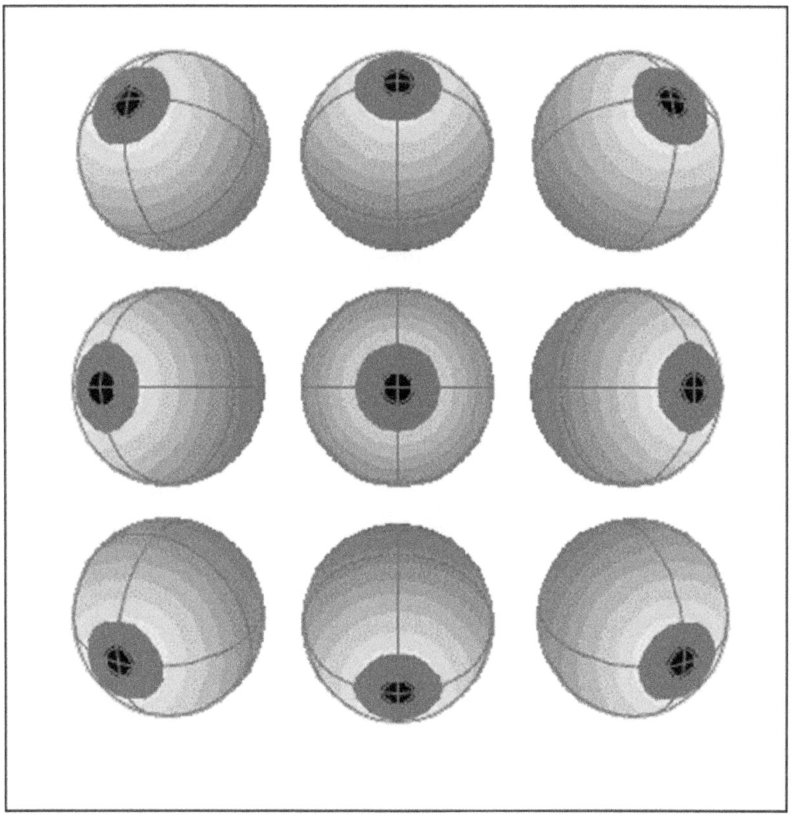

Figure 2.14
The eye in different gaze angles as per Donders's law.

response is what distinguishes simple and choice RT tasks. On any trial, there is just one stimulus and just one response. Yet the RT to make a response to a stimulus is longer when the stimulus and response are one pair of several that are possible. Through Donders's comparison of RTs in simple and choice tasks, it became possible to estimate the durations of mental events. That is precisely what you want to know if you hope to use RTs to study cognitive processes.

Donders's contributions have been celebrated. If you visit a city in the eastern part of the Netherlands called Nijmegen (pronounced Nigh-may-gun), you can see a scientific institute there called the Donders Institute for Brain, Cognition, and Behaviour, located at Radboud University. The Donders Institute specializes in brain imaging, a method not available to Donders in the nineteenth century, but true to Donders's interest in opening a portal to the mind and brain, it does just that.

Shepard Mental Rotation

Through his brilliantly elegant RT method, Donders helped us learn about the processes occurring in the mind and brain prior to action. Now we turn to another pioneering researcher who used RTs. Apropos of the earlier discussion of naturalistic observation, it is interesting to dwell for a moment on the naturalistic observation that this investigator and his coauthor (Cooper & Shepard, 1984) referred to in order to motivate their RT study:

> One of us witnessed a German Shepherd retrieving a long stick that had been thrown over a fence from which one vertical board was missing. The dog bounded through the gap, seized the stick in its mouth and plunged headlong back toward the narrow opening. Just as catastrophe seemed imminent the dog stopped short, paused and rotated its head 90°. With the stick held vertically it passed through the fence without mishap. The operation that took place in the mind of the dog in the moment before it turned its head presumably was not verbal. Might it not have been a preparatory mental rotation of the stick? (p. 106)

The text continued:

> Consider this question: how do you take a card table through a narrow doorway without folding up its legs? Most people report that they must envision the process of turning the table on its side, putting two of the legs into the opening, then turning and moving the table so that the other pair of legs pass through the door. (p. 106)

In experiments designed to track down the nature of such processes, Stanford psychology professor Roger Shepard was joined by Jacqueline Metzger, a graduate student at the time (Shepard & Metzler, 1971). They showed university students pairs of drawings like the

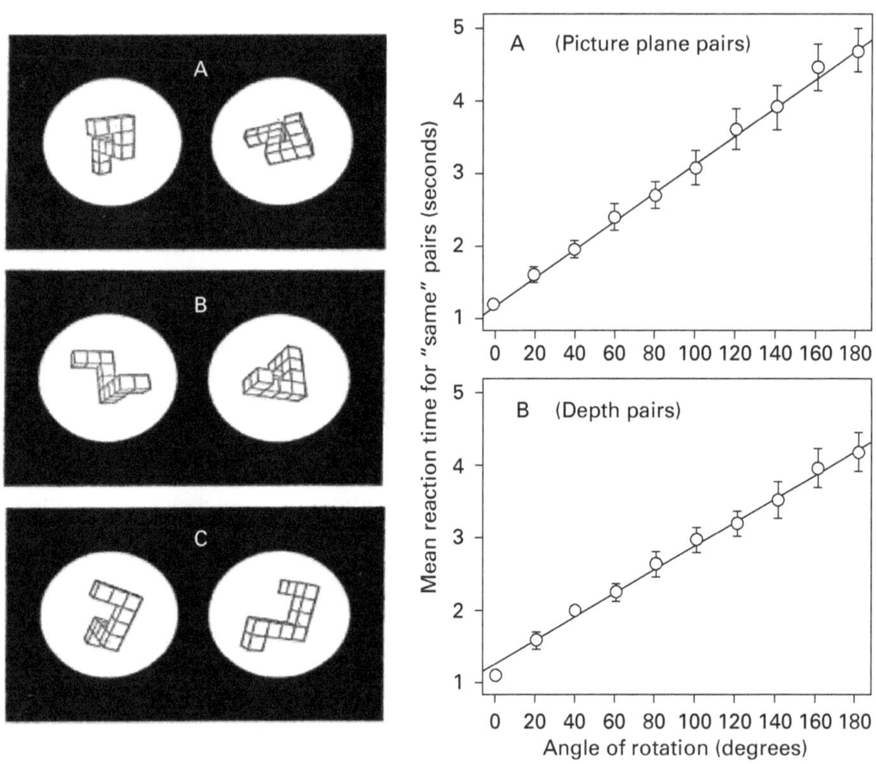

Figure 2.15
Stimuli (left panel) and RTs (right panel) from a study of mental rotation.

ones in Figure 2.15. For each pair of drawings, the question was whether the two members of the pair were the same block structure. In a random half of the trials, the two structures were different; there was no way to rotate one to bring it into alignment with the other. In the other trials, the two structures were the same. For those pairs, although the structures looked different, it was possible to turn one into alignment with the other. The ingenious feature of the experiment was that the turning was done mentally.

Shepard and Metzler designed the block structures so different degrees of rotation were needed to align the structures. In some cases, the degree of needed rotation was small; in other cases, it was large. The degrees of rotation needed for alignment were graded in steps of 20 degrees. The authors found something remarkable. As shown in Figure 2.15, the time to indicate that the two structures were the same increased with the angular separation between them. This was true up to an angular separation (number of degrees) of 180 degrees.

Another remarkable feature of the data was that the time to give positive responses (that is, to press the key signifying "the two structures are the same") grew at about the same rate if the rotation occurred in the picture plane (top left panel of Figure 2.15) or in the depth plane (middle left panel of Figure 2.15). If the rotation occurred in the picture plane, one of the objects could be turned like a disk *on* the screen (i.e., around an axis projecting out of the screen and perpendicular to it). But if the rotation occurred in the depth plane, one of the objects could be turned like a disk *in front of and behind* the screen (i.e., around a vertical axis in the screen). Either way, the larger the needed rotation, the longer the decision time. The rate of increase of the decision time was essentially the same in the two cases, as seen in the right panel of Figure 2.15. The bottom left panel is from a "different" trial.

This experiment had huge impact in cognitive psychology and neuroscience because it showed that it was possible to measure something that had not been measured before: the mental rotation of objects. Contributing to the impact of the result was the orderliness of the data, which have been replicated many times.

This experiment is a classic in the field of cognitive psychology, much as Donders's experiment was. Although Shepard did not get a stamp issued in his honor or have an institute named after him, he was awarded the National Medal of Science in 1995 by President Bill Clinton. The award was given to Shepard for his groundbreaking research on mental imagery and other topics.

Recording Brain Activity and Reductionism

Shepard's experiments revealed the inner dynamics of thought through an indirect measure, the time to determine whether two block structures matched. To an increasing degree, scientists and others have wanted to look more directly into the "black box" of the brain. The term "black box" comes from engineering and refers to processes whose inner workings are inaccessible to outside observers. For some investigators, it has been acceptable to rely on indirect measures like RTs to gain an understanding of the functional properties of the brain at work. For others, they have said that they prefer more direct measures. Among those in the latter, more-direct camp are some who declare they will only be content to rely on indirect measures like RTs after it has been found out what's really happening in the brain. Only by studying the brain itself, they say, can anything be truly understood about it.

I want to comment briefly on this view before turning to brain recordings per se because many students have this view as well. The idea that behavioral measures are acceptable only until one can open the brain's black box is understandable. Still, data from the brain itself aren't "more real" than data from the behavioral laboratory. Behavioral data are simply a different kind of data. The same rules of inference apply to both kinds of data. In addition, other comments can be informative.

First, to the extent that social recognition matters in the evaluation of both kinds of data, it's notable that the National Medal of Science was awarded to Roger Shepard, as noted above, and has been awarded to other behavioral researchers, too, as well as neuroscientists. Many investigators who have won Nobel Prizes and other top scientific prizes for their work have used both approaches simultaneously, so each kind of result can inform the other.

Second, arguing that brain data should be taken more seriously than behavioral data illustrates the view known as *reductionism*. This is the idea that, to understand something, you must break it down into its components. No one would question the value of componential analysis. If your car breaks down, you or a mechanic must determine which parts of the car need repair. If someone you know suffers a stroke—a sudden interruption of blood flow to some part of the brain—then of course that physical problem must be diagnosed and remedied by fixing the brain directly. On the other hand, refusing to take seriously large-scale functional measures is foolish. Many applications require large-scale functional analysis. Educators care about overall performance, for example; they can't spend years waiting for the details of the brain to be clarified before they move ahead with educational policies. Traffic engineers wanting to reduce accidents can't get stuck in gridlock, as it were, waiting for neuroscientists to figure out precisely what's happening in drivers' brains.

Third, the way the brain works depends on behavior itself. A brain may find itself, so to speak, in a speeding vehicle or in a quiet library. A brain may spend years on death row or in a monastery. What happens to the brain depends on the environment in which it finds itself, and that environment, in turn, is often a product of the behavior the brain itself produces. The brain depends on behavior no less than behavior depends on the brain.

Fourth and finally, if one adheres strongly to the reductionist agenda, it becomes difficult to defend whatever level of analysis one favors. Neuroscientists studying single neurons may be pleased with the progress they make until someone tells them they should stop looking at neurons and work out the chemistry of the system they're studying. But then, if they switch to chemistry, they may hear from some other supposed authority that they should actually switch to atomic physics. Having gone there, they might next be mocked for ignoring the mathematics their atomic structures actually reflect. Always trying to please others is an unhealthy approach to research (and life in general). No level of analysis is the best one or the only one; all are needed. Which one is best for you is a matter of your personal taste, ability, and—never to be forgotten—the practical problems you feel you need to address to improve the world as best you can.

Electroencephalograms

Recognizing that there are multiple paths to understanding the mind and brain in the control of action, consider measures of brain activity itself. The first measure is the electroencephalogram, or EEG.

As you may know, it is possible to place electrodes on the scalp and pick up electrical activity from the brain (Figure 2.16). The electrical activity is strong enough to pass through the *dura mater* (the tough membrane surrounding the brain) and then go even further, passing through the skull and scalp. The electrodes picking up brain activity needn't puncture the skin. Surface electrodes, as they're called, can sit on the scalp and pick up the electrical activity that emerges. The recordings, which typically

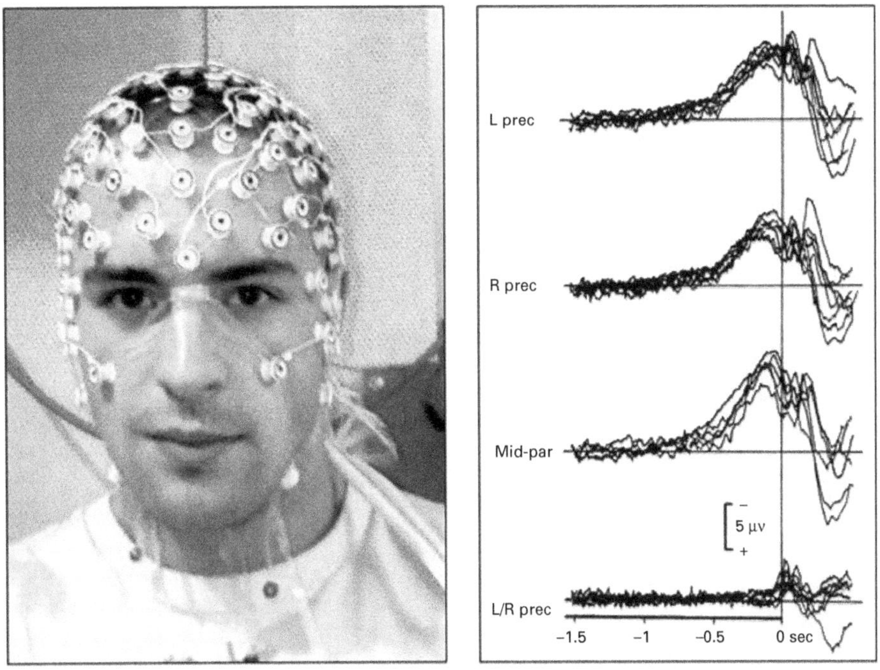

Figure 2.16
A subject wearing a cap with surface electrodes used for recording EEGs and (right panel) EEG waves showing growing readiness to move the right index finger at time 0 and the wave characteristics thereafter. The graphs show electrical activity starting 1.5 seconds before muscle activity in the right index finger, recorded with electromyography. The recordings come from the left precentral cortex, the right precentral cortex, and the mid-parietal cortex. The bottom graph shows the difference between the left and right precentral cortical activity. Each graph is the average of one subject's data tested eight times.

appear as continuously changing waves, have clinical as well as basic scientific importance.[17] Among the many clinical applications, EEGs are used to test the hearing abilities of newborn babies. For example, my granddaughter had her hearing checked a few hours after she was born. A technician attached an EEG cap to Greta's head and then played tones of various frequency to either ear over special headphones. The technician's smile suggested all was well as the resulting brainwaves were checked, and this was confirmed by the doctor. As this anecdote indicates, EEGs can be used to reveal what happens in the brain when other responses can't be relied on.

On the basic science front, EEGs have allowed for a multitude of discoveries. One is that most newborns exhibit greater specialization for language in the left hemisphere of the brain than in the right hemisphere. When speech is presented to neonatal subjects, the left hemisphere usually becomes more active than the right (Vannasing et al., 2016).

Another basic science finding is that EEGs help show what happens in the brain prior to the execution of voluntary actions. Shortly before voluntary movements are made, there is a distinct brainwave, called the readiness potential or *Bereitschaftspotential* (German for "readiness potential"). When this brainwave develops, muscle movement follows within a second and a half or so. Where readiness-potential waves arise in the brain provides clues about how the brain assembles actions (Deeke, Scheid, & Kornhuber, 1969).

A final discovery from EEG research to be mentioned here gives a sense of the enormous surprises in this field. In the relevant experiment, it was found that people had no awareness of the intention to act until *after* the act got started (Libet, Gleason, Wright, & Pearl, 1983). Participants watched a clock while having their EEGs taken. They pressed a key whenever they pleased, and after pressing the key, they indicated the clock time at which they thought they formed the intention to press the key. Curiously, the time they reported was *after* the brainwave appeared, not before.

This result might cause you to question a fundamental belief you have—that intentions lead to actions. The researcher behind this study, Benjamin Libet, of the University of California, San Francisco, questioned this commonsense assumption based on the result just mentioned. He reasoned that if one has an intention that leads to movement, then movement-related brainwaves should form *after* intentions become apparent in the brain, not *before*. Yet the order of events obtained in the experiment went the other way around. Could it be, Libet asked, that intentions are illusions? Could it be that, instead of forming intentions and then acting upon them, we act and then invent stories after the fact about why we did what we did? If you accept Libet's result and the reasoning just given, you might accept this surprising conclusion.

Positron Emission Tomography Scans

Other methods besides EEG exist for recording from the brain. Next consider brain imaging. One of the earliest technologies for brain imaging used positron emission tomography (PET). The outputs of such technology are called PET scans. The way PET scans work is that a radioactive isotope is injected into the bloodstream. For most people, the isotope is harmless. If you ever had a CT (computed tomography) scan, you may have been injected with an isotope similar or identical to the one used for PET scans. Any part of the body can be scanned once the isotope is injected. If the brain is the area of interest, the areas of the brain that are most active tend to "light up" in the scanner. This is because the more active areas absorb more of the radioactive isotope. Areas that are more active release more gamma rays, and these are picked up by the scanning device.

One of the first and most influential studies of PET scanning focused on motor imagery (Roland, Larsen, Lassen, & Skinhoj, 1980). People moved their fingers or just imagined moving their fingers while being PET scanned. As seen in Figure 2.17, when actual movements were made, two areas of the brain were especially active: the motor

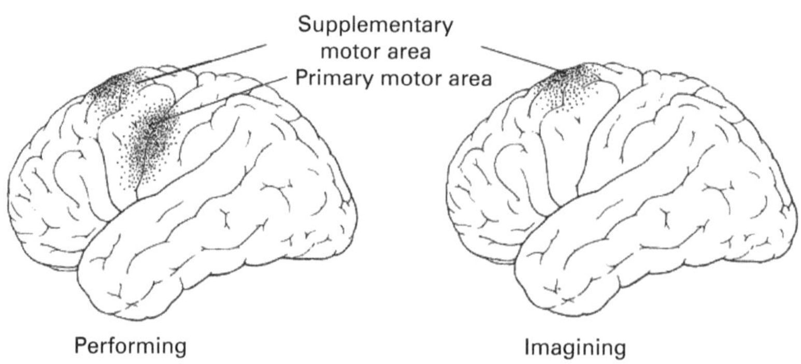

Figure 2.17
PET scan results of performing a movement sequence (left side) or imagining performing that same movement sequence (right side).

cortex and the supplementary motor cortex. By contrast, when the same people only imagined moving their fingers, just one of these areas "lit up": the supplementary motor cortex. From this result, it was suggested that the motor cortex is involved in moving the fingers while the supplementary motor cortex is involved in imagining the finger moving. Imagining moving the fingers wouldn't be expected to activate the part of the brain responsible for sending neural commands to the finger muscles, but actually moving the fingers would be expected to activate the motor-imagery *and* the muscle-activation regions. That story provides a reasonable account of the result.

Obtaining the dissociation between motor production and motor imagery caused tremendous excitement, for it showed the utility of PET scans for revealing functional specialization of different parts of the brain. In addition, it provided support for the general proposition that mental processes, including such elusive events as merely imagining movements of one's own body, could be scientifically studied and traced to identifiable sites in the brain. This conclusion complements the one given earlier based on the use of RTs to provide scientific evidence for mental rotation.

Functional Magnetic Resonance Imaging

Since the time of the PET study reviewed above, other methods have emerged. Among them is functional magnetic resonance imaging or fMRI. fMRI has better temporal and spatial resolution than PET but has lower temporal resolution than EEG. EEG, in turn, has lower spatial resolution than either PET or fMRI. fMRI does not rely on radioactive isotopes, but it picks up elevated activity in the parts of the brain that consume more blood-borne oxygen. The main measure in fMRI is a signal called BOLD, short for blood-oxygen-level dependent.

fMRI has largely replaced PET as a measure of functional activity in the brain. The reasons are twofold. First, radioactive injection is not required in fMRI, although it is required in PET. Second, spatial resolution is better in fMRI than in PET. There is a disadvantage of fMRI compared to PET, however. Temporal resolution isn't as good in fMRI, and movements, even small ones, made by the person in the fMRI scanner can cloud the fMRI signal, creating artifacts that can be tricky to remove. Identifying and taking out these artifacts is technically challenging, as is the task of picking up changes in activity levels in tiny portions of the brain and deciding how much extra activity per region counts as a statistically reliable change relative to background activity. Thousands of fMRI studies have been published since the development of the method in the early 1990s. Addressing the challenges just mentioned has forced researchers to sharpen their technical skills.

From the thousands of fMRI studies that have been published, I will report just one. It concerned a young woman who got into a terrible accident, leaving her totally paralyzed. By "totally paralyzed," I don't mean she was simply unable to move her arms and legs. She was unable to move any part of her body voluntarily. She couldn't speak. She couldn't move her eyes. She couldn't even blink, so she couldn't blink once for "yes" and twice for "no," for example. All she could do was lie there in her hospital bed. She was alive, her heart was beating, and her EEGs indicated that her brain was active, although it was impossible to tell from the EEGs whether she could hear or understand what was said to her or whether, indeed, she was aware of what was happening in her surroundings. There was distinct possibility she was "locked in"—fully conscious but unable to express herself.

Doctors attending to this young woman thought they might be able to find out whether she was locked in. After getting permission from the family, the physicians told the young woman that she would be put into a special machine called an fMRI scanner; that this was perfectly safe; that she would be hearing a loud, rhythmic noise while in the scanner, for that is how the scanner works; and that she needn't be afraid. Then they rolled her into the scanner, and once she was in place, the doctors told her to imagine playing tennis, which was something she had done before. They also told her to imagine walking through her house, which she had also done previously (Owen et al., 2006).[18]

When the doctors later examined her fMRIs in these two conditions (Figure 2.18), they found that different parts of her brain were activated depending on what she was asked to imagine herself doing. When she was asked to imagine playing tennis, her supplementary motor cortex lit up, but when she was asked to imagine walking through her house, a different part of her brain, the parietal cortex, lit up.

Was this outcome just a consequence of different sounds that happened to enter the young woman's ears when she was told what to imagine? To check this, the doctors tested age-matched neurotypical individuals in the same two imagery conditions. Those other individuals were also invited to image playing tennis and imagine touring their homes. Their fMRI results looked like the patient's (Figure 2.18).

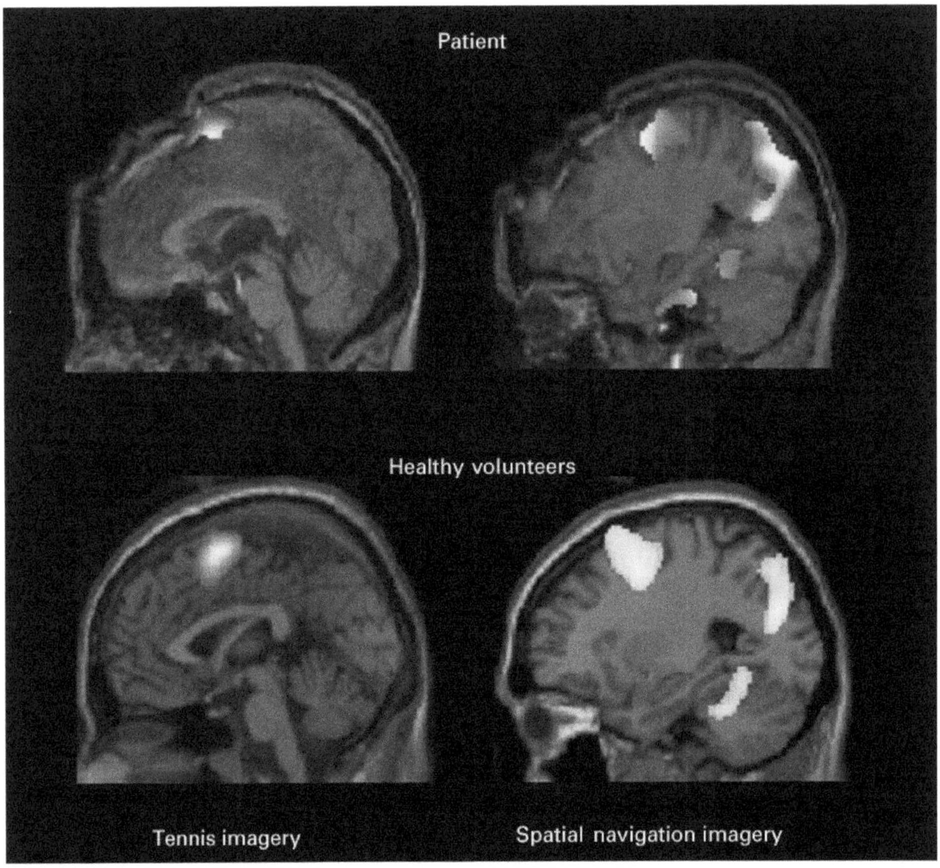

Figure 2.18
fMRI patterns of a patient in a vegetative state (top panels) and neurotypical controls (bottom panels) engaged in tennis imagery and spatial navigation imagery.

The implication was clear but chilling. The patient could understand what was said to her, and she could engage in different mental routines based on the instruction. The fact that her fMRIs were, for all intents and purposes, indistinguishable from the fMRIs of the control subjects meant that she understood the instructions and responded to them as her neurologically normal counterparts did. The young woman was mentally intact but unable to move. She was locked in.

How does this result make you feel? If you are like me and many of my students, you may find it unsettling. A number of students have told me they found it so. I have told them that I feel the same way. On the other hand, the finding may offer a ray of sunshine in an otherwise bleak scenario. Rather than seeing this patient and others like her as a "vegetable," these individuals can instead be viewed as fully cognizant human beings who can inspire the development of improved medical services. I have said to the students whom I've had the privilege to teach that I encourage them to become part of this vanguard.[19]

Magnetoencephalography, Optogenetics, and Diffusion Tensor Imaging

In addition to the methods of recording from the brain discussed above, others have appeared as well. One, magnetoencephalography (MEG), is a technique for picking up electrical currents in the brain using sensitive magnetometers. MEG has better temporal resolution than fMRI and is roughly on par with EEG when it comes to spatial resolution. Among its other uses, MEG, as shown in Figure 2.19, has been used to study infant brain activity in language learning (Ramirez & Kuhl, 2017; Zhang et al., 2009).

Another new technology is diffusion tensor imaging (DTI). This method addresses a limitation of fMRI not mentioned before. In addition to the low temporal resolution of fMRI and the vulnerability of fMRI to movement artifacts, fMRI does not reveal connections between brain regions. It shows how active a brain region is depending on what the brain is doing at the time of the scan—hence the term "*functional* magnetic resonance imaging" (fMRI)—but it doesn't show links, or strengths of links, between remote brain sites.

Seeing the strength of connections in the brain makes it possible to judge the extent to which brain regions communicate. In this regard, an expectation from neuroscience

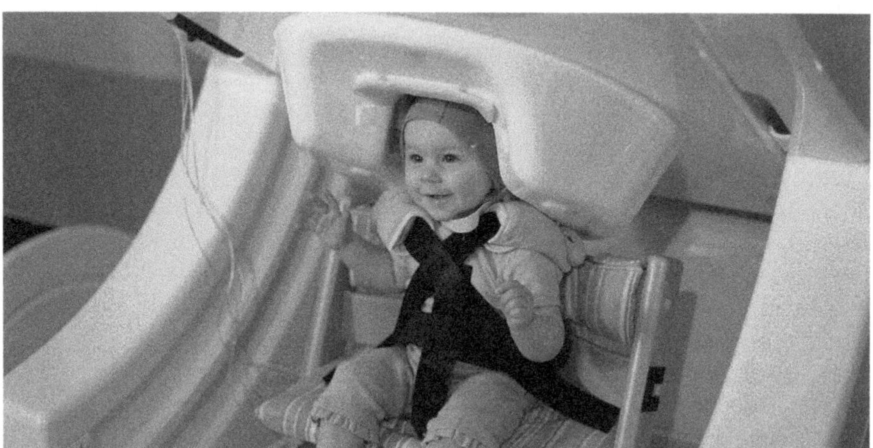

Figure 2.19
A baby in a MEG machine.

is that regions that often coactivate should develop strong connections. For example, if your hands and eyes work together for hours on end when you write, then strong bonds should form between your brain's hand-control and eye-control centers.

A catchy phrase captures this idea: "Neurons that fire together wire together." The sentence reflects a principle introduced in the 1940s by a Canadian psychologist named Donald Hebb, who suggested that the more often two interconnected neurons fire simultaneously, the stronger the connection between them. If two regions are weakly connected early in practice, Hebb (1949) proposed, they should get more strongly connected as practice continues.[20]

It would be helpful to have a brain-imaging technique that can reveal strengths of connections between brain sites. Such information would be useful for basic science purposes and also for clinical purposes. On the clinical side, for a case like that of the young woman described in the last section, it would help to know which brain connections were impacted by her accident. It's likely that connections in her brain that normally enable the organization or transmission of motor commands (neural signals to her muscles) were disrupted, whereas connections within her brain that permit comprehension of speech and the capacity for motor imagery remained intact.

In a moment, I'll tell you about a brain-imaging technique that shows connection strengths, but first I want to turn from the young woman in the hospital bed to a younger woman—or actually a twelve-year-old girl—sitting at a piano. The picture for you to look at (Figure 2.20) comes from an article in *Science* magazine, one of the top scientific publications in the world. The article I'm referring to was about the research of this twelve-year-old girl, or rather her grown-up self (though still young, in her thirties at the time the article appeared). When she was twelve, this girl was living in a household where her family's religion required her to wear a bonnet. Twenty years later, at the age of thirty-two, she donned a different kind a crown, receiving a MacArthur "genius" award.

Why did *Science* magazine show this particular photograph? The reason was to give the magazine's readers a sense of where this important scientist came from. To an increasing degree, publications like *Science* and its parent organization, the American Association for the Advancement of Science (AAAS or "triple A S" as it's called), seek to humanize STEM fields to open them up to people who have been less well represented in its ranks. I reproduce the photo here for the same reason, to help you see what an eminent scientist once looked like. My aim is to inspire you to see yourself as someone who can also make useful contributions. As I say to my students in every class I teach, "If not you, who?" I repeat that phrase often in my teaching, where I add another phrase as well: "Yoo-hoo! If not you, who?" I actually *sing* that phrase, or call it out as if to people at the other end of an open field, and I repeat the call here in print. Encouraging you is arguably more important than transmitting facts.

Who, then, is the girl in the photo? Her name is Danielle Bassett. She has done groundbreaking research with diffusion tensor imaging. Dr. Basset, a professor at the University of Pennsylvania at the time of this writing, did not invent DTI but was attracted to what it offered. She and others with whom she worked knew that DTI reveals strengths of connections between parts of the brain. Seeing those connection strengths, she reasoned, could be used to test Hebb's hypothesis.

The technical details of how DTI works aren't important for now. What *is* important is that DTI provides images of the brain that are unlike any seen before. An example appears in Figure 2.21, where you can see white matter (myelin) neural connectivity in the brain. Myelin is material that coats neural axons and allows for rapid transmission of signals from one neuron to another. More will be said about it in Chapter 3.

Figure 2.20
Danielle Bassett at age twelve.

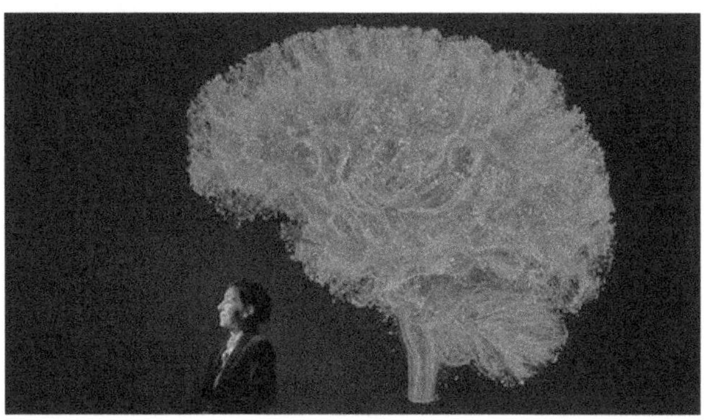

Figure 2.21
Danielle Bassett and a DTI image from her research.

Bassett and her coworkers asked how DTI images like the one shown in Figure 2.21 may change over time. In a 2011 article (Bassett et al., 2011), she and others focused on practice-related DTI changes. They had participants perform twelve-note piano-keyboard sequences in each of three sessions while having their brains scanned. Performers who improved more on the task as they went from one training session to the next were found to have greater flexibility in the connections formed in their brains. Consistent with Hebb's principle, the more the neurons wired together, the better the improvement in performance.

Several lessons can be learned here. First, new methods afford new insights. Second, the study of action can reveal much about brain function. Third, women no less than men can do cutting-edge research. (Even saying this feels chauvinistic. There should be no need to say such a thing.) Fourth, no one should look at a kid and say, based on their looks, "That kid can't grow up to do something important." Fifth and most important, look at yourself in a mirror and ask yourself, "If not you, who?"[21]

3 Bones, Muscles, Nerves

When I teach about action, mind, and brain and I want my students to start thinking about how they move and stabilize their bodies, I often ask them, "Who here has eaten muscle?" Typically, only a few hands go up. I pause a bit to see if the students get what I'm driving at. When it seems like the class has reached a point of collective puzzlement, I ask if anyone has eaten a hamburger. More hands rise. "And who here has had chicken?" I continue. Still more hands go up. My questioning goes on with a couple more examples: "Turkey? Ham?" Most of the students start to nod as they get the point. When I restate my first question, "OK, who has eaten muscle?" lots of students now raise their hands. More than once when I have gone through this exercise, students have cried out, "Oh my God! Meat is muscle!"

I share with the students that I didn't always know this either, that I reasoned it out for myself when I first came to understand that animals don't need meat for survival. They need muscles to move and to hold steady. "Oh my God!" I remember saying to myself when I figured this out. "Meat is muscle!"

Much as many of us don't know immediately that meat is muscle, most of us can't say much about our own muscles, bones, or nerves—or, for that matter, other structures in our bodies that relate to action: arteries and veins, other internal organs, sweat glands, and so on.

This chapter is about bones, muscles, and nerves and also about tendons and ligaments. Other physical components of the body are important as well, but the chapter will center on the structures just listed because they're the ones most commonly associated with bodily actions.[1]

As I prepared to write this chapter, I wasn't sure at first whether to discuss bones, muscles, ligaments, and tendons. I had no doubt about including nerves, but books about cognitive neuroscience don't typically cover the skeleton, muscles, tendons, and ligaments, so most students in the field aren't taught that tendons connect muscles to bones and that ligaments hold bones together. This is somewhat surprising given that many professors who consider themselves cognitive neuroscientists embrace the idea of embodiment, the notion that much of what we know is body based, as discussed in Chapter 1. It's ironic that so little is taught about the body itself when embodiment is embraced.

Several items in the news convinced me that bones, muscles, tendons, and ligaments are worth discussing. One article that caught my eye was the finding that during human fetal development, lizard-like muscles appear and then disappear in the thumb (Figure 3.1). This was reminiscent of the fact that human fetuses have gills. These fishy structures appear and then disappear in the human embryo, a result consistent with, and often used to justify, the adage *ontogeny recapitulates phylogeny:* individual development (ontogeny) replays species development (phylogeny). The appearance of temporary lizard muscles

Figure 3.1
Human hand muscles with lizard origins (left panel) and a modern-day lizard (right panel).

in developing human hands is a new source of evidence for this hypothesis. Beyond this, and of greatest surprise, the scientists doing this work suggested that the mobility of the human thumb—one of our most prized possessions, often cited as the most important anatomical bases for manual tool use—may be due to this vestige of evolution.[2]

Another article I found fascinating was a piece with the intriguing title "Ever Wonder What Dinosaur Meat Tasted Like? Try Eating This Bird" (Lauren Kent, CNN, October 4, 2019, https://www.cnn.com/2019/10/04/world/dinosaur-meat-taste-scn/index.html). The article described a study in which investigators studying bird development turned off the gene responsible for the formation of the birds' beaks. Birds evolved from dinosaurs, and in the course of evolution, the proto-birds (not-yet birds) lost their teeth and grew beaks. In the research, it was found that bird embryos with the altered genes developed teeth rather than beaks, consistent with the hypothesis that beaks became possible because of a genetic change from a toothier past. The research wasn't actually about the taste of dinosaur meat, but the journalist who wrote the piece succeeded in capturing my attention by posing the question about the taste of dinosaur meat. Nice job, journalist!

So what did dinosaur meat taste like? According to the article, dinosaur meat tasted like the meat of birds, which makes sense given the common ancestry of birds and dinosaurs. According to the article, dinosaur meat probably tasted like turkey meat. Turkey meat is dark meat, whereas chicken meat is both dark (legs and thighs) and light (breasts). It turns out that dark (red) muscle is specialized for sustained weight bearing, whereas light (white) muscle is specialized for rapid force generation. The dark meat of chicken legs supports the long-duration standing that chickens do, whereas the white meat of chicken breasts supports the intense forces that chickens need when they flap their wings to flee foxes or other predators. It turns out that long-distance runners have more dark muscle per cubic centimeter than sprinters do. Sprinters need to produce rapid bursts of intense energy, whereas long-distance runners need to go long (Costill et al., 1976).

Bones

What most people can say about bones is based on images they have of skeletons in haunted houses or experiences they've had with fractures. Few people who take psychology or cognitive-neuroscience courses have studied the skeleton in detail. The human body has 270 bones at birth, but this number shrinks to 206 over the life span. The fusion of bones takes a long time. Forensic anthropologists can estimate age of death well into adulthood based on the extent of bone fusion in the skull and other body parts.[3]

Another fact about bones that is little known is that bones are alive. They are made of living tissue that grows and reshapes. You can appreciate this fact upon reflection, for you know that bones lengthen and strengthen as you grow and that bones heal when they're fractured. Therefore, bones are dynamic, living structures. In fact, living bones are pale pink, not white as in haunted houses or in doctors' offices. The pinkness of living bones is understandable considering what's inside them—marrow, the source of blood.

Bones are hard enough to support the body but not so hard that they can't change shape based on usage. Such changes (Figure 3.2) have been demonstrated in studies of professional tennis players, in whom it has been found that forearm bones in the arm used to hold the racquet are significantly larger than in the other arm (Jokl, 1981). The difference between the two arms' sizes follows years of play. People who are destined to play tennis for years on end who hold the racquet with one hand rather than the other are not born with a Herculean arm on one side and a pip-squeak arm on the other. The difference comes from Nurture, not Nature.

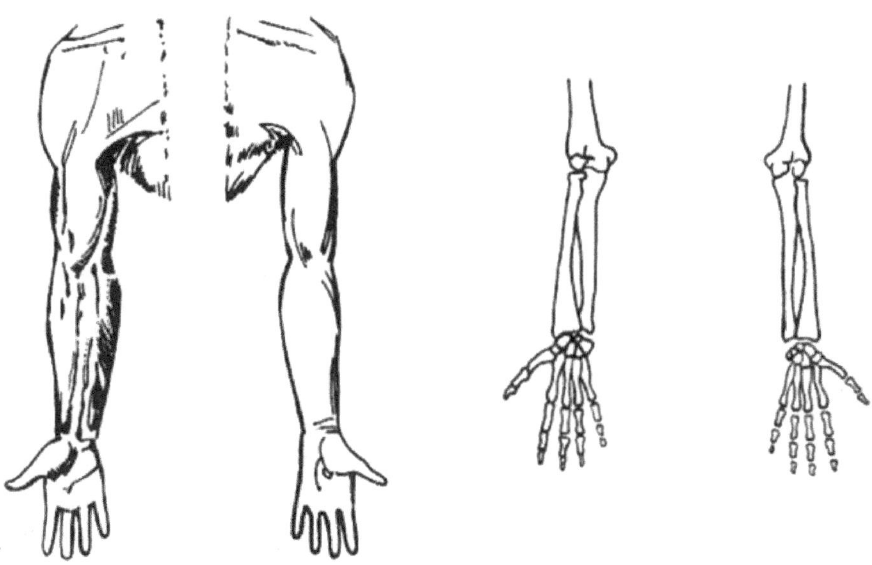

A B

Figure 3.2
Right and left arms of a Davis Cup tennis champion.

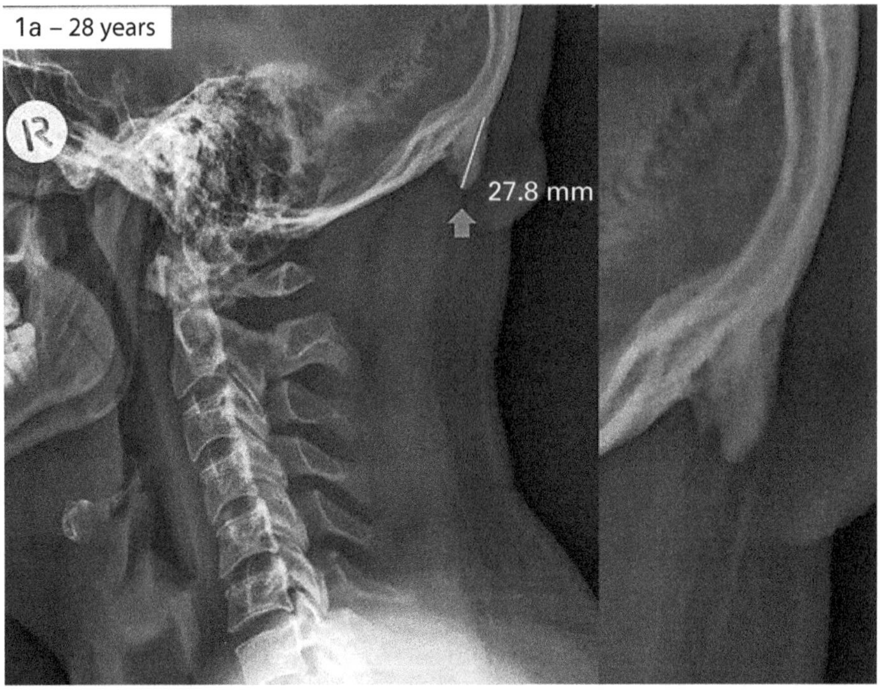

Figure 3.3
Bony bump on the back of someone's skull, with an arrow and white line highlighting the bump in the left panel, and the protrusion magnified in the right panel.

Another example of dynamic skeletal changes is one that got quite a bit of news coverage at the time this chapter was being prepared, although in the end, the matter turned out to be a will-o'-the-wisp (a fleeting illusion). To the surprise of many, it was claimed that young people who spend hours each day looking down at their cellphones develop bony protrusions on the backs of their skulls (Figure 3.3). So widely reported was the result that on more than one occasion, I was asked by friends and relatives who know about my interests whether I was aware of it and what I thought about it. I confessed I didn't know much and thanked them for asking. I said I'd get back to them when I learned more.

Getting back to them now via this book, here's what I found. In my subsequent reading, I learned that the claim was almost certainly wrong. Looking down for long periods doesn't cause bone spurs to grow on the back of one's skull, as confirmed by a noted authority on skeletal growth who was interviewed by the *New York Times* (Grady, 2019). Apparently, the individual who reported the connection between cellphone use and skull spurs failed to take into account that the spurs seen on X-rays were there *before* the patients started using their phones. The fact that a reputable media outlet, the *New York Times*, reported the finding and then retracted it (or said it should be retracted) is a good thing, for it shows that we as a society can be open to learning surprising things but should also be capable of disavowing strong claims once they've been scrutinized. A more general lesson is that skepticism is advisable, especially in connection with highly surprising results.

Another controversy concerning human bones swirled around possible differences between the lengths of the index finger and ring finger (Figure 3.4). *Science*, the

The long and short of it

John Manning and colleagues' 1998 study comparing the lenghts of index (2D) and ring (4D) fingers found a subtle difference between the sexes, which they and others attributed to hormones affecting early fetal development.

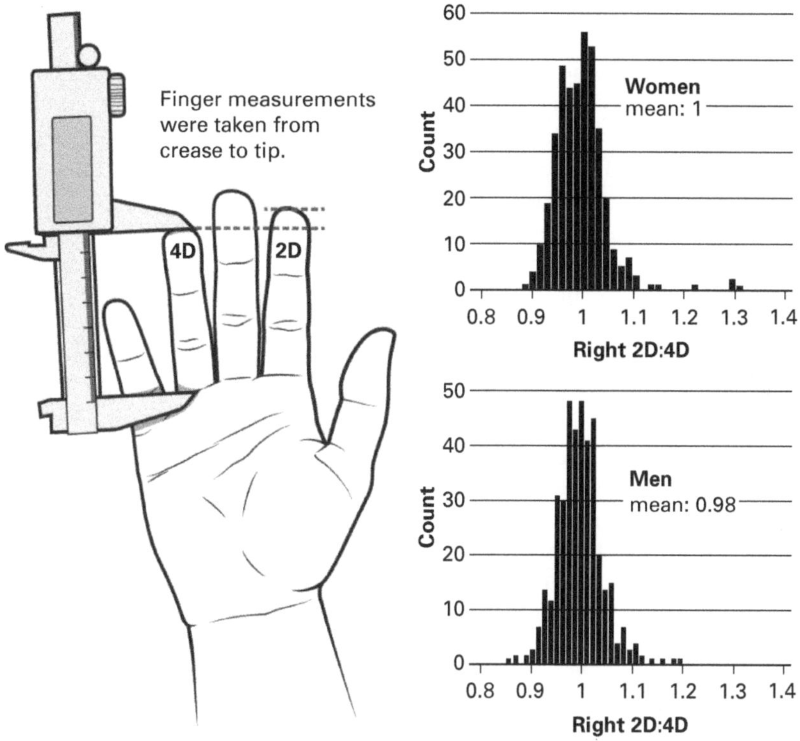

Figure 3.4
Number of observations of the ratios of the length of the second digit to the fourth in women and men. Each bar is the number of cases within a range of ratios.

high-profile magazine to which I referred earlier in connection with Danielle Bassett (Figures 2.20 and 2.21), devoted three full pages to this matter in an article published in 2019 (Leslie, 2019). The article addressed the hypothesis that the ratio of the lengths of the second digit (the index finger) to the fourth digit (the ring finger) may differ in men and women. Not only that, but it may also differ in homosexuals versus heterosexuals, between people on the autism spectrum and those who are not, and among people with different personality traits.

Why would anyone care if the 2D:4D ratio (the ratio of the length of the second digit to the length of the fourth digit) depends on the individual differences just listed? The interest stems from the possibility that if the ratio does differ, it is likely to be genetic, which could suggest a genetic basis for the more important individual differences listed above.

Consider the argument. Suppose there is a genetic basis for the 2D:4D ratio, and suppose, for the sake of argument, that the 2D:4D ratio is indeed different for homosexuals and heterosexuals, for people on the autism spectrum and people not, and for people with different personality traits. If all those suppositions are true, then a startling conclusion is possible: those individual differences have a genetic basis. This inference, if supported, could have profound implications. No longer would the differences be ascribed to environmental factors. Rather, people with the characteristics would be, or at least could be, said to be born that way. That possibility is what aroused the interest in the otherwise arcane topic of finger lengths.

To test the hypothesis that finger length differences are based on genetics, it's sensible to check for clear evidence of the difference in males versus females because the genetic basis for gender differences is clear. At least in individuals who are unambiguously male or female at birth, boys have X and Y chromosomes, whereas girls have X chromosomes only. Therefore, if 2D:4D ratios differ in males and females, there might be some basis for attributing the ratio difference to genes.[4]

What do the data say? The answer is not at all straightforward. The differences that have been found are extremely small. It has taken an army of statisticians to sort out whether the minute differences are real or not, and although at least one researcher has declared that he is convinced the difference is genuine, others are unsure.

I'm in the latter camp. I confess that I prefer to take a wait-and-see stance, wanting stronger evidence to be convinced of a real difference. We as a society should be hesitant to jump to conclusions about a possible hereditary basis for individual differences because such a conclusion could have unfortunate consequences, as occurred with eugenics, the practice of bettering stock—human or otherwise—by encouraging the transmission of desired genes and discouraging the transmission of undesired genes. Eugenics is one of the worst policies ever to beset humankind, as is all too clear from the Holocaust in World War II and other genocides. I raise this issue coming back to the Nature-or-Nurture question, which began with Plato, as discussed in Chapter 1. The discussion of Plato and the issues he raised was premeditated to set the groundwork for issues like the one just presented.

Five Fingers, Five Toes

There is another bone-related matter that we know little about: why exactly do we have five fingers per hand and five toes per foot? The answer, surprisingly, is that no one knows. Students are often stupefied by this fact. "You're kidding," they exclaim when I tell them this. Sometimes I see them glancing down at their hands, trying to tell the reason for the number of digits they see. In so doing, they seem to engage in a small rebellious act, which I appreciate. After hearing that the problem hasn't been solved, they inspect their own fingers, apparently thinking, "Well, *he* may say no one has solved it, but I bet *I* can figure it out!" I hope you can!

More background information may inspire you further. When I first learned that the basis for five fingers and toes is unknown, I thought, "Oh come on! This can't be so hard to figure out!" But I had no idea how to move forward on the problem. Soon after that, I had the good fortune to spend a week visiting Professor R. McNeil Alexander of the University of Leeds in England. Professor Alexander was one of the leading biomechanists in the world and also the director, at the time, of the London Zoo. At some point during my stay, I asked Professor Alexander if he had any idea why we have five fingers and five toes. I thought if anyone had figured this out, he had or would know who had. He smiled when I asked him; clearly, he had considered it before. In his deep English accent, he said simply, "I don't know." And that was the end of it.[5]

I have so much respect for Dr. Alexander, who passed away in 2016, that I feel that if *he* couldn't solve the five-digit problem, no one would be able to. I certainly haven't solved the problem myself but will share some facts that may help you think about it. The matter is interesting for a host of reasons, not least because our decimal system is based on our having ten fingers. The way we count is based on our anatomy. It would be nice to know why we count as we do.[6]

Begin by noting that certain anatomical arrangements have caught on in evolution and others have not. For example, all mammals have four extremities: four legs or, for us humans, two legs and two arms. There are no mammals with six extremities. The only ones that do are fictional: angels, Pegasus the flying horse, flying monkeys dispatched by the Wicked Witch in the Land of Oz, and others presumably. My reason for referring to these fanciful creatures is to point out that Nature favors economical solutions. Having *four* limbs is economical; having *six* limbs is not. Mammals that can fly—bats are the most common example—have two feet and two wings, as do birds. Whatever advantage might accrue to having six extremities is outweighed by the costs of having that many limbs. The extra potential costs might pertain to the physical weight of the extra extremities, the demands of nourishing six appendages rather than four, or, conceivably, the extra neural challenge of controlling still more limbs. I doubt the neural challenge is important, however, because plenty of animals have fine control of their tails, and elephants control not only their legs and tails but also their trunks with exquisite precision.[7]

No matter what the reason for four limbs, four is clearly the number of extremities that is "just right" for mammals, birds, amphibia, and reptiles. All animals that evolved after fish came along—amphibia, birds, reptiles, and mammals—are, in fact, members of a vast biological superclass called *Tetrapoda*, meaning animals with four limbs. The one apparent exception, snakes, had four legs but lost them (Held, 2014).

If four is the "right number" of limbs per body, why is *five* the "right number" of digits per limb? I put "right number" in quotes because, when speaking about evolution, there is no prescriptive right or wrong. The whole point of Darwin's theory of evolution is that there is no aim or purpose. There is just happenstance. Things happen to work out or not as measured by reproductive success. Five digits may be special because organisms with that many digits happen to have been more reproductively successful for whatever reason. What that reason is isn't clear, but having *more* than five digits per extremity seems to provide no extra benefit or possibly incurs cost. There are no tetrapods with more than five digits per limb. The one apparent exception is the giant panda, which appears, upon casual inspection, to have an opposable thumb. But giant pandas don't have true thumbs. They have five fingers and a bony protrusion that helps them grasp bamboo shoots, which is what pandas feed on. This bony protrusion—the *sesamoid* bone—enlarged over evolutionary time as pandas grasped bamboo shoots to eat them. Pandas that could hold onto the bamboo shoots had more offspring than pandas that couldn't. The story of the panda's "thumb" is a classic tale in the theory of evolution, illustrating the principle of *preadaptation*—how a structure or tendency that is present in latent form turns out to be useful when new challenges arise (Gould, 1980).

Returning to the question of why we have five digits, that question may not be the right one. A better question may be why don't we have *more* than five digits, for no tetrapod does. One way to progress on this problem is to ask when five-fingeredness emerged. As I learned from a *Scientific American* article, the five-digit limit took hold about 340 million years ago, when, according to the author, the fossil record shows that wrist and ankle joints came on the scene. Before then, there were some tetrapods with more than five digits per pod, but afterward, creatures with six or more digits disappeared (Coates, 2005). Why was that? Perhaps six digits weighed too much relative to whatever benefits they provided. Five was enough. In our case, being creatures with opposable thumbs, we can grasp tools, including wielding clubs for striking and grabbing stones for throwing. Being able to make and use tools is a great capability. Without that capability, we wouldn't be who we are. On the other hand, contrary to the view that only humans can use tools, that notion was disproved long ago when Jane Goodall observed chimpanzees inserting sticks into holes to fish out termites, a favorite chimp snack (Goodall, 1964). More recently, crows have been shown to insert wires into tubes to fish out mealworms for eating and even to bend the wires to turn them into usable hooks (Weir, Chappell, & Kacelnik, 2002). Tool use doesn't depend on having opposable thumbs, therefore, although having opposable thumbs is handy.

Ligaments

Bones support the body, but they couldn't do so if they weren't connected to one another. Ligaments serve this purpose. Without ligaments, we'd be bags of bones.

There's a ligament you've probably heard of: the anterior cruciate ligament or ACL. The ACL connects the lower leg bone (the tibia) to the upper leg bone (the femur) within the knee and behind the knee cap or patella. The ACL is anterior to ("in front of") the posterior cruciate ligament, and these two ligaments form a cross—hence the term "cruciate," which means "cross" (as in "crucifix").

You or someone you know may have suffered from a torn ACL. Your ACL might tear if you were subjected to a sudden, forceful twisting motion of your lower leg relative to your upper leg, as in landing the wrong way in basketball, turning too quickly in hockey, or taking a wrong step on a hiking trail. Health professionals who help patients with torn ACLs include physical therapists, orthopedists, and orthopedic surgeons. As helpful as these health professionals are, no less helpful—or maybe even more helpful—are those who help prevent such injuries in the first place. You can help reduce the chance of ACL or other injuries in yourself and others by being mindful of your surroundings and avoiding unnecessary risks.

Joints and Their Geometry

Ligaments tie bones together around joints. Different joints allow for movement about different numbers of axes (Figure 3.5). Some joints allow for motion about *one* axis. Examples are the hinge joint of the elbow and the pivot joint between the first and second cervical vertebrae of the spine. Other joints allow for motion about *two* axes. Examples are the plane joint of the foot's tarsal bones, the saddle joint of the wrist, and the condyloid joint of the wrist. Still other joints allow for motion about *three* axes. Examples are the ball-and-socket joint of the hip and the ball-and-socket joint of the shoulder. The ball-and-socket joint of the shoulder allows for movement of the upper arm bone (the humerus) within the clavicle (or, more specifically, the glenoid cavity of the clavicle). You can easily note the three axes of upper-arm movement for yourself by raising and lowering your upper arm, by bringing your upper arm toward or away from your body midline, and by twisting your upper arm.

The number of axes about which a limb segment can move corresponds to the skeletal degrees of freedom of the limb segment in that joint. A central problem in motor control is to understand how the degrees of freedom of the body are managed. That problem will be the focus of Chapter 8 and centers on the fact that there are typically more degrees of freedom in the body than in the description of the task to be

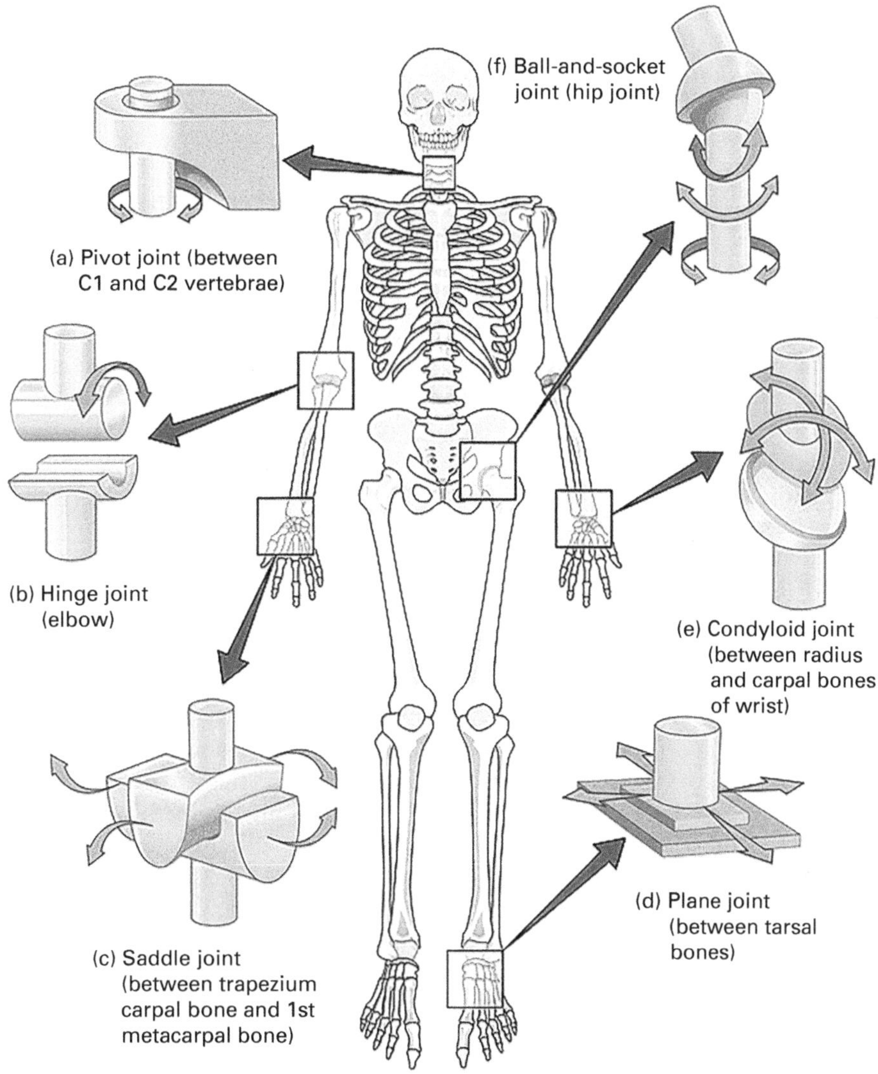

Figure 3.5
Skeletal joints. The shoulder's ball-and-socket joint happens not to be labeled here.

done. For example, if you extend your right arm and draw a rainbow, you'll probably change your hand's twist angle as you trace the arc. Your palm will probably be flat and downward at the peak of the arc and will go vertical at the bottoms. This pattern is reminiscent of Donders's law for the eye, discussed in Chapter 2.[8]

Joints and Joint Pain

A last comment about joints concerns joint pain. The most common kind of joint pain comes from arthritis, a disorder of the joints characterized by pain, stiffness, and reduced range of motion. Chances you know a senior citizen with arthritis (a grandparent, say), although younger people can get the disease as well.

There are two main kinds of arthritis: rheumatoid arthritis, which typically attacks the joints of the hands and wrists, although other parts of the body can be affected as well, and osteoarthritis, which tends to affect more parts of the body than rheumatoid arthritis does and tends to attack the cartilage and bones. Rheumatoid arthritis is an autoimmune disease, whereas osteoarthritis is a result of mechanical stress. Each year, millions of dollars are lost due to arthritis when people miss work or have difficulty getting their work done because of the pain or limited movement they suffer. Curing arthritis remains an immensely important medical challenge. Cognitive neuroscientists don't generally speak about arthritis and similar diseases, but the way the brain copes with arthritic pain and pain of other kinds is a topic worthy of investigation, not just for its clinical relevance but also for whatever basic science insights might be gained.

Arthritis isn't the only cause of achy joints. On the way to learning about other causes of joint point, look again at the joints shown in Figure 3.5. All of these joints are *synovial* joints—joints that have cartilage and contain a lubricant called synovial fluid. (You've probably seen cartilage on the ends of chicken bones.) Synovial fluid is encased in a surrounding membrane called the synovial membrane. If the synovial membrane becomes inflamed, the condition is known as synovitis.

Synovitis is associated with rheumatoid arthritis as well as *gout*, another type of arthritis. Quoting from a website, "Gout is a form of inflammatory arthritis that develops in some people who have high levels of uric acid in the blood. The acid can form needle-like crystals in a joint and cause sudden, severe episodes of pain, tenderness, redness, warmth and swelling."[9]

Joint pain also comes from *lupus*. Quoting from another website, "Lupus is a systemic autoimmune disease that occurs when your body's immune system attacks your own tissues and organs. Inflammation caused by lupus can affect many different body systems—including your joints, skin, kidneys, blood cells, brain, heart and lungs. Lupus can be difficult to diagnose because its signs and symptoms often mimic those of other ailments. The most distinctive sign of lupus—a facial rash that resembles the wings of a butterfly unfolding across both cheeks—occurs in many but not all cases."[10]

Notice that the suffix "-itis" keeps coming up here, as in *synovitis* and *arthritis*. You've probably heard of other "itises" as well: tonsillitis (inflammation of the tonsils), gastroenteritis (inflammation of the gastrointestinal tract), bronchitis (inflammation of the bronchi in the lungs), sinusitis (inflammation of the sinuses), and so on. Inflammation is the body's way of responding to pain and is often treated with steroids, which are often given to alleviate inflammatory pain. Steroids can work for a while but can have bad consequences if used too long. Finding a better way to treat inflammation is an important priority for medical research.

Tendons

When muscles move bones, they do so by tugging on the bones via connective tissue called tendons. You've probably heard of at least one tendon, the Achilles tendon (Figure 3.6). This term comes from a figure in Greek mythology. I'll say a little about the figure to draw another link between the material in this book and culture at large.

The tendon with the Greek myth connection is named after the great warrior, Achilles (Figure 3.7). Achilles could not be killed by arrows penetrating any part of this body except for his one weak spot, the back of his leg above his heel. Ironically, but perhaps inevitably given the nature of Greek mythology, Achilles was killed by an arrow that landed on just that spot. Today, if you speak of someone's Achilles' heel, you refer

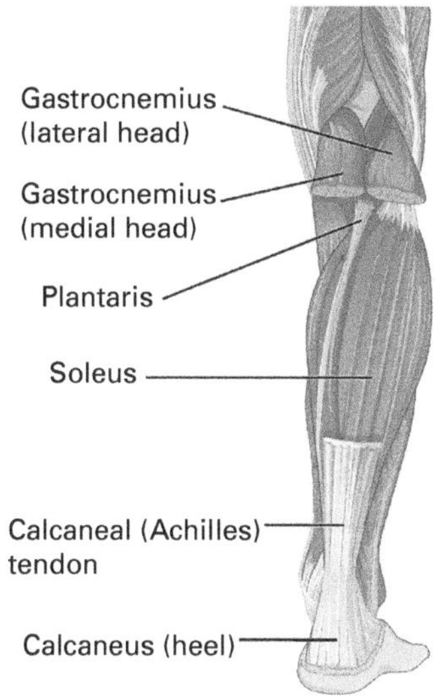

Figure 3.6
The Achilles tendon and other structures in the leg.

to their weak spot. In sports, in political campaigns, in legal suits, and in other contests, one side may say they're going for the other side's Achilles heel. You may refer to your own greatest vulnerability as your Achilles heel.

In case you're wondering why the Achilles heel tendon (the calcaneal tendon) (see Figure 3.6) was especially vulnerable, it got that way because, according to Greek myth, when Achilles was a baby, his mother held him by his heel when she bathed him (Figure 3.7, left panel). The water protected the rest of him, but because that one particular area above his foot wasn't dipped in the holy water, that spot remained vulnerable. So famous was this event that the great Renaissance painter, Peter Paul Rubens, memorialized it in a painting (shown in the left panel of Figure 3.7), *Thetis Dipping the Infant Achilles into the River Styx*. The painting was completed in 1625 and is displayed in a museum in Rotterdam. Another work of art, a statue on the Greek isle of Corfu (Figure 3.7, right panel), memorialized Achilles as a fallen warrior attempting to save his own life by dislodging the arrow that landed in—of all places—his one weak spot.

Whereas the Achilles tendon may have been a source of vulnerability for a warrior of Greek mythology, analogous tendons in humans and some nonhuman animals confer an advantage rather than liability. Tendons in people, kangaroos, and other animals provide a distinct mechanical advantage for hopping and jumping owing to the elastic properties of those connective tissues. The Achilles tendon may have made the Greek warrior vulnerable, but it also helped him crouch down and jump high, which no doubt contributed to his prowess on the battlefield. More will be said about elasticity in Chapter 8.

Muscles

We come now to muscles. These structures let you twist and shout, rock and roll, sit and stand. They let you breathe, swallow, blink, and move your eyes. Your heart, which

Figure 3.7
Achilles being held by his heel as his mother dips him in holy water (left panel) and his vain attempt as a warrior (right panel) to save his own life by trying to withdraw an arrow in his one vulnerable spot.

beats a billion times over a healthy human lifetime, has its own kind of muscle, so-called cardiac muscle, which is distinct from skeletal muscle, the kind that moves bones. Skeletal muscle has fibers that run parallel from one bone to another. Cardiac muscle fibers crisscross and are activated by the autonomic nervous system (part of the peripheral nervous system). Skeletal muscles, by contrast, are activated by the central nervous system (CNS).

A familiar quote in the study of muscle anatomy (muscle structure) and muscle physiology (muscle function) is that no engineer would ever design a machine with the crazy layouts that muscles have. The tongue-in-cheek meaning is that it is hard to imagine a more brilliantly rigged system than the muscle system. The arrangement of muscles may be dizzyingly complex, but clearly it works in wondrous ways.

Figures 3.8 and 3.9 show the main muscles of the human body. Looking at these figures, you may recognize some of the names of the muscles you see. The *gluteus maximus* is a familiar term for many (and a source of blushing for kids in schoolyards). The *deltoid* and *pectoralis major* often come up in connection with weightlifting. Psychologists interested in the control of facial expression often learn the names of the muscles used for smiling, frowning, scowling, and other facial movements (Ekman & Friesen, 1975).

Perhaps the most familiar muscles are the *biceps* and *triceps*. When you "make a muscle" by flexing your elbow, the bulge you see is your biceps. The biceps has two "heads," which is why biceps has the prefix "bi," meaning two. The two heads connect to different places on the scapula, also known as the shoulder blade.

The *triceps*, as its name implies, has *three* heads rather than two. The long head, medial head, and lateral head of the triceps attach to various parts of the humerus (the upper arm bone) near the shoulder and meet at the other end on a single tendon that connects to the ulna (one of the two bones of the forearm, the other being the radius).

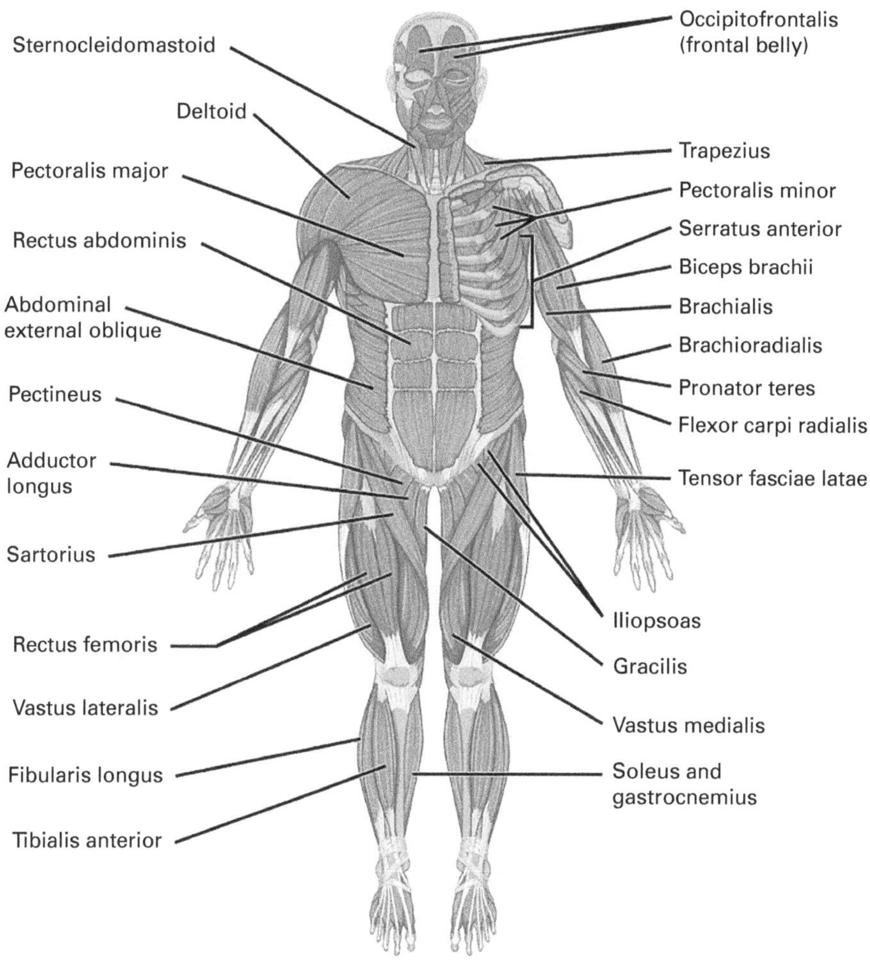

Figure 3.8
The muscles of the human body (anterior view).

How Are Muscles Controlled?

Muscles support each other, serving as agonists, and oppose each other, serving as antagonists. The triceps and biceps muscles illustrate this point. When the biceps muscle contracts, favoring flexion of the elbow (reducing the elbow angle, so the forearm moves closer to the upper arm), the triceps muscle must stretch for flexion to occur. Similarly, when the triceps muscle contracts, favoring extension of the elbow (increasing the elbow angle, so the forearm moves farther from the upper arm), the biceps muscle must stretch for extension to occur. How are these two simple maneuvers achieved?

There is no neural command to a muscle to make it stretch. The only way to get a muscle to stretch (to lengthen) is with neural commands to its *antagonist*, telling that opposing muscle to contract. The biceps–triceps pair illustrates this point clearly, although the same principle applies to all muscle sets. When the biceps contracts, the elbow flexes—provided the triceps relaxes (generates less force). And when the triceps contracts, the elbow extends—provided the biceps relaxes (generates less force).

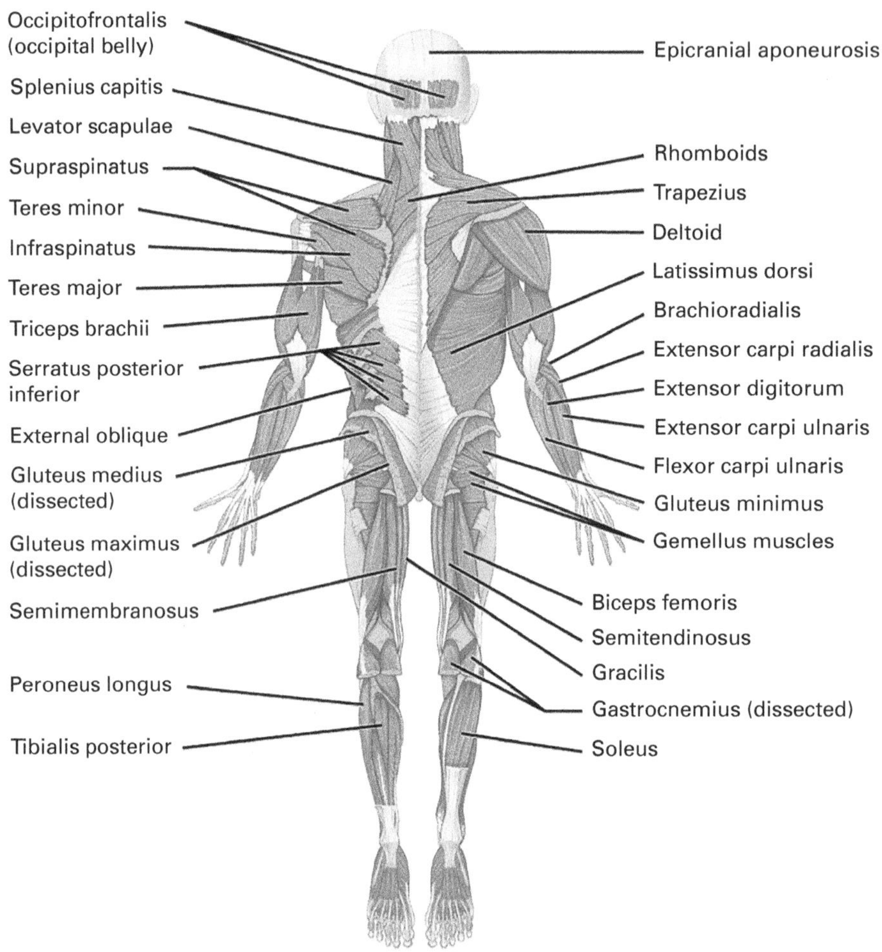

Figure 3.9
The muscles of the human body (posterior view).

Co-contraction is also possible, however, meaning that when one muscle contracts, the opposing muscle may do so as well. Full relaxation of one muscle in the pair isn't necessary. When both muscles are active and work together in the service of some purpose, they can be said to act as agonists.

What does co-contraction achieve? It achieves joint stiffening. Imagine a firefighter standing on the ground beside a burning building ready to catch a child or baby about to drop from a high window or ladder. Standing there expectantly, the firefighter may stiffen their elbows to catch the kid. Stiffening the joint comes from co-contraction of the muscle antagonists. Changing the degree of co-contraction of the muscles to the same degree allows the limb to stay in the same position while the opposing forces increase (or decrease). The limb can also be brought to a new position by changing the opposing muscle forces to different degrees. You may not know these principles at an articulate level, but you have been using the knowledge your whole life.

Are Muscles Balloons?

In the last section, I mentioned that muscles contract, but I didn't say how that fact came to be known. For a long time, it was thought that muscles act like balloons, shortening when inflated and lengthening when deflated (Figure 3.10). This notion originated in the third century b.c.e., when a Greek anatomist named Erasistratus proposed that muscles shorten by filling with fluid. According to Erasistratus, the fluid comes from nerves, causing the muscles to bulge and fill with fluid. When muscles relax, Erasistratus supposed, the fluid returns to the nerves; the relaxed muscles lengthen and narrow (McMahon, 1984).

This idea was widely accepted for hundreds of years. For example, René Descartes accepted the concept, as shown in Figure 3.11, where he sketched his own view of what happens when someone gets too close to a fire: fluid travels up to the brain via the spinal cord and then back down the spinal cord to the muscles, causing withdrawal from the flame.

The balloon theory of muscle function was exploded by two seventeenth-century investigators. A British physician, Francis Glisson (1599–1677), put an excised frog muscle into a tube of water and stimulated the muscle via the nerve connected to it. Glisson observed that the muscle contracted but the water did not rise as it should have if the muscle contraction were due to fluid entering the muscle. A Dutch contemporary, Jan Swammerdam (1637–1680), did something similar. He put a frog's muscle into a water-filled tube and stimulated the nerve running to the muscle. Like Glisson, he observed that the muscle contracted when the nerve was squeezed but did not see a bead rise up in a pipette extending from the tube. The bead should have risen

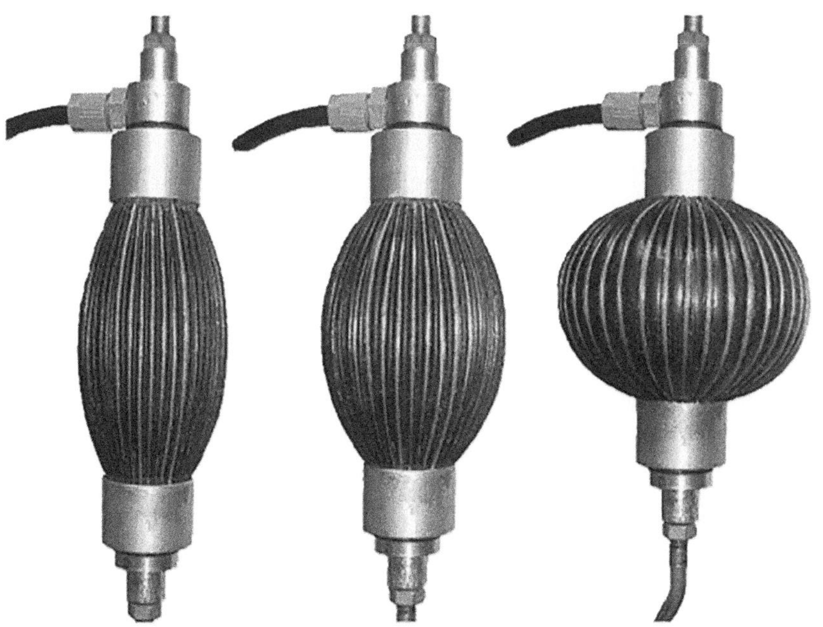

Figure 3.10
A pneumatic (balloon-like) actuator that lengthens when emptied of fluid (left panel) and contracts when filled with fluid (right panel).

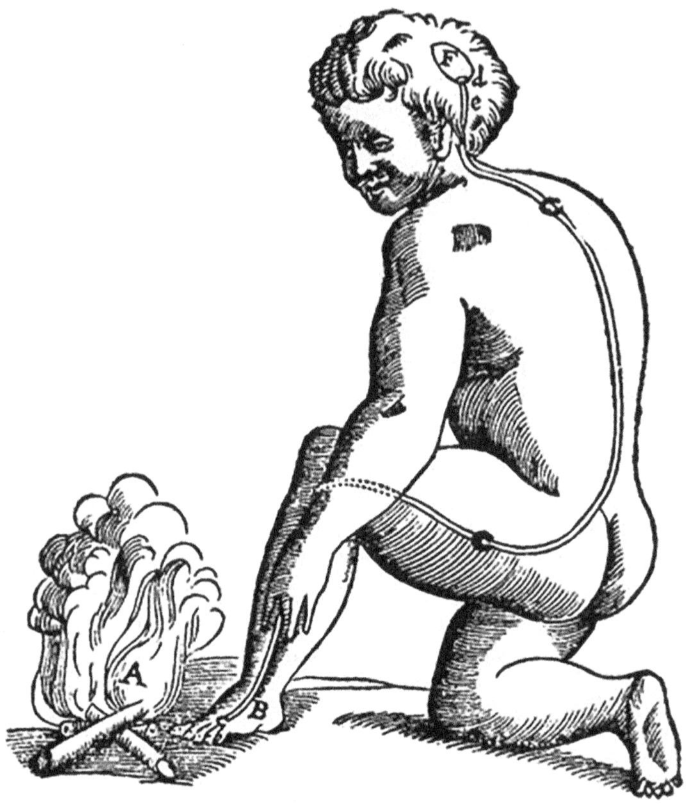

Figure 3.11
Descartes's conception of the reflex.

if the muscle contraction had been caused by the muscle filling with fluid. Glisson and Swammerdam rejected Erasistratus's hypothesis and, by implication, Descartes's (McMahon, 1984).

Muscle Cross-Bridges

Neither Glisson nor Swammerdam could say how muscles worked. That remained for later research. What did that later research show? How, indeed, do muscles work?

One hint is to note that some limb positions afford greater power than others. Consider this quote (Rothwell, 1987, p. 24) and the picture related to it (Figure 3.12).

> As every hero in a gangster movie knows, the way to make the villain release his grip on the gun is to flex the wrist forcibly to reduce the power of the finger flexors. The gun will thereupon fall dramatically to the floor, to be kicked neatly away by the hero's foot.

In this example, the wrist extends maximally and the muscles responsible for flexing the wrist stretch as much as they can.

The reason the gun-wielding villain finds himself or herself in a weak position becomes clear by viewing muscle fibers with a high-power microscope (Figure 3.13). Tiny

Figure 3.12
Disarming a gun-wielding opponent by taking advantage of the length–tension property of muscle: the more a muscle is stretched or contracted, the less force (tension) it can generate.

micro-bridges made of actin–myosin fibrils overlap entirely, not at all, or in between. At extreme positions—when the micro-bridges overlap entirely or not at all—the filaments can't grab hold onto each other; the resulting force is small. Greatest resistance is observed at the middle of the range of motion, when the filaments have the most give and take.

It may seem odd to talk about such microscopic features of the body in a book about cognitive neuroscience, but in our daily lives, we learn to anticipate the power capabilities of our limbs in different positions. Police officers are taught the disarming tactic shown in Figure 3.12, and ordinary people learn the underlying principle as well, as shown by the fact they (we) grab hold of objects in ways that afford most power when most power is needed (Winters & Kleweno, 1993). For example, when people reach out to grasp and rotate objects, the grasp postures they adopt when taking hold of the objects anticipate the power requirements of the grasp (Rosenbaum, van Heugten, & Caldwell, 1996). Hence, a fact about physical action capabilities that is rooted in microscopic features of biophysics comes to be known and exploited in everyday action control.

Nerves

A mystery still attends the story unfolding here. You have read that nerves don't disburse fluid into muscles or receive fluids from muscles. You have also read that muscles rely on tiny filaments that slide past each other and grab hold of each other with tensions that depend on the number of filaments that can be grasped. Something important has been omitted, however: How are muscles controlled? What causes muscles to contract and be stretched? The answer, as you almost certainly know, is through the action of nerves. But what is the nature of those nerve actions? How do we know that nerves affect muscles in the way they do?

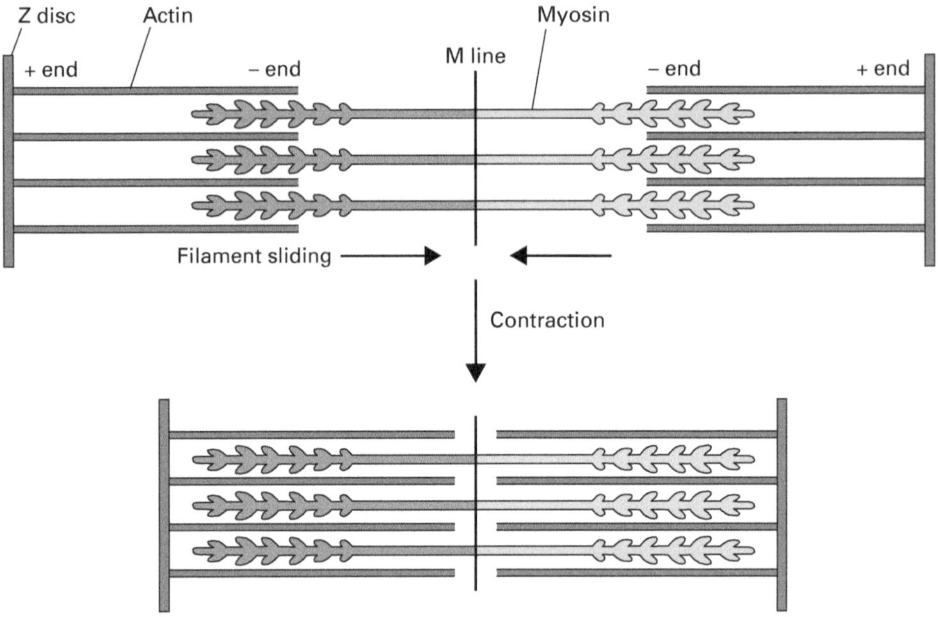

Figure 3.13
Less muscle contraction with less overlap of actin and myosin muscle filaments (top) and more muscle contraction with more overlap of actin and myosin muscle filaments (bottom).

To approach this question, it's interesting, and maybe a bit surprising, to consider, of all people, Benjamin Franklin, one of the signers of the U.S. Declaration of Independence, the first postmaster of the United States, and the founder of what would become the University of Pennsylvania, among other distinctions. The distinction of Franklin that is relevant here was his advancement of the understanding of electricity, a phenomenon people were enchanted by at the time (the 1700s). Chemical batteries had been refined in that period, and those batteries helped stoke people's fascination with static electricity—as in in rubbing a comb on a sweater and then seeing one's hair stand on end when the comb is brought near the scalp. People imagined that electricity embodies two different kinds of fluids, so-called vitreous fluid and resinous fluid. But Franklin doubted this idea. He proposed that electricity is just one kind of fluid whose constituents have positive and negative charges. Those opposing charges, Franklin surmised, balance out or cause flow if a potential difference is created by a battery or other generator. The concept of "plus" and "minus," visible on batteries to this day, comes from Franklin, as does the principle of conservation of charge: "The total electric charge in an isolated system never changes."[11]

Franklin's best-known research on electricity is his fabled experiment with a kite (Figure 3.14). Through this experiment, whose origin is unclear, Franklin allegedly showed that lightning is electricity.[12] The demonstration was achieved by attaching a key to a kite and taking advantage of the fact that the kite string, when wet during a thunderstorm, would conduct electricity down to the ground if the key were struck by lightning. Being careful to wear insulating material around his hands while holding the kite string and standing on an insulating surface during the storm, Franklin attached the ground end of the string to a Leyden jar, which could store whatever electric charge might descend

Figure 3.14
Benjamin Franklin pulling electricity from the sky in his mythic kite experiment.

from the sky. When lightning did strike, a huge surge of electricity charged the Leyden jar battery. Therefore, lightning was electricity, as Franklin surmised.

Based on this discovery, Franklin invented something so practical and common that it seems timeless: the lightning rod. Virtually every building in the world has one. A metal pole atop the building has a wire running down into the ground. If lightning strikes, the charge enters the ground and in so doing spares the building. Countless houses and buildings have been spared thanks to Franklin's invention.[13]

Franklin did his pioneering work on electricity in the 1740s and 1750s. A few years later, in 1780, two Italian investigators, Lucia Galeazzi Galvani and Luigi Galvani, demonstrated that electricity comes from nerves. The Galvanis applied electrical current to the legs of recently sacrificed frogs (Figure 3.15). Upon delivering the electricity, they saw that the frogs' legs contracted, and upon turning the electricity off, they saw that the legs relaxed.

This experiment is often called one of the most beautiful experiments in the history of science (Johnson, 2008). But in this connection, it is interesting to note that

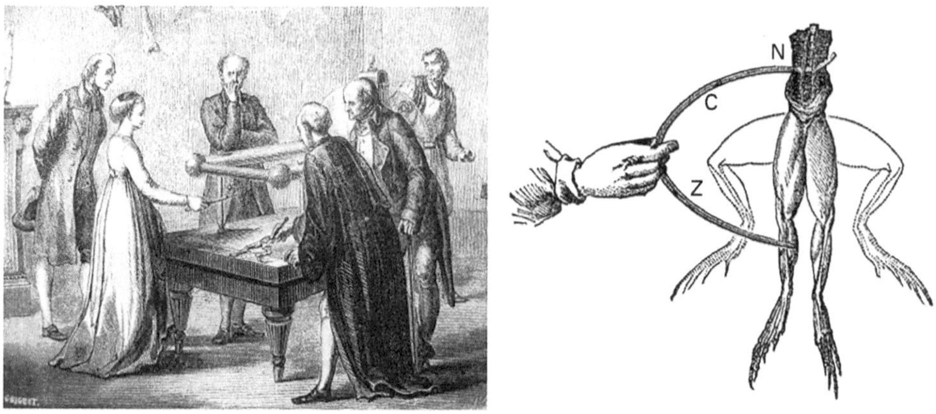

Figure 3.15
Lucia Galvani applying electrical current to a frog's legs (left panel) and closeup of the legs (right panel) in their resting state (hanging down) and after current was applied (lifted).

Luigi Galvani is typically credited with discovering bioelectricity. However, in preparing this text, I learned from Wikipedia that Galvani's wife, Lucia Galeazzi Galvani, "encouraged his independent research, and served as a counsellor, collaborator and guide for his experiments until her death. Due to the conventions of the time she wasn't credited for any scientific work she may have done in the lab. She grew up with science and her father was a prominent member of the Bologna Academy of Science." The same article said that "she additionally edited her husband's medical texts."[14] Based on this information, and in recognition of other cases where women scientists were shortchanged (e.g., Rosalind Franklin, who, arguably, codiscovered the helical structure of DNA along with Watkins and Crick), I have written this section in a way that acknowledges Lucia Galvani's role.[15] The picture of the application of electricity to muscle shown here (Figure 3.15) shows Lucia rather than Luigi providing the critical spark, perhaps in more ways than one.

Neural Gaps and Myelin

When a frog's muscles contract under normal circumstances, the stimulus to the muscle doesn't come from a wire controlled by human experimenters. Rather, it comes from nerves inside the frog's body. The same goes for other animals, too, of course, including us.

What happens when nerves excite muscles? The nerve fibers (Figure 3.16) connect to muscles at gaps called neuromuscular junctions.[16] Electricity doesn't pass from the nerve fiber to the muscle. Instead, a chemical is released from the end of the neuron into the neuromuscular junction.

Chemical neurotransmitters don't only cross neuromuscular junctions. They also cross gaps between nerve cells. These gaps are called *synapses*. The parts of neurons responsible for responding to neurotransmitters arriving via synapses are called *dendrites*.

A neuron tends to release a chemical neurotransmitter if it sends an excitatory burst of energy, called an *action potential*, from the neuron's cell body or soma to the end of a branch of the neuron called the *axon*. When the action potential reaches the end of the axon or *axon terminal*, it can cause the release of the chemical transmitter from the axon's end (the so-called terminal bud).

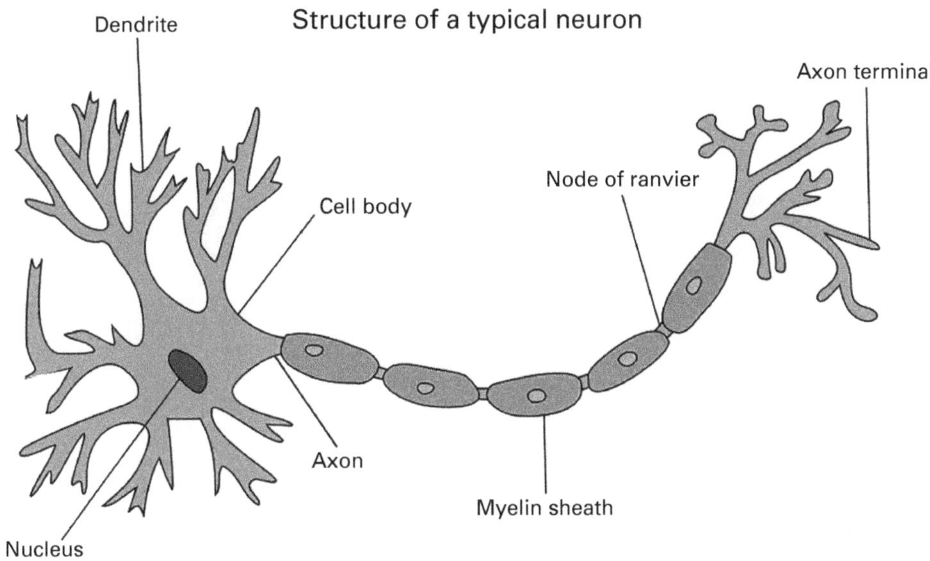

Figure 3.16
A typical neuron.

If the axon is coated with a fatty substance called *myelin,* the axon can travel more quickly. This is because the myelin is composed of small sausage-shaped bodies with gaps between them called *nodes of Ranvier.* Action potentials can "jump" from one node of Ranvier to the next, covering the length of the axon more quickly if myelin is present than if it is not.

Axons become sheathed in myelin over a number of years, commencing in infancy and continuing into puberty. In a disease called multiple sclerosis (MS, pronounced "Em Ess"), the immune system attacks the myelin, causing it to break down in multiple places. Those afflicted with MS tend to succumb in early adulthood and into middle age. As MS progresses, the patient finds it harder and harder to move in a coordinated fashion. Over time, there can be mental deterioration and sometimes, but certainly not always, death due to the disease.

One of the most famous MS patients of all time was Jacqueline du Pré, widely considered to be one of the greatest cellists ever (Figure 3.17). Tragically, she died of MS when she was only forty-two years of age.[17]

Motor Neurons, Motor Units, and the Final Common Pathway

Not all synapses are equal in terms of the strengths of the connections they provide. For some synapses, there is a high probability of a reliable postsynaptic response to a presynaptic input; those synapses are said to have strong connections. For other synapses, there is a lower probability of a reliable postsynaptic response to a presynaptic input; those synapses are said to have weak connections. Whether a connection is strong or weak depends on a host of factors, but the one of greatest interest here is experience. The more often a postsynaptic neuron fires when a presynaptic neuron does, the greater the strength of connection between them. This principle was discussed near the

Figure 3.17
Jacqueline du Pré.

end of the last chapter, in connection with DTI brain imaging. There, as you may recall, you encountered the phrase "neurons that fire together wire together."

Neurons can have strong connections even though the effect of one neuron on the other is inhibitory rather than excitatory. Inhibition is no less important than excitation for successful neural functioning. Whether one neuron excites or inhibits another depends on the interaction between the neurotransmitter emitted by the presynaptic neuron and the receptor properties of the postsynaptic neuron. The interaction may be such that transmission of a neurotransmitter from neuron A to neuron B excites neuron B or inhibits neuron B.

Because some synaptic connections are stronger than others, stronger links exist between some neurons than others, and that is true even if the neurons are widely separated. Similarly, there are stronger links between some neurons and muscles than others. The strongest neuromotor connections are ones in which neurons synapse directly onto muscles. There is a special name for such neurons. They are called *motor neurons*. Motor neurons exist in the spinal cord (Figure 3.18), where they occupy the ventral (belly) side of the spinal cord. Some motor neurons also occupy the brainstem, where they innervate

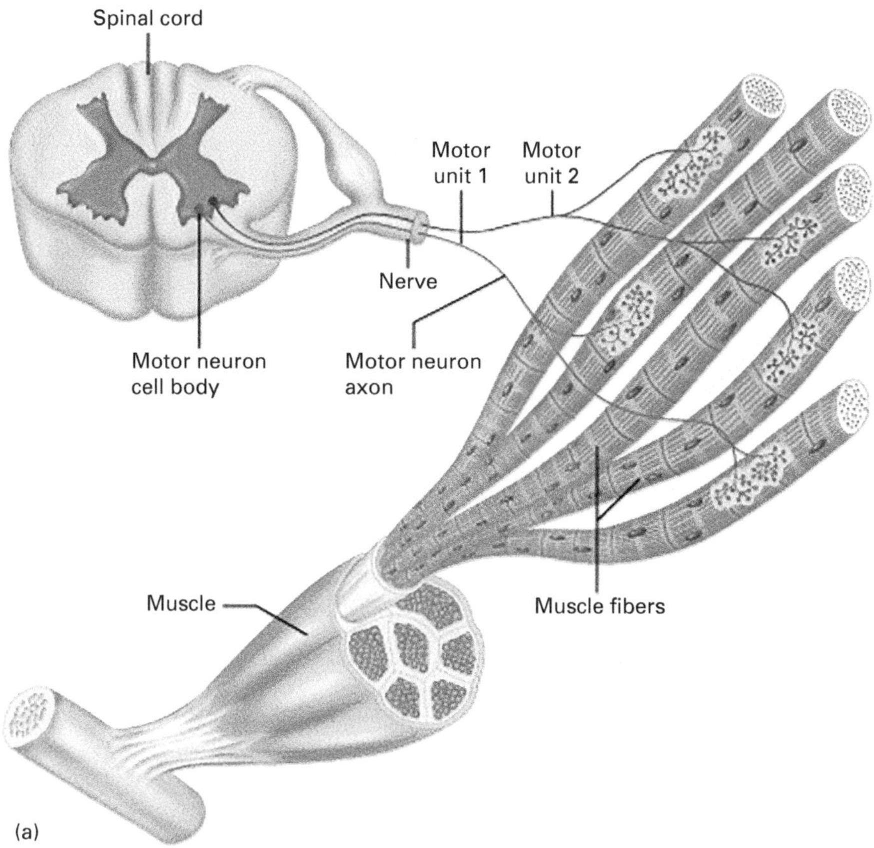

Figure 3.18
Motor units.

cranial nerves affecting muscles in the face and neck, as well as the torso. Motor neurons in the spinal cord only innervate muscles in the torso and limbs.[18]

There is a special term for a motor neuron and the muscle fibers it monosynaptically innervates. This ensemble is called a *motor unit* (Figure 3.18). A motor unit is hierarchical. Within a motor unit, a motor neuron innervates muscle fibers within one muscle, and those muscle fibers are innervated only by that one motor neuron. Because one motor neuron can activate multiple muscle fibers, and because every one of those muscle fibers is innervated by just that one motor neuron, the relation is strictly hierarchical.

Whereas motor units use monosynaptic connections, most connections in the nervous system are polysynaptic. Any given neuron is likely to be influenced by other neurons through multiple synapses. Similarly, any given neuron is likely to influence other neurons via a series of multiple synapses.

Because neurons connect with many other neurons (the theme of the last paragraph) but every muscle fiber is innervated by its own dedicated motor neuron, that one motor neuron can be viewed as the final common pathway to that muscle fiber.[19]

What a motor neuron does depends on everything that happened prior to its activation. Much as every instant of a radio program, television show, or movie is the culmination of a long, complicated process involving teams of writers, producers, directors, actors, and others, what you do at any given time is the culmination of a long, complicated process that brings your muscles to the state they're in at the time. This is one of the reasons why the study of action is so appealing: what you do at any instant is the culmination of everything that has happened to you up to that time. Indeed, at the risk of being overdramatic, your lifting your little finger or carrying out any mundane act is the culmination of everything that has happened to you, your species, and indeed the entire universe, up to that moment.

Motor Cortex

In 1870 (so ninety years after the Galvanis demonstrated that electrical stimulation of frogs' legs caused the frogs' legs to tense up), two German researchers named Gustav Fritsch and Eduard Hitzig did an analogous experiment in the brain. Fritsch and Hitzig's method was bold in two ways. First, they applied electrical stimulation to a species more like humans than frogs, namely, dogs. Second, they applied electrical stimulation to a site far removed from the muscles themselves (the brain), so if there were an effect of the electrical stimulation on the muscles, that would mean there were connections between the stimulated brain site and the muscles. At issue, then, would be the relation between the stimulation and the observed response. How specific or nonspecific would the effect be? Would parts of the brain that might classically be assumed to have nothing to do with physical action actually trigger muscle movements?

To carry out the experiment, Fritsch and Hitzig opened the skull of a dog and put an electrode in a part of the brain that was readily accessible, the cerebral cortex (Figure 3.19).[20] In anatomical terms, the site that Fritsch and Hitzig stimulated was in the *superior*

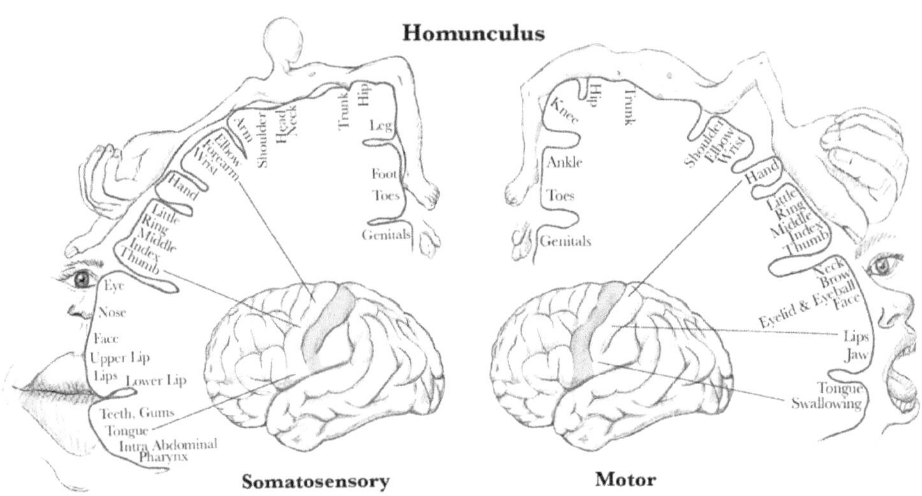

Figure 3.19
Overview of the human cerebral cortex highlighting the somastosensory cortex, located caudal to ("behind") the central sulcus, and the motor cortex, located rostral to ("in front of") the central sulcus. The cartoons show the amount of space (roughly the number of neurons) devoted to different parts of the body for touch (somatosensory cortex) and for muscle activation (motor cortex or, more specifically, primary motor cortex).

part of the brain (the top) rather than the *inferior* part (the bottom) and was more *rostral* (toward the front) than *caudal* (toward the back). When Fritsch and Hitzig applied current to the part of the brain of interest, a section that is rostral to a large "canyon" in the cerebral cortex—the *central sulcus*—the result was dramatic. As soon as an electrical impulse was delivered, the dog's leg twitched. Each time the stimulation was applied, the same thing happened.

Another outcome was equally important. The muscles that twitched depended on where the stimulation was applied. Shifting the position of the stimulating electrode by a small amount caused a different muscle to twitch. Moreover, the muscle that twitched when the electrode was moved by a small amount tended to be right next to the muscle that twitched before. Neighboring brain sites therefore controlled neighboring muscles.

This result was extremely important, for it showed that the investigators were not simply generating a diffuse electrical storm in the brain, eliciting movements anywhere and everywhere in the dog's body. Instead, because stimulation of particular sites led to twitches in particular muscles, it appeared that this part of the brain played a central role in motor control. If one had thought the cerebral cortex was just for thinking, one had to think again.

What Fritsch and Hitzig found was so influential that this part of the brain came to be called the *motor cortex* (Figure 3.19). The term "motor" means "related to movement." The motor of a car moves the car, the motor of a fan drives the fan, and so on. The phrase "motor control" is used in industrial engineering and related fields to refer to the management of mechanical motors. In areas like neuroscience, psychology, and kinesiology, "motor control" refers to the control of the body for movement and stability.

Calling a part of the brain the "motor cortex" reflects the conviction that this part of the brain is tightly linked to the control of muscles. The term was introduced soon after Fritsch and Hitzig made their discovery and it stuck. It's wise to remember, nonetheless, that if some part of the brain happens to act in conjunction with some aspect of experience or behavior, that doesn't mean that part of the brain exists *for* that function. Saying it does is to commit the *teleological* fallacy—equating a mechanism's effects with its purpose. (*Telos* is the Greek word for "purpose.")

If dogs have a motor cortex, does that mean humans do too? Thinking about this from a Darwinian perspective, you would be justified to say, "Yes, probably, because what's good for the goose is good for the gander." In other words, what worked for one species would be likely to work for others. Adaptive features are likely to be conserved in species facing similar selective pressures. Because dogs and people have a common ancestor, it's likely that our common ancestor had a motor cortex or some analogous structure. That structure or its genetic substrate was presumably passed down from generation to generation. Analogous cortical structures do in fact exist across all mammalian brains.

Two classic studies made the point that the kind of functional organization found for canine motor cortices was also found for human motor cortices. These studies justify showing a picture of a human brain in Figure 3.19, although the study referred at that point was carried out in the brain of a dog.

The first classic study entailed astute clinical observations by a neurologist. It beautifully illustrates naturalistic observation (see Chapter 2), albeit in a clinical context. In the mid-1800s, a British neurologist named John Hughlings Jackson observed spreading motor seizures in epileptic patients (York & Steinberg, 2011). Epilepsy, as you may know, is a disease in which there is a kind of electrical storm that spreads through the brain. When this occurs, the affected individual is said to have a seizure. In the case of spreading motor seizures, the seizure takes the form of muscle jerks, beginning in a limited area of the body and spreading further to other parts of the body.

Based on observations of spreading motor seizures, Jackson suggested that adjacent parts of the brain control adjacent muscles. He reasoned that if muscle jerks spread, the storm in the brain causing the jerks probably spreads through neighboring sites in the brain. Fritsch and Hitzig's electrophysiological study showed that this hypothesis was correct, at least in dogs.[21]

The second study pertaining to the functional organization of the human motor cortex came from a neurosurgeon attempting to help a (human) patient who, like the patients observed by Jackson, had epilepsy. The neurosurgeon was named Wilder Penfield. His fame became so great based on what he did (described below) that a major thoroughfare is named after him in the city where he did his pioneering work. The thoroughfare is called Doctor Penfield Boulevard. It is one of the main traffic arteries of Montreal, Quebec, Canada.

Penfield sought to alleviate spreading epileptic storms in a young man for whom epilepsy had become debilitating (Penfield & Jasper, 1954). At the time, in the 1950s, it was difficult to stem the tide of epilepsy with drugs; pharmacological agents available now were not yet available. Consequently, Penfield decided to perform surgery on the young man's brain. The idea was to create a kind of fire lane to stop the spread of the "forest fire" that sometimes ran rampant through the young man's cerebral cortex. Creating a fire lane would entail destroying neurons, and it was critical that neurons critical for the patient's basic functioning not be destroyed. Therefore, during the surgical procedure, the patient was awake. (It turns out, interestingly, that inserting a needle into the brain does not cause pain. The brain interprets noxious stimuli applied to the body outside the brain as painful but has no pain receptors itself.) Penfield and Roberts stimulated different parts of the patient's brain and asked the patient what he experienced at the time. On some occasions, the doctors did not apply stimulation but asked the patient what he was experiencing anyway. This was done to check that what the patient reported was in fact a result of actual electrical stimulation.

Besides asking the patient what he experienced when stimulation was applied to his brain, the doctors also observed what happened in the patient's muscles when stimulation was applied at different places within the brain. It was found that in the part of the brain that was anatomically analogous to the motor cortex of the dog, neighboring sites triggered muscle twitches in neighboring body parts. Shifting the position of the electrode step-by-step caused step-by-step shifts in the positions of the twitching muscle fibers. The result, then, was the same as that obtained by Fritsch and Hitzig.

Based on this work, it was possible to create a cartoon of the human body like the one shown in Figure 3.19. The cartoon is based on Penfield's finding that the hand and mouth are more amply represented in the motor cortex than other body parts are, suggesting that greater control of some body parts than others is due to larger representation in the motor cortex.

This same principle was supported by further research by Penfield and Roberts. Here, the neurosurgeons asked what responses in the brain were triggered by stimuli delivered to the patient. The stimuli of special interest were touches on different parts of the body. By seeing which brain sites were activated by touches on different body parts, Penfield and Roberts developed a functional map of the body surface. The map resembled the map of the body's muscles. Larger parts of the brain were responsive to touch on the hands and face than to touches on the feet, legs, or abdomen. These changes were recorded in a part of the brain caudal to the central sulcus. Because this part of the brain responds to sensory inputs to the body (the "soma"), it is called the *somatosensory* cortex (Figure 3.19).[22]

Population Coding

The cartoons shown in Figure 3.19 are just that: cartoons or caricatures. They capture large-scale features only. The cartoons reflect statistical features of the functions served by the depicted areas. They show which body parts are represented the most in adjacent brain regions, but they don't show the mix of body parts represented by any given region. Statistical considerations become important in thinking about the role of brain regions because neurons within regions work together in a way that reflects the statistical mixtures of their special properties.

Before I unpack this further, let me share with you that brain–computer interfaces rely heavily on the statistical characteristics of brain activity. A famous example came from a woman who lost the ability to move her arms and legs due to spinal-cerebellar degeneration. Messages from her brain could not get to her limbs due to the depletion of neurons in her spinal cord and cerebellum. The cerebellum is a structure on the back (caudal) and bottom (inferior) side of the brain. Among its other functions are the control of movement and balance. The woman, whose name is Jan Sherman, had been a highly energetic and ebullient (bubbly) person who once appeared as a contestant on the TV show *Jeopardy*. After losing her ability to move her arms and legs at age thirty-seven, she agreed to serve as a subject in an experiment at the University of Pittsburgh, the aim of which was to implant electrodes in her motor cortex and transmit the associated electrical activity to a robot. Although the procedure had challenges, it essentially worked. Jan was able to look at a cup and get the robot to move its arm and hand to bring the cup to Jan's mouth. She could look at a ball and pick it up via the robot. The way this transformation was achieved from an engineering perspective was to exploit statistical features of the firing of neurons in the motor cortex, treating the neurons as a population.[23]

What are those statistical features and why do I refer to a population of neurons? The answer comes from experiments done in the 1980s by Apostolos Georgopoulos, along with his colleagues, at Johns Hopkins University (Georgopoulos, Schwartz, & Kettner, 1986). Georgopoulos and others, including Andrew Schwartz, the scientist at the University of Pittsburgh who later set up the brain–computer interface used by Jan Sherman, rewarded monkeys with sips of juice if the monkeys made specified hand movements. The monkeys held a lever that could slide in or out and left or right. In a typical trial, a light appeared in the center of the workspace and the monkey brought the handle to the lit location. After a short wait period, another light appeared somewhere around the central start location and the monkey had to move the lever to the target and hold it there to get another juice reward. The special feature of the experimental arrangement was that the researchers could record the position of the handheld lever at each moment in time while also recording activity in the monkey's motor cortex. The question was how the activity of the motor cortex was related to the movement of the hand. By having each trial begin with the handheld lever at the same central position, the researchers could vary the direction of the movement outward from that center position to each of a number of surrounding targets. In the simplest version of the experiment, the distances of the targets from the center position stayed the same.

The discovery made by Georgopoulos and his colleagues is shown in Figure 3.20. They found that individual neurons fired more when the hand moved in some directions

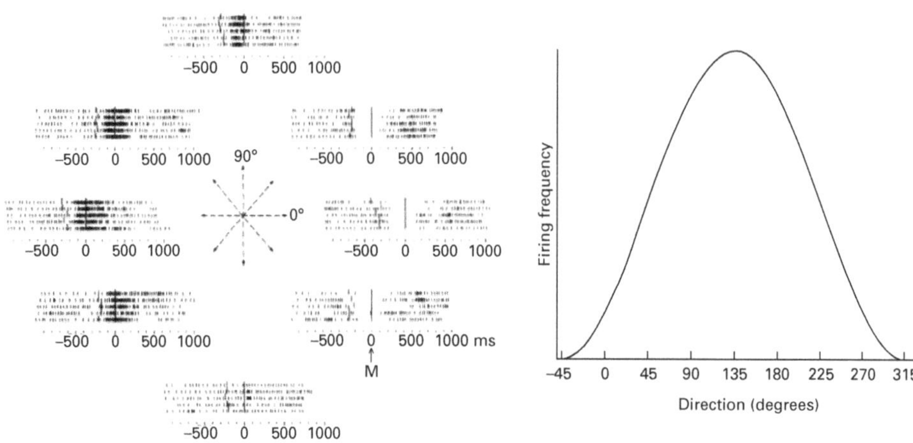

Figure 3.20
Preferred direction of one neuron in monkey motor cortex. The left panel shows individual trials in which the monkey moved straight out to the "north" (90°), to the "northwest" (135°), to the "west" (180°), and so on. The numbers are times in milliseconds relative to the start of the movement, which occurred at time 0. The neuron's preferred direction is 135°. The neuron fires the most for that direction. The right panel shows the tuning curve for this neuron.

than others. A given neuron responded most vigorously if the monkey moved the hand to the east, another neuron responded most vigorously when the monkey moved to the west, and so on. For any given neuron, there was a preferred direction of movement. The neuron featured in Figure 3.20 became most active for movements in the 135° direction. But this neuron, like all the neurons studied, still fired if the movement direction departed from the neuron's favorite direction. The greater the departure from the favorite direction, the less vigorously the neuron fired. This meant that the neurons had broad tuning curves. Thus, even though the neurons had preferred directions, they weren't completely fussy, so to speak, firing only if the required direction exactly matched the direction the cell preferred the most (the direction to which it was optimally tuned).

At first, the presence of broad neural tuning curves might suggest that the motor-control system is extremely messy. However, it turns out that having broad tuning curves with different preferred directions has advantages. One is that the activations of the neurons can be summed and the contribution of each neuron's activation to the sum can reflect its relative degree of activation. The result is a weighted sum. The weight of a given neuron's activation is equal to its activation relative to the sum of all the activations of the population. The entire population "votes," and the resultant direction corresponds to the summed weighted outputs of all the neurons in the group.

The other advantage of this method is that the system is resilient to neural loss. If one neuron happens to die or become momentarily inactive, the remaining neurons can still contribute to the net output. It is still possible to go anywhere in the space.[24]

By relying on this neural coding scheme, Andrew Schwartz, who had been a student of Georgopoulos, was ready to help Jan Sherman. When Jan Sherman fell prey to her disease, Schwartz and his team used her brain signals to move the robot arm in particular directions and also over particular amplitudes. They did so by capitalizing on the principle of neural control described above, which is called *population coding*. The term refers to the fact that the entire population contributes to the decision. As stated above, the neurons that are most active contribute the most, while the neurons that are least active contribute the least. The principle of population coding is one of the most important concepts in neuroscience.[25]

Population Coding and Mental Rotation

If the population code from the motor cortex directs the arm to move in some direction at some time, that doesn't mean the arm must move in that direction at that moment.

Recall from the last chapter that Roger Shepard showed that mental rotation is a systematic process that can be inferred from reaction times. Shepard demonstrated that the greater the angular separation between two objects (up to a maximum of 180°), the longer it takes to compare them. Shepard's data were obtained from human subjects pressing either of two keys, one for "same" and one for "different." From his data, Shepard inferred that the object comparison process drew on an internal process akin to turning objects in space.

Georgopoulos, Schwartz, and others realized that they might be able to track the mental rotation process itself within the brain (Georgopoulos, Lurito, Petrides, Schwartz, & Massey, 1989). Their idea was that, by computing the movement direction implied by the combined activity of many neurons at once in the motor cortex, they could tell, at each moment in time, which direction was favored by the neurons in the population as a whole. Accordingly, they developed a technique that combined their reaching task with Shepard's mental rotation task.

Each monkey tested in this experiment was shown a target light, as in the original experiment, but now the monkey had to move to a different location than the one occupied by the target. The new location was shifted 90° clockwise relative to the original. If the target appeared in the northwest direction, say, the monkey had to move to a target in the northeast direction.

Intuitively, you might expect the monkey to first move its hand toward the northwest position and then veer to the northeast position once it realized it was going the wrong way. There are experiments with human subjects showing that such changes of motion direction do in fact occur (Song & Nakayama, 2009; Tanenhaus, Spivey-Knowlton, Eberhard, & Sedivy, 1995). More will be said about these experiments in Chapter 9. In those experiments, participants can report on their change of mind. They can tell what they expect and do to correct their first movement tendencies. Monkeys can't do that, of course, but their brains can be studied to reveal what's happening "under the hood."

What happened in the monkey experiment was illuminating. The initial veering toward the first target did in fact occur, but it occurred in the motor cortex before much movement was observed. As seen in Figure 3.21, the population-code vector pointed toward the displayed target at first, but it was weak (had very small amplitude), indicating that the neurons contributing to it were not fully on board for the move. Over time, the vector lengthened and turned toward the required location until an actual movement occurred. Here then was neural activity that directly bore out mental rotation. The rotation that occurred was reflected in the continual change of the neural population vector.

When this discovery was published, it was heralded in the popular press. The media reported that brain scientists had discovered the basis for thinking. In some sense, they had, or at least they had opened a window into events within the brain that correspond to thinking. It was a truly exciting advance.

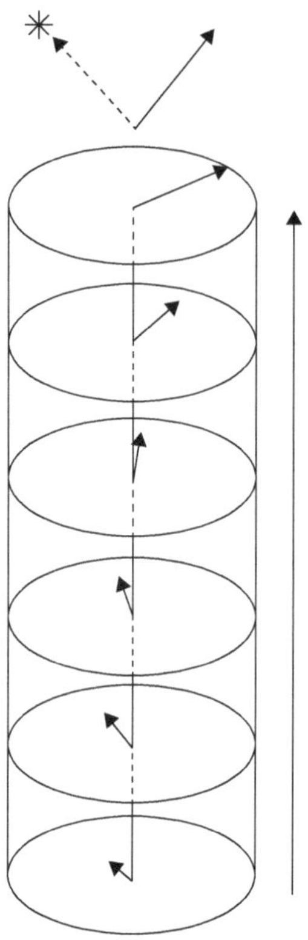

Figure 3.21
Rotation and magnification of the neural population vector from an initial small impulse toward a visual stimulus (dashed arrow toward asterisk on top) toward an actually required direction where no visual stimulus appears (solid arrow on top). The population vector grows and redirects toward the required direction in successive time intervals between 90 ms and 190 ms following presentation of the visual stimulus.

The finding had another important implication, too. The implication was that the motor cortex is more involved in thinking than might have been expected before. If the motor cortex were "purely motor" in the sense of being nothing more than a launch pad for muscle commands, it would have only gotten activated when all the relevant muscle commands were assembled. That is not what happened, however. Instead, the motor cortex was involved while the decision making continued. This points to less of a clear division between cognition and action than might have been expected otherwise. In fact, it suggests that action, mind, and brain are tightly interwoven, not just in a metaphoric sense but in a literal, neurally implemented sense that can be recorded in the lab.[26]

Some Final Words

The last few sections of this chapter touched on a tiny fraction of what is known about nerves and neural control, just as the earlier sections merely scratched the surface of what is known about bones, joints, ligaments, and tendons. The dearth (lack) of information about these topics is justified to the extent that an introductory text can't (and shouldn't) go into everything in tremendous detail. Relative to what's normally said in psychology texts about bones, joints, ligaments, and tendons, this text has said vastly more than usual. In psychology, these mechanical structures of the body are viewed, at least implicitly, as outside the scope of what's relevant or important. Yet it's untrue that they're unimportant psychologically, because, leaving aside that of course we need these structures to carry out our physical activities, even if psychologists don't explicitly discuss these parts of the body, all of us, in our daily lives, actually know an enormous amount about them. Body knowledge is a core part of the knowledge that we put into practice every day.

In case you're not sure of this or want more convincing, recall that you spent the first part of your life learning how to control your body. You learned how to look and reach, as well as how to control your spittle, bladder, and bowels. As a baby and toddler, you (and I) wore diapers and went through potty training. We wore bibs to catch the drool coming from our mouths and noses, and we got our noses wiped by our caregivers. People at the other end of the developmental spectrum, or people who have lost or never developed normal (typical) control of body functions, revert to or continue to require such assistance. Only out of prudishness do psychologists and cognitive neuroscientists spend less time discussing these matters than they should. In fact, only recently have neuroscientists begun to study urinary and rectal representation in the somatosensory cortex. Bladder and bowel control is an extremely important topic from a medical standpoint, of course. Just because people blush at the mention of rectums and the like, that doesn't mean those topics shouldn't be part of psychological research, including research on adult cognition.

In terms of bones, joints, ligaments, and tendons, we tend to forget what it's like not to know what we can and can't do anatomically. Hindsight bias, the tendency to forget what it was like not to know something, extends to physical action just as it extends to other forms of knowledge. In this connection, it's interesting to ask, in keeping with the debate between Nature and Nurture, what aspects of body knowledge are available at which points in development. It turns out that babies know they can't touch their left elbow with their left hand and can't touch their right elbow with their right hand, as recently demonstrated by Jeffrey Lockman and colleagues at Tulane University (Leed, Chinn, & Lockman, 2019). By simply applying a brief, harmless stimulus to babies at different places on their skin, Lockman and colleagues showed that the babies immediately reached for the points of stimulation with the hand best suited for the job, not the one that couldn't reach the site. This discovery was published in 2019. That it wasn't published in 1919, although it could have been, shows that psychological science has only recently woken to core questions that can be asked about action, mind, and brain.

4 Moving in Space and Time

How do you know where you are as you move, and how do you time your actions? What kinds of neural and mental representations are used for spatial and temporal orientation? Why are some people better than others at spatial navigation or timing? These questions, as you might imagine, are basic to an understanding of action.

To start the discussion of moving in space and time, I will share a personal fact: I am seriously deficient when it comes to spatial navigation. If I'm in a city where I haven't spent much time, once I get turned around for a short time, I'm goners. I have to hunt for landmarks, ask for directions, or get help from a map (electronic or paper) to figure out how to get to my destination. It's not unusual for me to retrace my steps and go back to my starting point to begin my trip all over again, making sure in the next journey to follow the route I know. I'm profoundly spatially challenged.

This spatial deficiency is likely genetic, for males are generally good at spatial navigation, but my father had the same problem. He was not the only one who took me on trips when I was a child, though. I got spatial training from my mom, who was spatially skillful and tried her best to help me. An added source of support for my belief in a genetic basis for my spatial incompetence is that my wife, who is spatially gifted, and I have nonidentical (female) twins, one of whom is severely spatially challenged as I am, but her sister is spatially able, like her mom. The two girls were raised in the same household. Some feature of Sarah's brain has made her spatially capable, like her mom, whereas some feature of Nora's brain makes her spatially inept, like her dad. Enough has been learned about the neural basis of spatial navigation to make some educated guesses about what the neural difference might be.

Route Maps, Survey Maps, and Cognitive Maps

Consider two ways of navigating. One is to rely on *route maps*. These are paths taken from one place to another, as in the example of returning home and following a familiar series of turns to get to a desired destination. When a route map is used, the memory representation guiding the trip is often captured by verbal instructions such as "Drive .5 miles, then turn right at the school, then drive 1.5 miles and turn left at the bowling alley." Services like Google Maps use such directions.

Another way of navigating is to use *survey* maps. These are representations that are more abstract and less tied to specific landmarks or to specific behaviors such as turning right, proceeding for .5 miles, and then turning left. When you look at a survey map, you see a symbolic rendering of thruways, streets, and the like so you can get situated and, in principle, reach your destination through any of a number of routes.

Survey maps reflect more cognitive sophistication than route maps. Creating a survey map requires integration of information gained through on-the-ground travel or, if the map maker is blessed with an airplane or satellite, a few aerial photos. Creating a survey map from on-the-ground travel takes intelligence of a kind that comes more easily to some than others. But even spatially challenged people like me and one of my daughters can eventually form survey maps in our minds, or what are also known as *cognitive* maps.

How would you know that someone has a cognitive map? The main way is to show that they can easily find new routes in previously explored areas. This way of testing for cognitive maps was pioneered in one of the most influential experiments in the history of psychology. Edward Tolman, a professor of psychology at the University of California, Berkeley, used the reasoning just mentioned (Tolman, 1948). He tested maze-running rats, letting the rats explore the same maze repeatedly. The method of testing rats in mazes became so well known that it became a trope for cartoons in magazines, newspapers, and, more recently, the internet. One of the cartoons is shown in Figure 4.1.

In Tolman's experiment, during initial training, the rats always started from the same release point in the maze and could find food at the same place every time. The rats' travel times got shorter with practice. Tolman then wondered whether the rats had merely learned the route they were taking or, perhaps, had formed cognitive maps. To test for cognitive maps, he released the trained rats at new spots in the maze. He observed that the rats still got to the target site quickly. Tolman took pains to ensure that the

Figure 4.1
A cartoon based on rat maze studies of the kind pioneered by Tolman. Such cartoons are a staple of our culture, reflecting one of the many ways that research related to action, mind, and brain has influenced society. This particular cartoon plays on the distinction between route maps and survey maps.

animals were not simply responding to ambient smells or other cues, and he went so far as to test the rats such that the muscle movements they made differed from what they had used before. He flooded the maze and saw that the rats now swam to the target, having only walked or run there before. This outcome showed that the rats hadn't merely learned muscle movements to get to the target location. The fact that the rats could even swim to the target when released into a suddenly flooded maze from a different starting point than the one they had previously experienced showed that the rats had formed abstract spatial maps, not just memorized patterns of muscle movements.

If rats have cognitive maps, do people? Indeed, we do. Experimentally, this conclusion was demonstrated with procedures modeled on Tolman's. People have been brought to areas they previously explored and had to start from unfamiliar positions or go around barriers that weren't there before. They could still find their way around without much time cost, provided they had sufficient experience.

A related method, not used with rats but compatible with the general approach, is to have participants point to named sites when they are stopped at various points in regions they have experienced repeatedly. They can do so quite well. Chances are, you can point to locations on your college or university campus when you stop at various places on campus, provided you've had adequate prior exposure.

In another procedure, subjects are asked to estimate distances between pairs of points separated by barriers. The estimates are often very good, indicating that the participants have formed mental overviews of the areas they've been in (Hermer & Spelke, 1996; Kosslyn, Pick, & Fariello, 1974).[1]

Smaller Spaces

Questions about the mental representation of space needn't be limited to large spaces traversed by walking, running, biking, or driving. Even simple acts like directing your hand from one position to another can rely on a spatial representation. Research on the control of hand movements has shown that hand moves are actually controlled with respect to external spatial coordinates, not just with respect to internal muscle-based coordinates. Relevant evidence comes from studies of hand trajectories.

As seen in Figure 4.2, a handpath can be linear in external space, resulting in a curved path in joint space, or linear in joint space, resulting in a curved path in external space (Hollerbach & Atkeson, 1986). The data generally support the first hypothesis. Observed handpaths generally look fairly straight in external space but are more curved in joint space (Morasso, 1981). This suggests that abstract spatial representations are available for hand movements, just as they are for whole-body movements.[2]

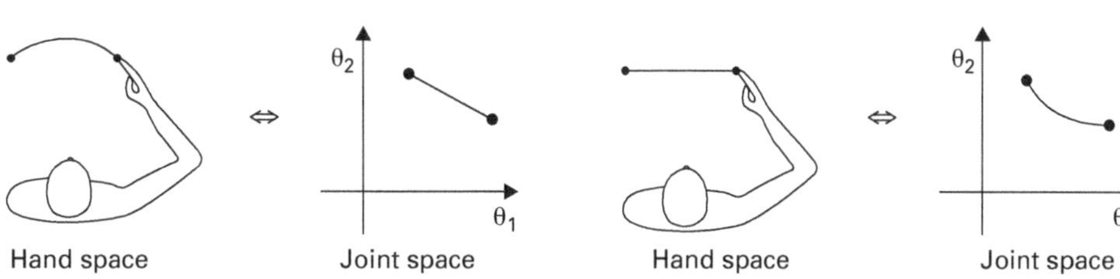

Figure 4.2
Simple paths in joint space (two left panels) and in external space (two right panels).

Neural Representations of Space

How is space represented in the brain? What neural substrates let you know where you are in the world? Answering this question might give hints about why some individuals are more spatially challenged than others.

A remarkable discovery in the 1970s showed that neurons in the brain become tuned to particular locations. These neurons are called *place* cells (Figure 4.3). The experiments that led to their discovery were done with rats. It was found that when a given rat occupied a given place in an area it explored repeatedly, a cell tended to fire preferentially when the animal was at that place. When the rat was at another location, another cell fired preferentially, and so on. Even after removal of a host of sensory cues that could have possibly caused the neural response, the cells fired the most when the rats were in particular locations. So it was the places themselves, their addresses in the

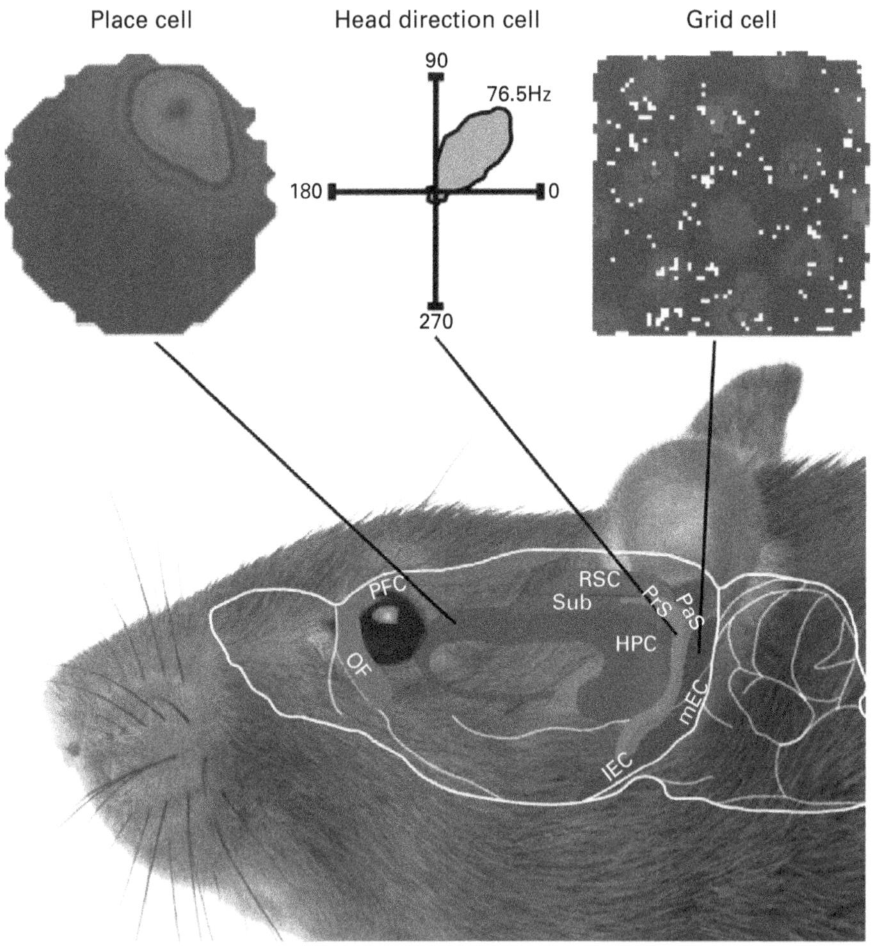

Figure 4.3
Place cell, head-direction, and grid cell locations in the rat brain. Place cells are located in the hippocampus (HPC), head-direction cells are located in the dorsal presubiculum (PrS), and grid cells are located in the medial entorhinal cortex (mEC).

world, so to speak, that were encoded by these cells. Place cells later were demonstrated in macaque monkeys and are believed to exist in humans.[3]

As mentioned above, place cells code locations rather than sensory cues such as sounds, sights, or smells corresponding to those locations. The cells also ignore (fail to respond systematically to) movements made to get to those places or the body positions animals are in at the time they occupy the places. Another kind of cell does this, however. *Head-direction* cells (Figure 4.3) code body positions in the external environment. Which way the head is oriented in the outer world is coded by these cells, so they act like compasses for the brain: "That way is north," one head-centered cell signals the brain, or "That way is south."[4]

Having place cells and head-centered cells is helpful for indicating where one is at a given time, but neither type of cell conveys how successive positions are related to one another. Each cell provides information along the lines of "You are here, at such and such address," without indicating how that address relates to others in a larger map. As a result, these cells might be useful for providing information about where you are at the moment but might not be useful for telling you how to get from place to place.

Other neurons provide more information about relative positions. These neurons are called *grid cells*. They occupy the medial entorhinal cortex (Figure 4.3) and fire most vigorously when an animal is in a *neighborhood* rather than in a specific *place*. A given grid cell tuned to a given neighborhood fires no matter where the animal is in that neighborhood, no matter which direction the animal is facing (unlike head-direction cells), and no matter how quickly the animal is moving. Grid cells have that name because of their arrangement in the brain. Distances between grid cells are roughly constant, which means that they occupy vertices of equilateral triangles rather than corners of squares or rectangles. Points on corners of squares or rectangles cannot all have the same distances between them, but points on vertices of equilateral triangles can. Because of the triangular grid occupied by these cells, they have the name they do (Moser, Kropff, & Moser, 2008).

Having grid cells, place cells, and head-direction cells puts animals at an advantage when it comes to spatial navigation. With place cells, you can tell which spot you're in. With head-direction cells, you can tell which way you're pointing. And with grid cells, you can tell which neighborhood you're in. Having access to all these kinds of cells can keep you oriented.[5]

Neural Representation of Manually Reachable Space

The last section concerned whole-body motion through the environment. When it comes to representing the space immediately outside the body, other parts of the brain are mainly involved, namely, the parietal cortex and premotor cortex (Figure 4.4).

Patients with damage to the parietal cortex may suffer from spatial neglect (Figure 4.5). They may ignore the side of space opposite the side of the damaged parietal cortex, drawing clocks with half the face missing, shaving just one half the face, and so on. Patients with this syndrome don't consciously downplay the neglected side of space; they are simply oblivious to it. The neglected part of space doesn't register at all. For all intents and purposes, it doesn't exist.

Spatial neglect is cross-modal, spanning touch, vision, and action. Neurons in another part of the brain, the premotor cortex (or ventral premotor cortex, more specifically) are multisensory as well (use many senses). As shown by two researchers at Princeton University (Graziano & Gross, 1994), neurons in the ventral premotor cortex are tuned to

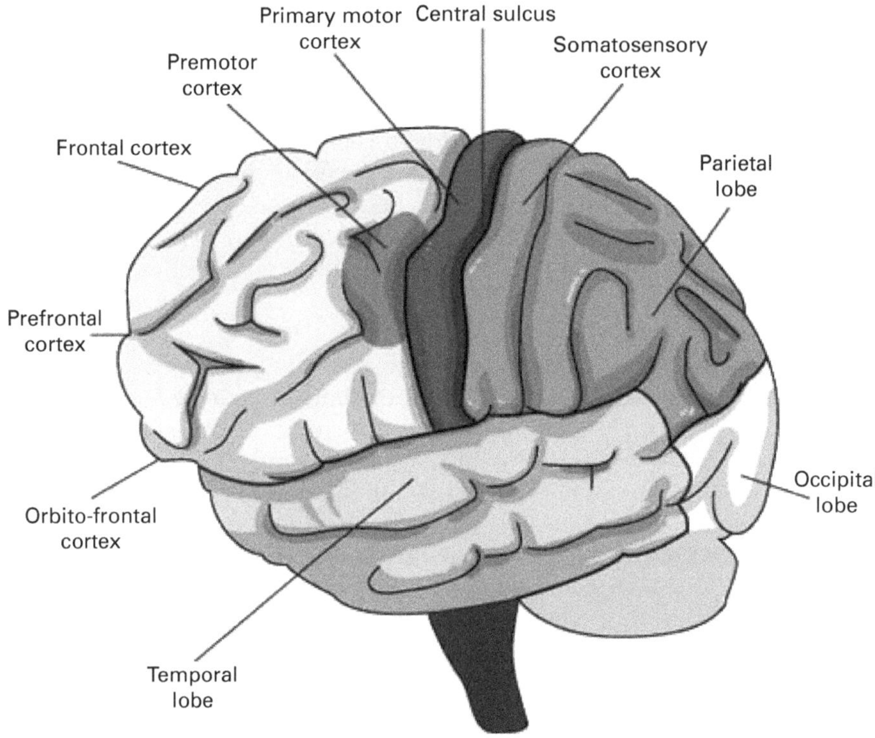

Figure 4.4
The human cerebral cortex.

the proximity of objects to the body (Figure 4.6). If a monkey sees an object approaching its extended hand, the closer the object gets to the hand, the more vigorously the neuron responds. The same neuron responds to touch on the hand. Critically, it doesn't matter to the neuron where the hand is located over a wide range of positions. What does matter is how close the object is to the hand itself. Because the hand's position determines the cell's response, the cell's receptive field is said to be *hand-centered*.

You can easily imagine why there would be cells responsive to the approach of objects to the hand. If the object poses a danger, it is important to withdraw the hand as quickly as possible. Similarly, if the object is attractive—an apple to be grasped, for example—it is useful to get neural feedback about the hand's approach to the object, expressed relative to the hand itself in hand-centered coordinates.

The face also has neurons of this kind (Figure 4.6), which makes sense insofar as the face, like the hand, needs to be protected. If an object hurtles toward the face, it's critical to get out of the way quickly. Perhaps this is why these neurons are close to the primary motor cortex, and likewise for the hand-centered ells.

Individual Differences in Spatial Navigation Abilities

Find your way back now to individual differences in spatial navigation, the topic broached near the start of this chapter. A number of studies have confirmed that there are differences in people's ability to navigate external environments. In one of those

Figure 4.5
Model images to be copied (left side) and copies made by a patient with hemispatial neglect (right side).
Source: Public domain

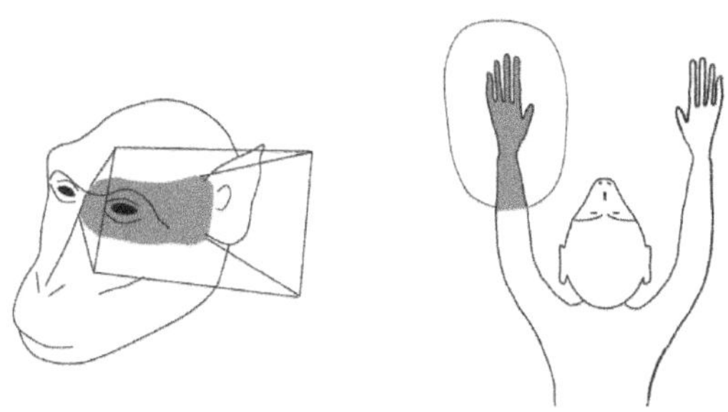

Figure 4.6
Regions of the hand and face with neurons in the ventral precentral cortex that respond to touch on those regions or to sight of objects approaching those regions.

studies, people were escorted through an area once a week for ten weeks. At test time, the participants were brought to one spot or another in the toured area and were asked to point to various target sites. Pointing accuracy was very high for some people but very low for others (Ishikawa & Montello, 2006).

What distinguished the good pointers from the bad ones? Sex differences played a role. Males on average were better at spatial cognition than females, as implied earlier in this chapter. Experience played a role as well. Those with more training in the area did better. But the most interesting finding was that the amount of neural "real estate" devoted to spatial representations of the area, as tested with brain imagery, differed for the good pointers and bad (Wolbers & Hegarty, 2010).

Does it follow that only Nature (genes) rather than Nurture (experience) affects the neural representation of space? This might be possible in the case of the study just reviewed, because amount of training did not fully account for the different levels of pointing proficiency. Still, amount of experience can have an impact, as shown in a different study involving London taxi drivers. These London cabbies were found to have more gray matter volume in the mid-posterior hippocampus than did London bus drivers who drove the same number of hours on average. Although the driving times for the two groups were comparable, the taxi drivers had to wend their way through the streets of London, going to new destinations, exploring and updating their spatial knowledge on a daily basis. The bus drivers, by contrast, just drove the same routes over and over again. Exploring space apparently led to the formation of more gray matter for spatial updating (Maguire et al., 2000; Maguire, Woollett, & Spiers, 2006).

Timing

Move now from moving in space to moving in time. In practice, this division is more important for didactic reasons (for teaching) than for real performance, because passing through space takes time. The brain may represent space and time in one format, at least when it comes to spatial navigation (Buzsáki & Llinás, 2017). Even so, it is reasonable to think about timing per se, just as it was reasonable to considerable spatial navigation on its own. Often timing requires very little movement, and timing may pose some of its own challenges. A prime example is stuttering, which is a problem of speech timing. Stuttering has confounded speech scientists and speech pathologists for years. People have stuttered at least since the Bible. Moses, some scholars say, may have stuttered (Brosch & Pirsig, 2001).

What makes stuttering such a confounding problem? A distinguishing feature of stuttering is that it occurs persistently in some individuals but not others. Another is that it appears in speech but not in other activities. Also, it tends to occur at the beginnings of words and is more likely to occur with certain sounds than others. Stop consonants can be especially challenging, as in "Papa" or "Memphis."

The word "Memphis" sticks in my mind because of a video of a stutterer trying to say that word.[6] In the video, a young man tries over and over again to say "Memphis" as he attempts to name the university on whose basketball team he played. The video comes from a leading speech pathologist demonstrating a method for helping stutterers. The method involves playing delayed auditory feedback. With the method, the speech that is heard is postponed by a fraction of a second. In the video, the young man's stuttering abates to some degree when he is given delayed auditory feedback of his own ongoing talking. Thus, the delayed auditory feedback method can be helpful.

Still, the question remains: Why doesn't the method help the young man learn to speak more effectively *without* the artificial delay? If the method is so helpful, why isn't it used all the time to eliminate stuttering completely? More generally, and stepping away from delayed auditory feedback alone, what causes stutterers to stutter?

For some time, it was thought that stuttering is caused by emotional or psychiatric problems, but that view has changed. Now it is believed that something about the neural aspects of speech timing goes awry in these individuals. In the text to follow, I will present information that relates to that view. I will offer information about the timing of action that leads to a model of neural timing that may suggest how stuttering arises. Along the way, I will talk about basic aspects of behavioral sequencing and timing.

The Problem of Sequencing and Timing

Consider the question of how you control the sequencing and timing of your actions. Begin by considering the name of a well-known ice cream company: *Ben and Jerry's*. When Ben and Jerry named their company, they could have called it *Jerry and Ben's*, but that sounds odd. Likewise for *Barnes* and *Noble*. That order sounds better than *Noble* and *Barnes*. Why?

Thousands of years ago, an Indian linguist named Panini noticed that short words tend to be followed by long words when the order of the words is otherwise arbitrary.[7] If Panini were alive today, he (or she?) would be unsurprised that we have *Ben and Jerry's* rather than *Jerry and Ben's* and that we have *Barnes and Noble* rather than *Noble and Barnes*. Linguists have pondered the basis for this preference ever since Panini's observation. A currently popular hypothesis is that the short–long ordering frees up cognitive resources for planning the second word while the first word is being said (MacDonald, Koranda, Bulgarelli, & Weiss, 2020).

Regardless of the exact reason, the broader question is this: How does the brain sequence and time the elements of behavior? When you dance the Hokey-Pokey, a dance done by young children at birthday parties and other events, how does your brain represent the requirement that "you put your right leg in, you put your right leg out, you put your right leg in, and you shake it all about?" You and the others in the dance can all carry out these acts in the order specified. What happens in the brain to allow this? How, most fundamentally, do you sequence and time your behaviors?

Tapping

To tap into the underlying mechanism, consider a simple experiment on finger tapping. Press one computer key over and over again at a comfortable pace and imagine that as you make the taps, the times between them are recorded. The series of intertap time delays might reveal the basis for their generation.

That simple experiment was done in the mid-1800s, using a telegraph key rather than a computer key. The experiment showed that times between finger taps had a characteristic pattern. If one intertap time was long, the next one tended to be short, and if one intertap time was short, the next one tended to be long. Expressed in the language of statistics, successive intertap times were negatively correlated.[8]

How can this result be explained? The researcher who made this discovery suggested that participants waited for feedback from the preceding tap before making the next tap. If the average time between taps stayed the same, then if one tap interval happened to be shorter than average, the next tap interval would be longer on average. Similarly, if one tap interval happened to be longer than average, the next tap interval would be shorter on average.

Waiting for feedback for successive responses seems reasonable, but is that what happens? In an article about this line of research by a graduate student at the time (Alan Wing) and his graduate adviser (Alfred Kristofferson), it was shown that the wait-for-feedback explanation didn't have to be accepted (Wing & Kristofferson, 1973). Wing and Kristofferson showed that times between finger taps could be set up in advance and regulated with an internal clock (Figure 4.7). According to the model, the time between forthcoming taps is established in advance. A command is issued to generate the first finger tap. Then, after the prescribed time, a command is issued to generate the second tap. Next, after another prescribed time, a command is issued to generate the next tap, and so on. If there is variability in the times to move the fingers given the motor commands they are sent, then if by chance the first response happens to be generated more quickly than usual, the time for the second response will be longer than usual. Similarly, if by chance the first response happens to be generated more slowly than usual, the time for the second response will be shorter than usual. Through statistical tests, Wing and Kristofferson showed that this model did a reasonably good job of explaining finger-tapping data.[9]

Subsequent work extended Wing and Kristofferson's analysis. One of the most interesting extensions of the model was that the planning of forthcoming responses can specify the timing of entire *groups* of responses (i.e., more than two taps). Thinking along the same lines as Wing and Kristofferson, other investigators predicted that in activities like playing the piano, the brain prespecifies the times between starts of entire *groups* of responses (key presses for musical measures). Timing of individual notes within the measures is controlled at lower levels.

Why should one care how finger taps are controlled? One reason is to see that sophisticated analytic techniques can be used to dissect mechanisms of control. Even the simplest of measures, like times between key presses, can be wrung for riches. No scalpels were used to dissect the underlying mechanism. The analysis boiled down to thinking and drawing on knowledge of statistics. I introduce this example to share the power of analytic work that can be brought to bear in behavioral research.[10]

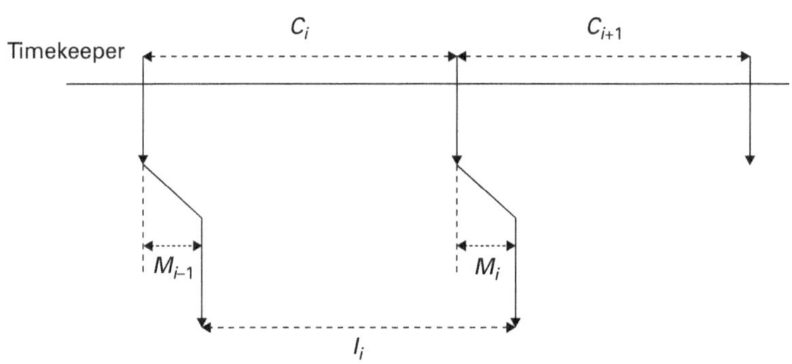

Figure 4.7
Clock model for finger tapping. An inner time keeper is set to run for C_j ms during the jth interresponse interval. When the interval starts, a motor command sent to the finger activates the finger after a delay M_{i-1}. The observed ith interval, I_j, reflects the combined effects of the clock delay and the motor delays leading up to the responses before and after the interval.

Another reason to care about the control of finger tapping is that it turns out that even this incredibly simple act relates to a much larger debate about how behavior is controlled. On one hand, behavior may be controlled by responding to feedback. On the other hand, behavior may be preplanned, at least over some span of forthcoming behavior. This contrast is analogous to the distinction drawn earlier between acting and reacting (Figure 1.9).

Sir Charles Sherrington, Reflexes, and the Bell–Magendie Law

The idea that feedback is the basis for timing and sequencing was advanced by a giant of neuroscience whose name has already been mentioned here: Charles Sherrington (Figure 4.8). Sherrington argued that the nervous system is made up of distinct neurons. He also named the synapse and clarified that activation of one muscle is accompanied by relaxation of that muscle's antagonist, as discussed in the last chapter.

Sherrington's main focus was the reflex arc (Figure 4.9). You may have learned about the reflex arc in other contexts. As a reminder, think about a medical exam in which a doctor taps your patellar tendon (below your knee) with a rubber mallet to see if your leg kicks out. If your leg extends quickly and vigorously, that's a good sign. It means that the events shown in Figure 4.9 occurred properly. When the patellar tendon is tapped, it causes the muscle above your knee, the *quadriceps femoris*, to stretch. Within that muscle (and in all others), there are stretch receptors known as muscle spindles. A spindle is a rod with twine wrapped around it, like a top with a string wrapped around its axle. If the muscle fibers stretch, the stretch receptors send action potentials to the spinal cord, where they enter the dorsal ("back") side of the spinal cord and synapse onto other neurons. Some of the neurons receiving these inputs are *interneurons*. They are called that because they receive signals from some neurons and send signals to others. When interneurons receive inputs from muscle stretch receptors, they send action signals to other parts of the spinal cord if the inputs are sufficiently intense. Some of the action signals go up to the brain, from which neural commands may come back down, reaching motor neurons that signal the quadriceps femoris to contract. When that happens, the stretch is relieved. All of these events occur quickly, in about 110 milliseconds (ms).

That time, 110 ms, is impressively quick. Nonetheless, that delay is nearly twice as long as a more direct connection that involves just *one* connection, a so-called *monosynaptic* connection between the stretch receptor and motor neuron whose activity

Figure 4.8
Charles Sherrington (left), B. F. Skinner (middle), and Karl Lashley (right).

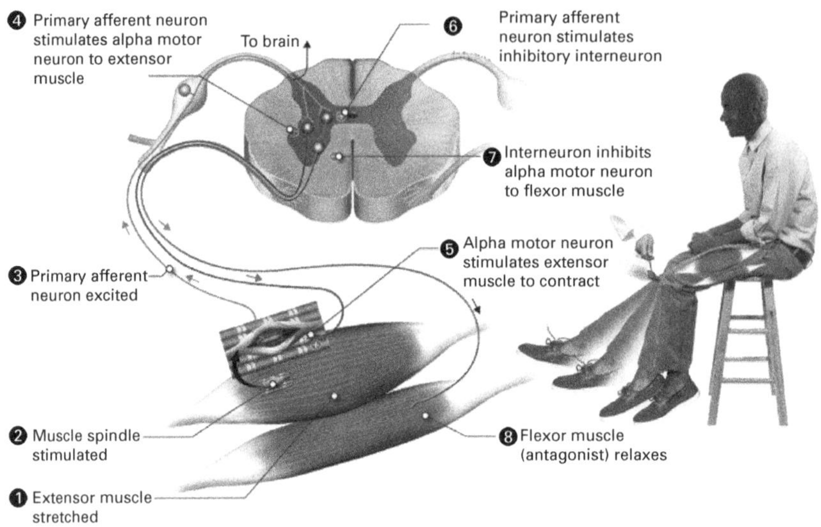

Figure 4.9
The muscle stretch reflex.

leads to contraction of the muscle. Such a connection is shown in Figure 4.9. The time for this connection to have an effect on the stretched muscle is about 55 ms.

Looking at Figure 4.9, you can see that the fiber providing sensory input to the spinal cord enters the *dorsal* side of the spinal cord, as mentioned above. The fiber providing motor output to the muscle exits the *ventral* (belly or anterior) side of the spinal cord. The fact that sensory nerves enter the spinal cord's *dorsal* side while motor nerves exit the spinal cord's *ventral* side was independently confirmed in the 1800s by a British researcher named Charles Bell and a French researcher named François Magendie. The *Bell–Magendie law*, named after these investigators, holds that sensory signals are transmitted into the spinal cord via dorsal (or posterior) inputs, whereas motor signals are transmitted out the spinal cord via ventral (or anterior) inputs. The Bell–Magendie law is a cornerstone of neuroscience. The fact that it is called a *law* reflects its importance.

Charles Sherrington, who identified the monosynaptic and polysynaptic pathways of the reflex arc, became convinced that these pathways and the entire nervous system work in a highly integrated way. He summarized this view in a book called *The Integrative Action of the Nervous System* (Sherrington, 1906/1947). A key idea that Sherrington promoted is that a simple dynamic underlies all behavior. According to Sherrington, behavior is ultimately explainable in terms of stimulus–response pairs: a stimulus is encountered and a response is made to it, and then another stimulus is encountered and that stimulus triggers the next response. Some of the stimuli may be produced by the organism, in which case an animal's own autogenic (self-produced) stimuli can arouse more behavior.

B. F. Skinner

Sherrington's idea was expanded in a critical way by B. F. Skinner (Figure 4.8) in a school of thought known as *behaviorism*. Here a key construct is *reinforcement*. The idea is that the probability of a response to a stimulus depends on the organism's history of past rewards and punishments. Skinner, a psychologist at Harvard University, was the chief proponent of this view, which he advanced with a simple experimental apparatus that he built to study the effects of reinforcement. His famous Skinner box was (and is) a device in which a pigeon, rat, or other subject can press a lever, tap a key, or make some other response and get reward or punishment for doing so. Skinner developed the apparatus to test the hypothesis that if a rat, say, presses a lever after a stimulus and is immediately and reliably rewarded for doing so, that response will reliably recur when the stimulus appears again. If the reward is withheld, the response rate will become less likely.[11]

According to behaviorists, all the world is, in effect, a Skinner box: the environment is a place where behavior is continually reinforced one way or the other. Everything we do is a learned pattern of behavior shaped by our histories of reward and punishment. Underlying that pattern is the fundamental mechanism of stimulus–response (S–R) chaining, envisioned by Sherrington.[12]

Karl Lashley

Is the S–R view correct? Is your life a chain of stimulus–response events? Do you accept the idea that everything you do is just a response to an immediately preceding stimulus? Regardless of your personal answer to this question, you already got one answer in connection with the simple task of tapping a key over and over again. Wing and Kristofferson, whose analysis of response timing was discussed above, showed, in effect, that Sherrington's reflex-based idea did not necessarily explain the timing of successive finger taps. Instead, Wing and Kristofferson showed that times of forthcoming finger taps are specified in advance. The time between one finger tap and the next isn't just the time to get feedback from the first tap. Rather, the delay between the first tap and second is planned in advance.

Before Wing and Kristofferson came along, other thinkers questioned Sherrington's reflex-based account of behavioral sequencing and timing. One was John Dewey, introduced in Chapter 1, whose disquiet was partly philosophical. Another was Karl Lashley (Figure 4.8), now introduced in this book for the first time. Lashley was a professor of physiology at Harvard University. He had been a medic in World War I and while treating soldiers in that war—the Great War as it was called before World War II—noticed that some soldiers he treated had lost the capacity to feel their limbs but could still move their limbs reasonably well (Lashley, 1917).

Subsequent reports confirmed this observation. For example, as shown in Figure 4.10, a patient with complete sensory neuropathy, unable to feel their hand and arm, could nonetheless draw circles, figure eights, and squares without seeing the action unfold. This patient's behavior was recorded after Lashley's life ended, but in his own time, Lashley (1951) noted that the ability to carry out extended movement sequences without feedback posed a problem for Sherrington's view and, by extension, for the radical behaviorist view of Skinner that behavior is governed entirely by S–R chains.[13]

What view did Lashley (1951) endorse? He proposed that behavioral sequences are planned in advance. The elements of behavioral sequences aren't just triggered by stimuli that happen to come up in the midst of ongoing behavior, Lashley said. Behavior isn't just a series of *reactions* but instead are *actions* based on plans reflecting intentions. In effect, Lashley resurrected Dewey's pronouncement, covered in Chapter 1.

No matter how obvious and intuitive it is to say that actions are based on plans, the idea was radical at the time (the early 1950s) because behavioral and neural scientists (especially in the United States) were wedded to the S–R view. The dogma at the time, due especially to Skinner, was that successive behaviors are initiated in response to immediately preceding stimuli. By challenging this view, Karl Lashley opened the door for what would come to be called the *cognitive revolution.*

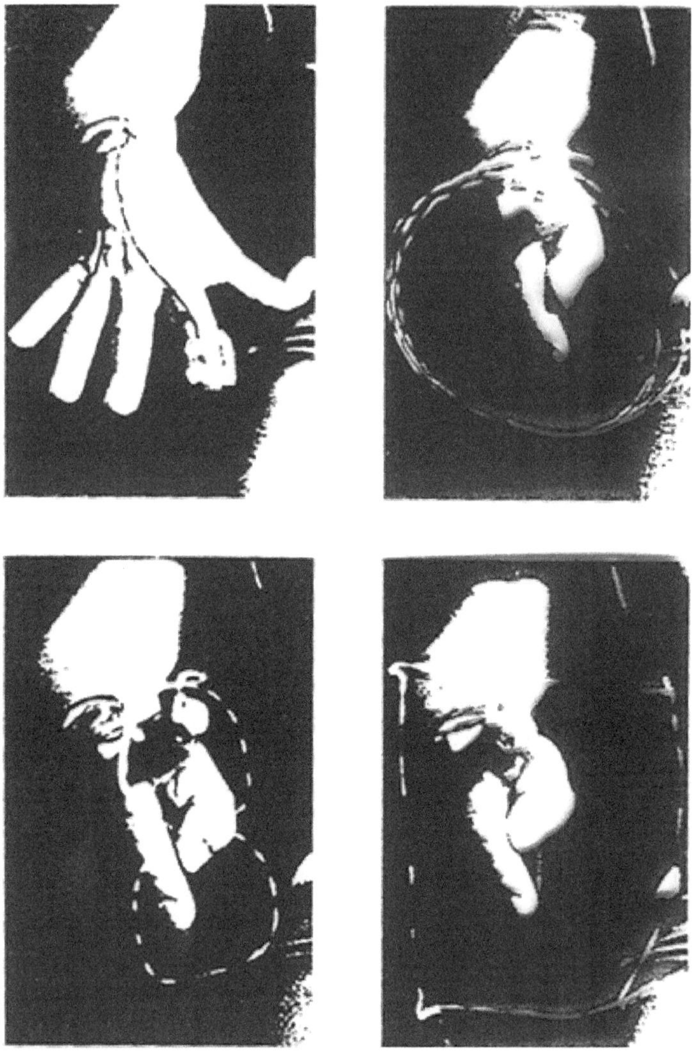

Figure 4.10
Movements made in the dark by a patient who was unable to feel arm or hand movements. The top left panel shows the marker worn on the finger as well as the wire attached to the marker to record the patient's movements. The remaining panels show a series of circles, a figure eight, and a square all drawn at will by the patient.

It is worth reviewing Lashley's arguments for the existence of behavioral plans. There were four (Lashley, 1951). One was that at least in some cases, it is possible to carry out movement sequences even without feedback. This is the argument that was just covered.[14]

Lashley's second argument was that some response sequences occur too quickly for each movement to be triggered by feedback from the movement before it. Here, Lashley described piano players who made successive keystrokes with delays that were less than feedback delays. Typists, too, can type at very high speeds (Figure 4.11), making it unlikely that they wait for feedback from each keystroke to start the next one.[15]

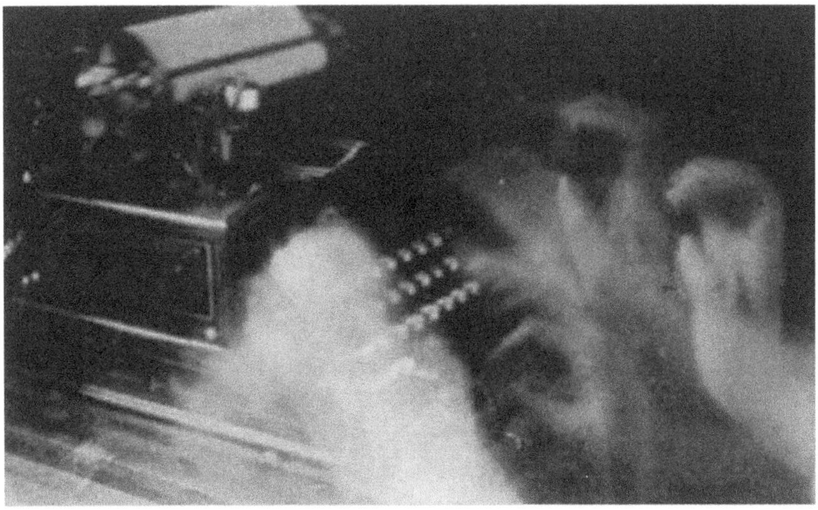

Figure 4.11
Time-lapse photograph providing a vivid image of the rapidity of typing.

Third, Lashley noted that the same response can be followed by different responses on different occasions. For example, after saying "Karl," you can say "Lashley" or "Marx." Nothing in the pronunciation of "Karl" predicts what word will follow. The feedback from "Karl" isn't different enough to trigger "Lashley" or "Marx" afterward. Therefore, the basis for saying one name or the other is in the plan, not in the feedback.

Lashley's fourth argument concerned meanings and mistakes. Regarding meanings, he offered a sentence consisting of a series of homophones—words that sound alike but mean different things:

> The mill-wright on my right thinks it right that some conventional rite should symbolize the right of every man to write as he pleases. (Lashley, 1951, p. 116)

Lashley (1951) said of this sentence,

> The individual items of the temporal series do not in themselves have a temporal valence in their associative connections with other elements. The order is imposed by some other agent. (p. 116)

What was that other agent, according to Lashley? It was the part of the brain that handles higher-order thinking, including intentions, meanings, and plans. Without such a brain center, Lashley argued, there is no way to understand sentences like the one concerning the millwright.

Because this book is about action, it is important to observe that Lashley did not limit his argument to language. He extended it to nonverbal behavior as well, as in this quote from the article covered above:

> The gaits of a horse, trotting, pacing, and single footing, involve essentially the same pattern of muscular contraction in the individual legs. The gait is imposed by some mechanism in addition to the direct relations of reciprocal innervation among the sensory-motor centers of the legs. (Lashley, 1951, p. 116)

Going on, Lashley (1951) wrote,

> The order in which the fingers of the musician fall on the keys or fingerboard is determined by the signature of the composition; this gives a set which is not inherent in the association of the individual movements. (p. 116)

In other words, plans underlie voluntary acts whether they are verbal or not.[16]

The other part of Lashley's fourth argument concerned mistakes. The analysis of errors is so rich and evocative that it deserves an entire section.

Errors Reveal Plans

In his 1951 paper, Lashley described mistakes made while typing. Sometimes he doubled the wrong letter, as in typing *lokk* instead of *look*, or in typing *iil* instead of *ill*. Such errors he took to mean that typing uses an abstract doubling operator. If the doubling operator is applied to the wrong letter, then that letter, rather than the right one, is repeated. This is such a clear, intuitive conclusion that it's easy to gloss over it. Lashley realized that abstract operators would never be sanctioned by S–R chaining theorists. Anything smacking of mentalism, such as abstract operations, would be anathema to them. Perhaps knowing this, Lashley offered a pithy (meaty, deep) remark to crystalize his hypothesis: "The order is dissociated from the idea," he wrote. In other words, there is the idea concerning the word to be expressed and, separately from that idea, yet another abstract representation of the need to repeat some of its elements. Lashley (1951) went on to suggest a similar principle for language:

> The polyglot, who has become proficient in a secondary language, who thinks in it and even dreams in it, may still tend to use the grammatical structure of his native tongue. If, as in French, that tongue applies gender to inanimate things, the English pronouns referring to them may take the gender of the French equivalents, though the French nouns are not thought. The German postponement of the verb or the Magyar use of the past infinitive may be incorporated in the new language. In such cases, the structuring seems to be dissociated both from the content and from the simple associative connections of the words themselves. (p. 118)

Extending this way of thinking, Lashley also focused on pig Latin, or what he called hog Latin. Before giving the quote from Lashley on this topic, let me mention what pig Latin is, because I have found that some students I've taught don't know about it. Pig Latin is one of a number of recreational languages that you can speak, often in a few minutes. This is what makes it so entertaining and also so useful for conversing privately with others in front of others who won't understand you, although they typically have a vague glimmer of understanding because the fake language you're speaking is so closely related to the real language on which your recreational language is based.

The defining feature of pig Latin and other language games (or argots, as they're also known) is that one or more rules transform certain sound patterns into others.[17] An example is

> Ifway ouyay owknay igpay atinlay ouyay ancay oolfay ouryay iendsfray,

which is the translation of "If you know pig Latin, you can fool your friends."[18]

Here is what Lashley (1951) had to say about this:

The ease with which a new structure may be imposed on words is illustrated by the quickness with which children learn hog Latin. The form which I learned involved transposing the initial sound of each word to the end of the word and adding a long a. Thus—at-thay anmay oes-gay own-day e-thay eet-stray. Some children become very facile at such inversions of words, and re-structure new words without hesitation. From such considerations it seems to follow that syntax is not inherent in the words employed or in the idea to be expressed. It is a generalized pattern imposed upon the specific acts as they occur. (p. 119)

Lashley here again points to abstract rules or cognitive operators, akin, one might say, to functions in mathematics. It is historically significant that Lashley made the observations he did several years before the world's most famous linguist, Noam Chomsky, presented his theory of language. According to Chomsky, there are abstract structures for syntax. Lashley came up with the idea before Chomsky did.[19]

Freudian Slips

Errors have long fascinated students of psychology. In fact, the study of errors helped launch the development of clinical psychology and psychiatry. Sigmund Freud, the de facto inventor of psychiatry, found speech errors enlightening. He argued that speech mistakes reveal pent-up urges. If someone substitutes a sexual word for a nonsexual word, the accidental utterance betrays what's going on "beneath the surface."

Students love hearing about Freudian slips, and for good reason. One of the most alluring aspects of psychology is that with its methods, one can see more than meets the eye. One can peer into the mind, so to speak, and claim to understand what's brewing beneath the surfaces. Getting a degree in psychology provides one with certification that one can do this in a more educated way than someone without the degree. The material in this book should help give that certification some basis.

Freudian slips are fascinating to learn about. When I teach about them, students blush. By contrast, the kinds of errors that Lashley was interested in were less sexy. Hitting the wrong letter twice on a keyboard is unlikely to titillate (arouse passions), yet such errors offer insights into the workings of the mind and also point to unconscious mechanisms.[20]

Action Slips and Abstract Representations

Doubling of keystrokes in typewriting isn't the only kind of action error that people make. You may have thrown your dirty socks into a toilet instead of a hamper when you were tired or distracted, or you may have poured ketchup rather than milk into your coffee when your mind was on other things. Such errors have been studied by psychologists, whose analyses of these sorts of mistakes—*action slips*, as they're called—have supported the hypothesis that actions rely on plans with abstract representations, as Lashley suggested. For example, if you throw your dirty socks into a toilet instead of a hamper, it's not because you have bad aim, assuming the toilet and hamper are far apart. Rather, at the moment, you have a goal in mind of throwing something into a receptacle, and the particular object assigned to the receptacle slot in your mind is, by chance at that moment, the toilet rather than the hamper. Similarly, if you pour ketchup rather than milk into your coffee, you have an abstract representation of something to be poured. The ketchup bottle rather than the milk bottle happens to take that place in your mental representation.[21]

Saying that action plans have abstract representations is nothing that a radical behaviorist would sanction, for abstract mental representations are, or were, outside the scope of respectable science for these individuals. Today, abstract mental representations are seldom questioned, thanks to arguments by Lashley and others. Even if mental representations can't be observed directly, they can be inferred, which is true as well of abstract constructs from other scientific disciplines. A force has never been seen, for example, but everyone has seen the effects of forces. Trees sway because of the force of winds, and apples fall because of the force of gravity. The fact that forces have never observed doesn't cast doubt on their existence. Neither should plans or other abstract representations.

Two further aspects of the abstract-representation concept are worth mentioning. One is statistical. The other is practical. On the statistical side, allowing for abstract representations allows scientists to look for statistical regularities in behavior with respect to aspects of the abstract representations under consideration. When action researchers develop models of action planning, they invoke probabilities of some action features being associated with others, as in the examples above—the probability of the toilet rather than the hamper being the aiming target, or the probability of ketchup rather than milk being the thing to be poured. It only makes sense to compute such probabilities if they bear on internal processes one finds interesting or important.

Here's another example. While writing the paragraph above, I first typed "if you poured ketchup rather than mild into your coffee." I typed "d" instead of "k," so I misspelled "milk." The letter "d" is often typed instead of the letter "k" by touch typists (of whom I'm one), and the reverse mistake happens, too: "k" is often typed for "d." A way to interpret this mishap is to say that there is an abstract representation for "middle finger" to which a hand feature "left" or "right" is assigned; on some occasions, the wrong assignment is made. This suggests that there is a middle-finger feature per se and an index finger feature per se. Touch typists apparently determine that they need to use a middle finger and then determine which middle finger to use—the left or right. With some probability, the hand that is picked is wrong. Perhaps, then, keystrokes are assembled on-line out of their constituent features. There may not be prepackaged routines for pressing the "d" key with the left middle finger or pressing the "k" key with the right middle finger. Instead, the features may be brought together on the fly.

This idea is reminiscent of an influential model of visual perception developed by the late Anne Treisman, winner of the National Medal of Science (Figure 4.12). Treisman argued that visual perception works through feature integration (Treisman & Gelade, 1980). According to her model, when visual stimuli are encountered, they are decoded by low-level mechanisms into their features, and then those features are reintegrated to create representations of the objects from which the visual stimuli came. An indication that this is how vision works is that on occasion, the wrong feature goes with the wrong object: a guy in a red cap standing near a blue van may be remembered as a guy in a blue cap standing near a red van. Finding similar errors in output suggests an analogous process operating in reverse for generating actions.

Evidence for this view about assembling features for action was obtained in reaction-time experiments where participants got advance information about some, none, or all of the features of actions they would be called upon to produce (Rosenbaum, 1980). The data indicated that the features could be pulled together in any order to define the action that would ultimately be required.

Figure 4.12
Anne Treisman with President Barack Obama when she won the National Medal of Science in 2013.

Bow Bloopers

I'd now like to tell you about a kind of action slip that provides straightforward evidence for the view that features are hooked together to assemble forthcoming actions. I discovered the action slip at a rehearsal of a community orchestra. Along with the other amateur violinists in the orchestra, I played a passage consisting of a quarter note followed by two eighth notes (Figure 4.13). We played the passage over and over again at a rapid pace and spontaneously played it with a down bow for the quarter note followed by two up bows for the eighth notes. We played the passage well. Then, for reasons unknown to us, the conductor asked us to play the passage "as it comes," which means with regularly alternating down and up bow strokes. Switching between down and up bows is generally easy. It's what every beginner does. A child playing "Twinkle, twinkle, little star" plays with a down stroke followed by an upstroke followed by a down stroke, and so on. There's nothing particularly hard about switching between up and down bow directions. However, when we violinists in the orchestra tried to play that particular quarter-eighth-eighth passage with regularly alternating bow strokes, we couldn't do so. We fell apart and broke into laughter because of how impossibly hard it was. The conductor then relented and told us to play the passage as we had before.

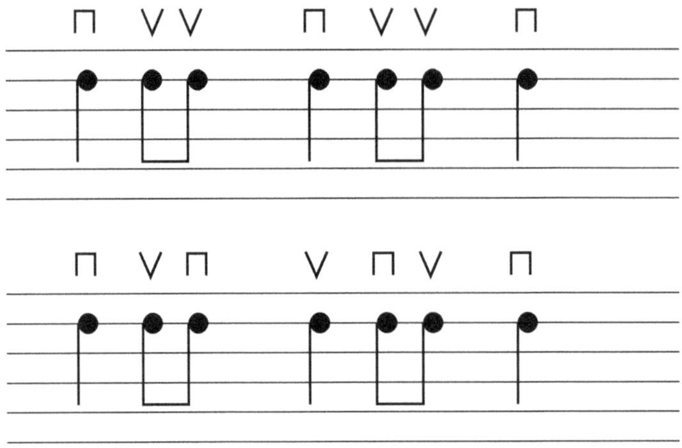

Figure 4.13
Music that is easy for violinists (top staff) or hard (bottom staff). Downbow strokes are denoted with rectangular marks. Upbow strokes are denoted with triangular marks.

After the rehearsal, I wondered why it was so hard to play this simple passage with regularly alternating bow strokes. It occurred to me that the bow stroke pattern we spontaneously adopted had an important feature that the regularly alternating pattern did not. In the down-up-up pattern, the relation between bow stroke directions and durations stayed the same, whereas in the down-up-down pattern, the relation kept changing. In the down-up-down case, all down strokes had long durations and all up strokes had short durations, but in the up-down-up-down case, the relations kept changing: long down, short up, short down, long up, and so on. I hypothesized that these changes made things harder.

Why would it be harder to play patterns whose associations between bow directions and bow durations kept changing? The action sequences might be controlled by changing features distinguishing successive renditions of the sequence. If that were true, it would imply that after elements of action sequences are performed, their defining features are maintained in working memory. Changing those features would be an economical way to control extended action sequences. Instead of creating a plan from scratch for each forthcoming sequence, the plan would simply be to change the features that need changing. If changing action features were time-consuming and prone to errors, mistakes would be expected when many changes were needed. A bow stroke pattern requiring frequent remappings of bow directions to bow durations (or vice versa) would be harder than one requiring none.

Finger Fumblers

Based on this reasoning, it would be reasonable to expect problems to arise no matter what kinds of actions require feature remapping. Not just in violin bowing but in other domains, too, there should be a *parameter remapping* effect—a cost of changing the mappings of action features to one another (Rosenbaum, Weber, Hazelett, & Hindorff, 1986)..[22]

To test this prediction, three colleagues and I tested two different finger-tapping sequences. One had the property that each time a finger was pressed, it was always pressed the same number of times. The right middle finger, for example, was always pressed twice, the left index finger was always pressed once, and so on. By contrast, the other sequence had the property that each time a finger was pressed, it was never pressed the same number of times. The right middle finger was pressed *twice* the first time it was used and *once* the next time it was used, the left middle finger was pressed *once* the first time it was used and *twice* the next time it was used, and so on (Rosenbaum, Weber, Hazelett, & Hindorff, 1986).

When participants attempted to produce the two kinds of sequences over and over again at rates specified by a metronome, they generated fewer correct renditions when the sequences had variable mappings than when the sequences had fixed mappings. In addition, the effect of the nature of the mappings—variable or fixed—was present even at the lowest required response rate (three finger presses per second). This outcome confirmed that people would have greater difficulty with finger sequences that have changing features. We called these sequences *finger fumblers*. In predicting and observing finger fumblers based on the earlier evidence for *bow bloopers* (another term we introduced), we felt more confident that this was a general phenomenon. As indicated above, we called it the *parameter remapping* effect.

Tongue Twisters

Would the parameter remapping effect also apply to speech? It would if the phenomenon were general. To test this prediction, we also asked college students to say the first *n* letters of the alphabet over and over again as quickly as possible for 20 seconds under the stipulation that they always alternate between stressed and unstressed pronunciations (Figure 4.14). For *even* values of *n*, the task was easy, exactly as predicted because the mapping of stresses to letters stayed the same. On the other hand, for *odd* values of *n*, the task was hard, also as predicted, because the mapping of stresses to letters kept changing.

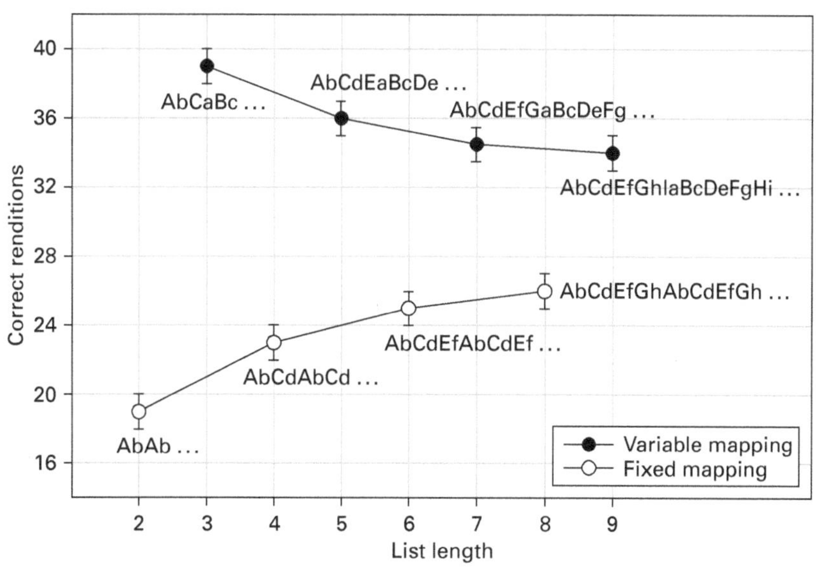

Figure 4.14
The parameter remapping effect for speech. Uppercase letters denote stressed pronunciations. Lowercase letters denote unstressed pronunciations.

As seen in Figure 4.14, odd-n conditions were generally harder than even-n conditions, but there was also a systematic effect of the value of n. The hardest condition of all was $n=3$. Here the hypothesized rate of change was highest. The more often changes had to be made (the smaller the value of odd n), the higher the cost (the larger the damage wrought by the parameter remapping effect). Presumably, the strength of the association between a letter and its stress level diminished over time. With longer delays between successive productions of a given act (saying a letter with some level of stress), the easier it got to undo or inhibit the previous association and replace it with the other.

Lessons Learned

Lessons can be learned from the work described above. One is methodological. A great deal can be learned by observing behavior in the everyday environment. Just as Lashley drew useful inferences from observing his own typing errors, my colleagues and I were able to draw useful inferences from observing violin-bowing errors and related phenomena. Naturalistic observation is at the heart of science, although it tends to get less play or advocacy than it might in psychology. Textbooks on psychological methods generally downplay or even omit naturalistic observation, but such observation can spur systematic observations illuminating phenomena of interest.

The second lesson concerns the picture of the cognitive system underlying action planning and control. In a word, the system is *conservative*. It tends to conserve (preserve) features of performance that were just performed, changing features that distinguish what needs to be done from what just was done. This view implies several things. One is that there must be a storage medium where action plans are maintained and manipulated. The storage medium is likely to be working memory. Second, plans are likely to include as few changes as possible. The reason is that if it is difficult to change plans, it is better to have few of them. Repeating what has just been done, either in whole or in part, is easier than "flipping all the switches." This is presumably why toddlers say "mama" or "dada" before saying phrases with many elements. And presumably it is also why individuals with brain damage are more likely to rock or murmur the same phrase over and over again than to produce more varied behaviors. Repeating actions and changing as few elements as possible makes sense from a physical standpoint, too, because physical systems have inertia. So too do cognitive and neural systems.

Yet another implication of the conservative view just offered concerns the spans of behavior that are planned. If sequences are repeated, there must be sequence starts and stops. Those starts and stops must bound spans of behavior with some duration, measured in time or number of behavioral elements. A sequence can't be a day long or as short as a millisecond. Typically, sequences that are repeated last a second or less.

An attraction of thinking about repetitions is that they provide a way of identifying planning spans. The planning span for a behavior is definable as the length of the behavior that benefits from repetition. Beyond the span, it doesn't matter whether elements are repeated or not. All of this is another way of saying why there are sequences at all, or series of acts with starts and starts. A priori, there needn't be. All of behavior could be one continuous chain, but just as there are gaps between neurons, there are gaps in behavior, functional junctions between subsequences when computations are done to prepare the next bit of behavior by drawing on bits that were just completed.

Handpath Priming

The view that action planning is conservative leads to the expectation that there should be carryovers from behaviors that were just produced. Those carryovers can lead to surprising behavioral effects, as confirmed from another naturalistic observation that I happened to make (Figure 4.15).

Once at a dinner party I got into a discussion with someone and listed to him with such rapt attention that I found myself doing something funny. While listening to this person who sat across from me at the dinner table, I absentmindedly munched on olives, taking one olive at a time from a dish beyond a wine bottle that stood between my plate and the dish. I reached around the bottle to get each olive, then brought my hand back to my mouth to pop the olive into my mouth, then reached out for the next olive, and so on. At some point, another person took the wine bottle to help herself to a glass of wine, after which she set the bottle down somewhere else. Perhaps because I was only dimly aware of the bottle's removal, I kept reaching for the olives as if the bottle were still there. My hand followed a more curved path than needed, and this went on for several out-and-back hand trips. When I noticed that my hand was behaving as if an obstacle was still in the way, I straightened out the handpath but made a mental note about what had just happened. "Cool!" I remember thinking. "My hand kept moving as it had before even though it didn't have to!"

This sort of behavior is an example of perseveration, the tendency to continue doing what was being done before. Perseveration is well known in clinical contexts. It is a common symptom of Parkinson's disease.[23]

My hand may have kept moving in curved paths because I simply failed to notice that the wine bottle had been removed. I was only dimly aware that the bottle had been taken away. To check the observation under more controlled conditions, a graduate

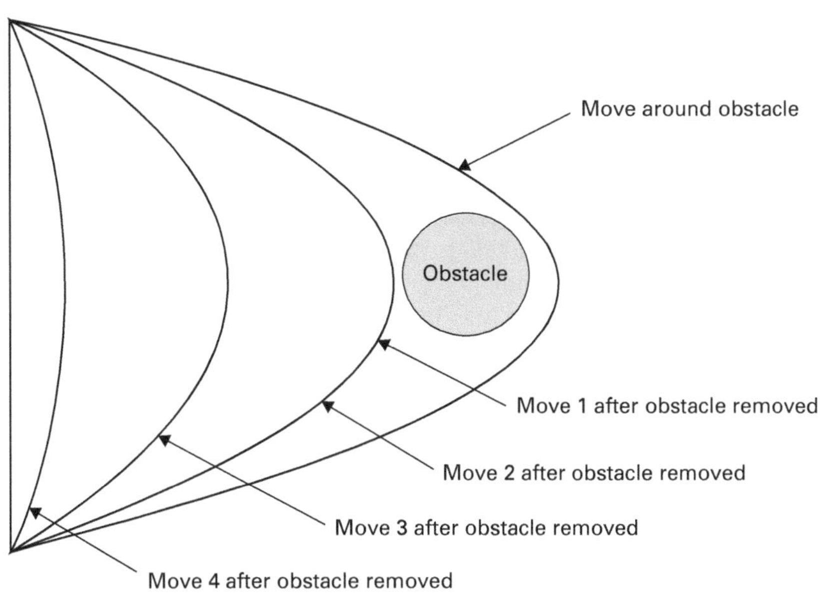

Figure 4.15
The handpath priming effect.

student of mine at the time, Steve Jax, and I brought the observation to the lab. Our participants laid their preferred hand on a table top and saw a stick-figure image of the hand on a screen before them. A target appeared on the screen and the participants moved the hands to the target. Once the target was reached, another target appeared and the participant moved the hand to that target. This went on for a while. All the movements were made between a central target to a peripheral target and then back again. In the conditions of most interest, Jax and I included a big spot between the central target and peripheral target; this was an obstacle to be avoided. After several trials with the obstacle in view, we removed it. Would the observed handpaths be more curved than needed, as occurred at the dinner party? The answer was yes. Soon after the obstacle was removed, the handpaths had high curvature, but eventually they straightened out, becoming as straight as in the control condition, when no obstacle ever appeared.

These results are shown in schematic form in Figure 4.15 and confirmed what I had observed at the dinner party. There is a tendency to persist with handpaths that have just been used. This *handpath priming* effect, as we called it, wasn't due to missing the disappearance of the obstacle. The obstacle and its disappearance were as conspicuous as possible, and we even conducted an experiment in which we described the handpath priming effect to participants and urged them to resist the temptation to keep making needlessly curved handpaths after the obstacle disappeared. We still found handpath priming in this condition.

How can handpath priming be explained? Using the same plan repeatedly, only changing it as needed, reduces cognitive costs (Jax & Rosenbaum, 2007; van der Wel, Fleckenstein, Jax, & Rosenbaum, 2007).[24]

Hierarchies

The theme of the last sections was that forthcoming actions tend to preserve features of just-completed actions. For this to be true and for series of actions to be goal directed, there must be an overarching plan concerning the identities and linkages of the action elements. Changes occurring at local levels must make sense for the larger purpose. Said another way, plans must be hierarchical. Small-scale goals must be organized for larger ones, and high-level goals must be organized so small scales can be achieved.

This idea is intuitive, but can it be tested? Albrecht Inhoff, Andrew Gordon, and I reasoned that if action plans are hierarchically structured, it should be possible to obtain data consistent with that view when participants choose between actions (Rosenbaum, Inhoff, & Gordon, 1984). We invited college students to associate distinct action sequences with distinct stimuli and then, on each experimental trial, to do whichever sequence was called for. The participant's task was to minimize the time between presentation of the stimulus and completion of the corresponding sequence. Our reasoning was that we could vary the properties of the sequences, keeping the discriminative stimuli the same when we did so, meaning that difference in the times to start and complete the sequences could be attributed to the sequences themselves—to their composition and the relation between them.

The stimuli we used were the capital letters "X" and "O," two stimuli that are very easy to tell apart. We communicated to subjects before each experimental block of trials which action sequence to perform if they saw "X" and which action sequence to perform if they saw "O." The mappings of sequences to stimuli were counterbalanced.

Figure 4.16 shows the model we developed and the times we observed, plotted as a function of the number of operations assumed in the model to produce each response

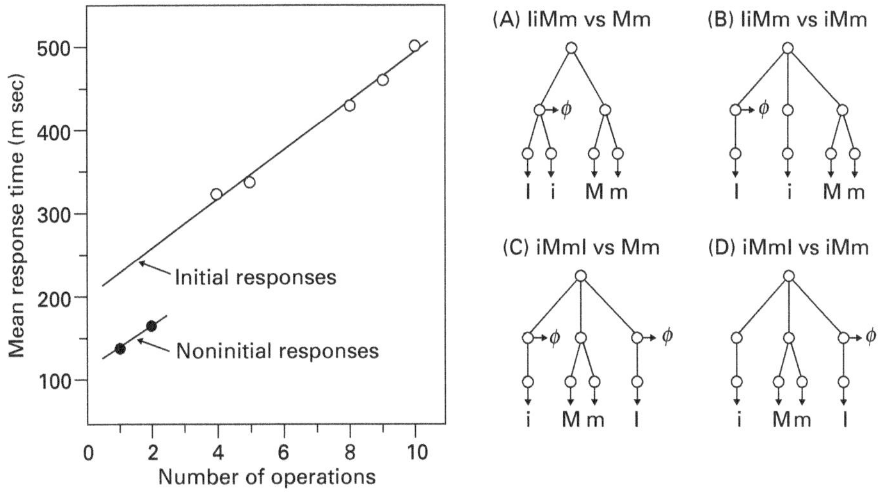

Figure 4.16
Times to initiate initial and noninitial responses when choosing between action sequences (left panel) and the choices to be made as well as the theoretical steps required to choose them. The letters "I," "i," "M," and "m" denote keypresses made with the right index finger, left index finger, right middle finger, and left middle finger, respectively. So "raised" letters are for the right hand and lowercase letters are for the left hand. The null sign, ø, denotes the option to cancel keypresses.

(each keypress in the sequence to be performed); error rates were low and essentially uniform over conditions. The model assumed a hierarchical plan with choice points where choices were needed. The model made exact predictions about the number of operations needed to produce each keypress. The operations could be conceived in terms of stepping through a hierarchy. Each element was specified and implemented, and then control returned to the higher level once all the commands beneath it were done.[25]

When I tell students about these results, and especially the finding that the time to start an action sequence increases with the complexity of the sequence, I point out that besides having theoretical implications, the results also have a practical application. "If you are late submitting an assignment," I tell the students, "you can tell your professor you have big plans."[26]

Neural Networks

Drawing trees like the ones in Figure 4.16 makes it easy to visualize control, but it's easy to fall into a trap when considering such models. In fact, I fell into that trap, or chose to enter it, in the last section, when I drew trees and said that control passes through the trees. It's convenient to do this, but it begs questions. What exactly are the "top" and "bottom" of the tree, and what does "left" and "right" mean?

It's conceivable that neural representations in the brain are literally read from top to bottom and left to right but it's not obvious why the brain would rely on such a process or how it actually could. Keeping track of where the "inner reader" is in the path would require a mechanism that, in effect, would need to control sequencing and timing, but discerning how that mechanism works is the aim here; relying on the mechanism itself can't be the explanation.

A model that avoids these problems was proposed by William K. Estes, one of the founders of mathematical psychology. Estes (1972) envisioned a simple neural network for controlling the sequencing of elementary behavioral acts such as typing the letters in a word. Suppose the word to be typed is "very" (Figure 4.17). The *v* node inhibits all the other nodes, the *e* node inhibits the *r* and *y* nodes, and the *r* node inhibits just one other node, the *y* node. When the *v* node fires, causing a keystroke on *v*, it inhibits itself, causing it to stop inhibiting the other nodes. At this point, one node is sufficiently activated to fire: the *e* node. When that node fires, it causes the *e* keystroke to be made and it inhibits itself. This lets *r* reach threshold, and that node fires, allowing *r* to be typed, causing *r* to auto-inhibit, whereupon the *y* node, which had been inhibited all along, is finally uninhibited, and it fires, generating *y* at the very end.[27]

Estes's proposal is elegant because it relies on simple neural functions to control sequencing. Moreover, it doesn't the beg the question of how sequencing is controlled. Unlike a tree diagram, Estes's model doesn't use spatial position as a proxy (stand-in) for serial order. In principle, Estes's model can be scaled up, with small cell assemblies like the one in Figure 4.17 nested within other cell assemblies, allowing for hierarchical control.

Estes's network might also provide a clue to the source of stuttering, the problem raised earlier in this chapter. Imagine that auto-inhibition does not work properly, and suppose that auto-inhibition is particularly vulnerable for nodes at the starts of sequences. If the amount of inhibition coming into the network as a whole is too great relative to the amount of inhibition distributed over the nodes, then multiple nodes will fire. To combat this overabundance of activation, the inhibition from the first

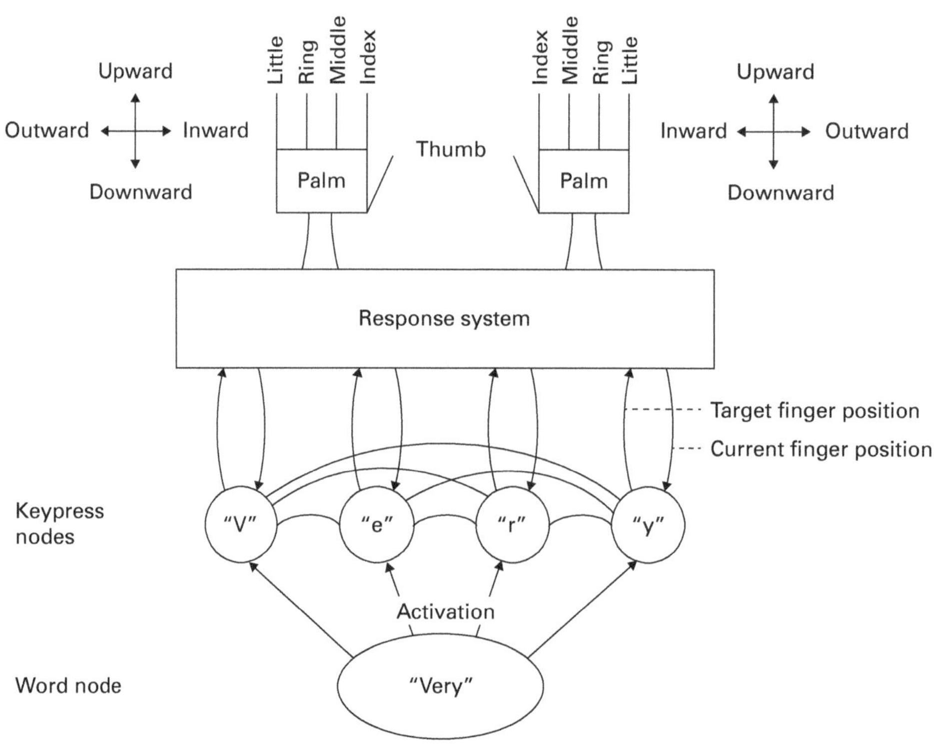

Figure 4.17
Neural network model for typing the letters in the word "very."

node to the others might need to increase, but if increasing that inhibition to the other nodes makes it harder to get the first node to inhibit itself, the result could be repetitions of sequence-initial elements. It might be very difficult to stop saying the "M" of "Miami," for instance, and a stutterer could get stuck on that sound.

This account is speculative and incomplete. I introduce it because no one, as far as I know, has been able to figure out how a neural network could cause stuttering. My suggestion doesn't explain why stuttering would be more likely for some sounds than others, nor does it explain why stuttering arises specifically in speech. Stutter-like motor patterns are manifested in some neurological disorders, such as Parkinson's disease, in which patients falter before initiating movements for reaching or walking, although not, curiously, before initiating speech movements. Meanwhile, stutterers stutter in speech but not in other behaviors. These patterns are puzzling to be sure. Much remains to be learned about sequencing and timing. Basic mechanisms, like the one suggested by Estes, may hold the key.

Edward Taub, Constraint-Induced Therapy, and the Ethical Treatment of Animals

A final topic for this chapter is a body of work that had a tremendous effect on society. Recall that one of the core arguments that Lashley offered for plans concerned feedback-free performance. Lashley observed that people can do things like draw circles and figure eights even when deprived of feedback. Such performance should be impossible according to S–R chaining theory. Lashley based these reports on wartime injuries. Controlled studies had not been done to check them.

Figure 4.18 shows a monkey from a line of experiments that was designed to do so (Taub & Berman, 1968). When I gear up to present this work in my Action, Mind, and Brain class, I preface the description with a warning. "This may a bit difficult," I tell the students. "It's a pretty tough study in terms of what happened to this monkey and others involved, but the research was done in the name of science and useful information came from it." I then go on to say, "Some of what I will tell you may be hard to hear. If you want to, you can cover your ears or leave the room for a while." I explain

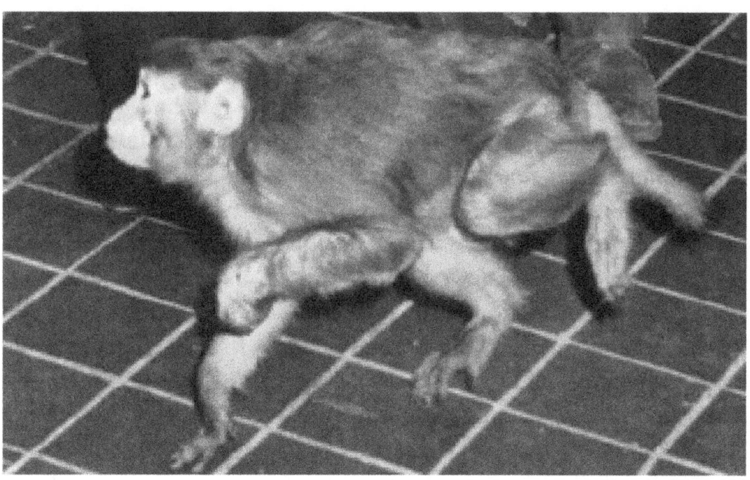

Figure 4.18
Monkey deafferented on one side refrains from using the arm on that side.

that I won't relish the opportunity to report the results but that it's important for them to know about it both as students and as citizens. I pause to see if anyone gets up and leaves. If you, dear reader, wish to do so, you can skip to the next chapter and check in with a friend or your professor to get a delicately worded summary of what's to come.

In Figure 4.18, the monkey holds up one of its arms. The monkey was a subject in an experiment on the effects of loss of sensory feedback. The kind of sensory loss was similar to the kind experienced by the soldiers who sustained gunshot wounds to the dorsal region of the spinal cord. In accord with the Bell–Magendie law, the soldiers couldn't feel their limbs, but they could still move their limbs as desired.

The injuries the soldiers sustained were variable, making it hard to be sure what happened, anatomically, to any given combatant. Years later, in the study described now, this matter was brought under experimental control. The main investigator, Edward Taub, sought to explore more systematically than Lashley had what occurs when various neural deficits are imposed. Taub cut the spinal dorsal roots on one side of a monkey's spinal cord but not the other. The monkey with the surgical procedure refrained from using the arm whose sensory fibers had been cut.[28]

If Sherrington and Skinner had seen this result, they would have said this result is exactly what they predicted. From the perspective of S–R chaining theory, monkeys deprived of sensory feedback would be unable to move, and that's what happened here. The monkey was apparently unable to use its deafferented arm.

There was an alternative interpretation, however, as Sherrington and Skinner surely would have recognized. The alternative interpretation was that the monkey that underwent the surgical procedure may have kept its arm still because it couldn't feel its arm, not because it literally was unable to move the arm skillfully. The monkey may have simply *preferred* not to use the arm it couldn't feel. Losing feedback from a limb is disconcerting. Think about getting up from your seat in a theater after you've sat for a long time and find that your leg is "asleep." You may refrain from using that leg as much as you normally would, although you can still move it if necessary.

Taub and Berman conducted another experiment that bore out this possibility. They severed the dorsal nerve roots of *both* sides of the spinal cord, not just one side or the other. With both sides deprived of incoming sensory signals, there was less of a chance that one arm would remain unused simply because it couldn't be felt.

The result following the bilateral surgery was dramatic. With sensory feedback absent in both arms, both arms were deployed. The monkey could walk again, suggesting that the reason for disuse of the singly deafferent arm was that the monkey *preferred* not to use it, not that the monkey *couldn't* use it.

This striking result implied that the S–R chaining account was wrong, just as Lashley had said. Contrary to S–R chaining theory, sensory feedback was *not* necessary for motor acts of the kind Taub and Berman's monkeys performed. By implication, sensory feedback may be unnecessary for movement sequences generally.[29]

Other lessons were also learned from this research. One was based on the inferred reluctance to use one limb. As just said, there can be unwillingness to use a limb when the limb can't be felt, at least when the other limb *can* be felt. Based on this result, Taub suggested that a useful form of therapy for patients with trouble moving one arm following stroke would be to constrain the "good" arm and force the use of the "bad" arm. This form of therapy is called *constraint-induced* therapy (Taub & Uswatt, 2006). The motivation is that reluctance to use the bad arm will cause its muscles to atrophy and its neural control centers to wither or be taken over by other functions. By constraining the "good" arm and requiring use of the "bad" arm, the intervention should prevent degeneration of the "bad" arm and possibly allow for partial or even complete regeneration of function.

Constraint-induced therapy is widely used, thanks to Taub's research. The therapy is difficult because the good arm is tied to the torso (with the consent of the patient). Patients find this frustrating; they want to use the arm that works, not the arm that doesn't work, and they tend to be hesitant to use the therapy when on their own. Someone generally needs to be on hand to urge the patient on (Kwakkel, Veerbeek, van Wegen, & Wolf, 2015).

A second lesson from Taub's experiments concerns the ethical treatment of animals in scientific research. A great brouhaha (emotional storm) arose after it was learned that the monkeys in Taub's research were hurting themselves. The monkeys cut themselves on wires sticking out from the cages and often developed infections. Feeling nothing, the monkeys experienced no pain when brushing up against the cage wires. An even more alarming outcome—and now we come to the gruesome part—was the allegation that some monkeys actually ate parts of their own arms.

When word got out about these allegations, there was much ado about it. Newspapers carried stories about it and so did TV news. Taub's lab was closed down, and the monkeys were relocated to safer lab facilities. Taub was prevented from getting research funds from the U.S. government for some time, and he faced legal proceedings, although ultimately was acquitted.

What became clear from the tumult around these events was that there needed to be explicit guidelines for the safe and ethical care of animal subjects. Prompted by the controversy, such guidelines were developed. Now, all researchers planning to conduct experiments with animals must go through rigorous review by their institutional review boards (IRBs) to ensure that the animals will be treated humanely to the fullest extent

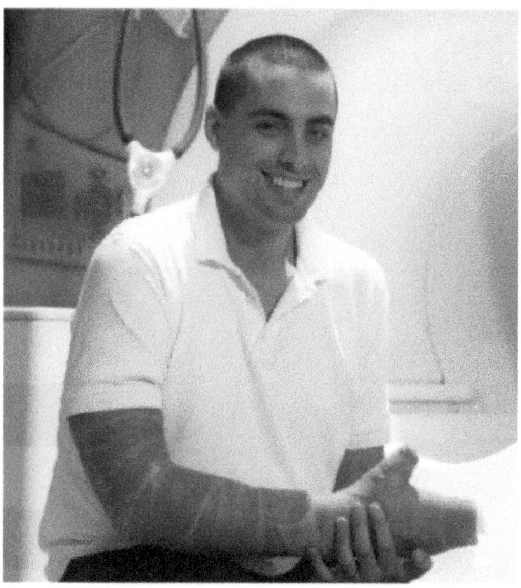

Figure 4.19
Dr. Nico Dosenbach in the cast he had put on his perfectly healthy arm to study the effects of arm immobilization on his own brain.

possible. When researchers submit applications to IRBs for planned studies involving animal or human subjects, they must defend the proposition that whatever risks may be incurred by the subjects will be outweighed by the potential benefits of the findings. Increasingly, scientists proposing research with animals are asked to consider the necessity of testing living (or once living) subjects. Could the research question be addressed by reviewing previous literature or by developing mathematical models, computer simulations, or other artificial devices? These developments are partly due to the Taub affair.

Although Taub's name will always be associated with these matters, his scientific and clinical contributions have been acknowledged. By deafferenting monkeys and observing the effects, he showed that sensory feedback is not absolutely necessary for carrying out voluntary physical actions even though loss of sensory feedback does cause movements to be made in a less finely tuned way than is possible otherwise. This conclusion confirms what Lashley had suggested earlier based on his observations of soldiers in the Great War.

Constraint-induced therapy, the form of therapy derived from Taub's research, has recently been brought under a happy spotlight, so this is a nice way to end this chapter. Figure 4.19 shows a young doctor, Nico Dosenbach, who decided to study brain changes associated with immobilization of one of his arms (Hamilton, 2020). He immobilized one arm by getting it put it in a cast—a pink one, as it turns out, whose color he chose to entertain and assuage any concern in his young daughter. While wearing the cast for an extended period of time, Dosenbach got his brain scanned at various times. By doing so, he and his colleagues could see dynamic changes in neural connections within his brain in regions responsible for moving the restrained and unrestrained arms. Characterizing these changes may provide a basis for recommending effective treatments for patients with actual rather than feigned mobility problems.

5 Learning

Imagine someone who can learn to do things like no one ever has. She watches a cellist perform a solo and picks up a cello for the first time, playing as well as the soloist did. She looks at paintings by a leading artist and paints as well the painter did the first time around. She watches a world-class dancer and steps out onto the dance floor, never having danced before, and dances as well as the star dancer did, immediately and with no hesitation.

No one can achieve mastery all at once. This scenario is science fiction. It's also rare for someone to become a top expert in unrelated skills. Why does mastery take so long, and why is the range of skills that can be mastered by one person so limited?

A quick answer is that the time to learn a skill depends on the components the skill relies on. To play the violin, for example, you need to hold the instrument under your chin, move your right hand the right way to move the bow, place your left fingers on the fingerboard to get the right notes and vibrato, and modulate your phrasing and dynamics to give the tune the emotional tone you want. Only in science fiction can all these components come together instantly. But why exactly does skill learning take years and not minutes?

10,000 Hours

Consider a fact that has attracted a great deal of attention. In 1993, Anders Ericsson, Ralf Krampe, and Clemens Tesch-Romer reported that people need about 10,000 hours to become experts. The area of expertise the researchers studied happened to be violin playing, although the authors made clear that they viewed violin playing as a stand-in for skills generally (Ericsson, Krampe, & Tesch-Romer, 1993).

Before making the 10,000-hour discovery, Anders Ericsson, along with another colleague, William Chase, had reported that someone who was tested extensively on memorization could expand his memory span for one type of materials but not another. The kind of material the subject could memorize more effectively was the kind he practiced (numbers). The kind he could not learn better was a kind of material he hadn't practiced (letters) but was tested on occasionally (Chase & Ericsson, 1981).

The kind of materials the subject practiced was, as just indicated, numbers. Each day, over a two-year stretch, he was given lists of new numbers to memorize. The lengths of the lists increased based on how well he did. By the time the study ended, the subject, who enrolled in this study for pay, was able to recall an astonishing eighty digits per new number list.

The other kind of material the subject was tested on occasionally was lists of letters. In each test session, he was given a few letter strings and tested on those. With much less practice on letters than numbers, his memory span for letters remained where it was at the beginning, about seven items, which was also his initial memory span for digits.

The fact that the participants' digit span could get up to eighty indicated that he had developed expertise for encoding and recalling numbers. However—and this was the most important point—he hadn't strengthened an overall ability to memorize. If he had, he would have also improved with letters. The broader implication was that for this individual and, by extension, the rest of us to the extent the subject in the study was typical, there is no memorization talent per se. Some people may be better memorizers than others, but a general talent for memorizing of any kind can't explain improvement for one kind of material rather than another.[1]

Returning to the 10,000-hours study, Ericsson et al. found that violin students who put in about 10,000 hours of deliberate practice were more likely to become concert-level performers than students who didn't put in that many hours. Those in the latter group may have been labeled "talented" to begin with, but that label didn't get them very far. Being talented—having a knack for the skill, having great potential, showing all the earmarks of being destined for greatness—guaranteed little about actual success. What mattered instead—what increased the chance the skill would flourish to expert level—was hard work. Sustained, attentive practice over a span of approximately 10,000 hours was needed for high-level achievement.[2]

The figure of 10,000 hours wasn't a number that Ericsson and his colleagues predicted on theoretical grounds. Instead, it more or less popped up on its own in studies that Ericsson et al. did. The figure could be rationalized (made sense of after the fact) in terms of the normal time for training. Ten thousand hours is equivalent to forty hours a week for fifty weeks a year for five years. That duration is a typical stint for a music conservatory student. So too is it a typical training period for a PhD student, a medical student, and so on. (Most of these curricula last four years rather than five, but a post-doc of one sort or another is usually tacked on to go from competence to expertise.)

So enticing was the 10,000-hour figure that it became the subject of a best-selling book, *Outliers: The Story of Success*, by Malcolm Gladwell (2008). In *Outliers*, Gladwell emphasized that the discovery made by Ericsson et al. overturned the idea that people can do great things based on innate potential alone. What made Mozart (Figure 5.1) such a great composer, for example, wasn't just that he was a genius—a useful term, perhaps, for conveying enthusiasm about him—but rather that he spent an enormous amount of time (virtually his whole life) honing his musical skill. From a very young age, he received music instruction from his father, spending most of his waking hours around the piano and other instruments. Mozart's musicality flourished because he was steeped in music, plus he had the temperament and drive to take advantage of that experience. Motivation matters. Without it, arduous practice would not occur.

Bill Gates (Figure 5.1) was another protégé aided by the sheer amount of time he got to spend in his field: computers. As Gladwell explained, Gates was around computers all through childhood, in his teenage years, and as a young adult. Had he not been around computers as long as he was, he wouldn't have had the spectacular success he did in creating (or co-creating) MS-DOS, the operating system that set the stage for Microsoft.

A similar tale was told about Tiger Woods (Figure 5.1), the great golfer, who, like Mozart, had a dad who trained him from the time he was a tyke. By the time Tiger could stand, he was putting.

Similar upbringings applied to Venus and Serena Williams (Figure 5.1), the great tennis stars whose dad got them on the tennis court when they were very young,

Figure 5.1
Outstanding individuals who spent their entire lives developing their skills. Top left panel: Wolfgang Amadeus Mozart. Top right panel: Bill Gates. Bottom left panel: Tiger Woods. Bottom right panel: Serena and Venus Williams.

encouraging them to put in long hours on the game, enabling them ultimately to become the stars they did.

The bottom line, then, as Ericsson et al. showed and as Gladwell recapitulated, is that sheer talent is a myth. Although some may have it, the amount of time spent deliberately practicing is the best predictor of extraordinariness. This is what makes one an outlier (the title of Gladwell's book). Extensive focused practice is needed to reach the extreme right side (the best side) of the skill distribution.

But why 10,000 hours in particular? Why *that* number? The question can be reframed: What kinds of skills are the ones that take 10,000 hours to develop? What distinguishes those skills from ones that take less time to learn? If there are skills that take *longer* to learn than 10,000 hours, what distinguishes them from their quicker-to-learn counterparts?

Suppose that skill learning is incremental and that the nature of the learning is such that the amount learned per unit time (per hour, say) is some proportion of the amount that ultimately needs to be learned. Under that scenario, you would get the relation shown in Figure 5.2, which is a typical learning curve.[3]

In situations where the learning curve applies, the amount of improvement is large at first and diminishes over time. With each new hour of practice, it gets harder and harder to see improvement. Saying this another way, it become more difficult to discern the effects of more practice. It may be, then, that 10,000 hours is the largest number of hours for which there is realistic hope of picking up differences in skill levels related to amount of practice. Ten thousand hours may simply be the most hours for

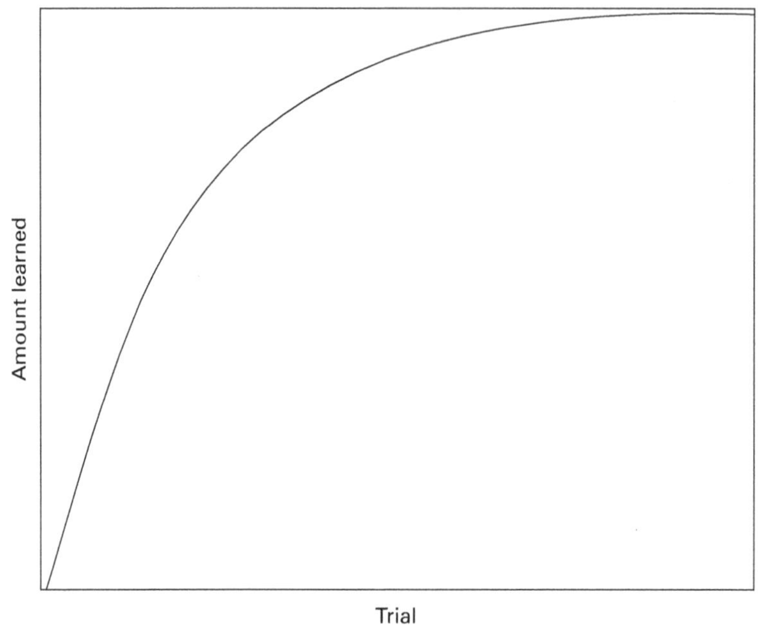

Figure 5.2
The learning curve of Rescorla and Wagner.

which it is possible, practically speaking, to detect time-based improvement. Perhaps this is why graduate school, medical school, and the like take four to five years. Society may have timed the duration of training to be the period beyond which it is hard to tell whether further training has a noticeable effect. There would be little sense in requiring students to spend ten years in school if it takes that many years to tell whether ten years is better than five.

YouTube Overconfidence

Next consider another discovery about skill learning that has received a fair amount of recent attention. The result is related to watching people engage in activities on YouTube. The activities require skill: throwing darts; juggling bowling pins; pulling tablecloths out from under dishes.

Watching YouTube has become a favorite pastime of millions of people. I'm one of them. I enjoy watching YouTubes related to my hobby, playing chamber music. I'm an amateur violinist, as mentioned before, and I enjoy watching highly accomplished musicians play their instruments, especially in string quartets, piano trios (violin, cello, and piano), and violin and piano sonatas. I play in all these contexts and like to watch them perform fabulously because—I confess—they engender in me the feeling that I can play as well as the great performers I watch. There is something about seeing them play as well as they do that gives me the feeling I can play just as well. Truth be told, I can't—not in a million years, let alone 10,000 hours.[4]

In 2018, Michael Kardas and Ed O'Brien of the Booth School of Business at the University of Chicago showed that such overconfidence is widespread (Kardas & O'Brien,

2018). As background for the study, the authors drew on earlier work showing that people who watch others engage in various activities heightens the chance that the observers will engage in the same activities themselves. In the most famous example of this earlier work, children watched a film of another child strike a Bobo doll. After observing the other child do this, the observing kids were more likely to punch the doll (Bandura, Ross, & Ross, 1961). The researcher who led this study, Albert Bandura, concluded that modeling others is an important basis for skill learning. The theory he developed around this idea, social learning theory, challenged the view from B. F. Skinner that the only way to learn is to get direct feedback about one's own actions. Watching others is also important, Bandura showed.

Kardas and O'Brien asked participants to watch YouTubes of experts performing various skills such as pulling tablecloths out from under a flower pot, tea cups, and a tray (Figure 5.3). Next they had the participants give ratings of how well they thought they could perform the skills themselves. Crucially, the participants had never tried the skills before. The result was that those who watched the YouTubes gave higher confidence ratings than subjects who did not. The YouTube viewers didn't lose all connection with reality, however, for when they were given the chance to try the skills themselves, they conceded that they needed more training.

This state of affairs—watching others and feeling inspired to do what they do—may have evolutionary benefits. Birds who see their moms and dads fly from the nest may finally venture forth, emboldened by what they see their guardians doing. Children who can't yet hunt or gather need to feel confident enough to strike out on their

Figure 5.3
Pulling a tablecloth out from under a table setting.

own. Much as play is likely to be an important vehicle for exploring action possibilities, watching others may serve a similar role.[5]

The Case of H.M.

Acquiring skill relies on the brain, and the brain's role in skill learning has been elucidated by studying the effects of brain damage on the way skills are learned (Scoville & Milner, 1957). An important advance was made on this front in the 1950s when a young man, who until his death came to be known only by his initials, H.M., suffered from severe epilepsy. At the time, no drugs were available that could curtail H.M.'s seizures, so he agreed to undergo surgery, much as the young man did who went under the knife in the surgery room of Wilder Penfield, discussed in Chapter 2. In the case of H.M. (not one of Penfield's patients), the surgery involved removal of the hippocampus, a region discussed in the last chapter in connection with place learning and episodic recall. The rationale for removing the hippocampus was that it occupies a central place in each cerebral hemisphere (so there is one on the left and one on the right), making it a possible hub for the generation of epileptic seizures. If this site were surgically removed in whole or in part, the surgeons thought, the storms might subside. To double up on the hoped-for effectiveness of the procedure, the surgeons decided that both hippocampi should be removed. H.M. and his family agreed.

The surgery was successful. H.M.'s seizures subsided. However, there was an unforeseen side effect. After the procedure, H.M. had great difficulty learning new facts. He was unable to acquire the names of streets in the new neighborhood his family moved to, and he couldn't learn the name of any new U.S. president past Eisenhower. There were many such examples, including one that struck close to home for the neuropsychologist who visited H.M. every day to test H.M.'s memory. "Have we met before?" she asked him. "No," he said. "This is the first time."[6]

The neuropsychologist to whom I refer is Brenda Milner. Besides testing H.M. on his ability to learn new facts, she tested him on his ability to learn new *skills*. One of those skills was mirror tracing (Figure 5.4). The goal here is to draw a line from a start position to an end position, staying between the borders on the way. What makes the task challenging is that the drawing is done without direct sight of the path. Instead, the drawing is viewed through a mirror, causing everything to be seen in reverse. Moving the pencil to the right causes the image to move to the left, and moving the pencil to the left causes the image to move to the right. People get better at this with practice. The number of times they cross the line borders declines as they do the task more and more, and the time to complete the task declines as well.

Neurotypical individuals—that is, control subjects with intact hippocampi—did better and better on this task as they practiced it, and to no one's surprise, they also recognized the individual who administered the task. Not so for H.M., however. As indicated in the made-up dialog above, he claimed never to have met the doctor who gave him the mirror task. Yet remarkably, he got better on the task, just like the control subjects.

How could this be? Brenda Milner argued that the hippocampus is essential for learning new facts but is inessential for learning new skills. Some other part of the brain, she suggested, makes it possible to learn mirror tracing.

Milner's finding paved the way for a new picture of the neural basis for memory (Figure 5.5). In this picture, distinct neural systems are thought to underlie the acquisition of declarative knowledge, on one hand, and procedural knowledge, on the other.

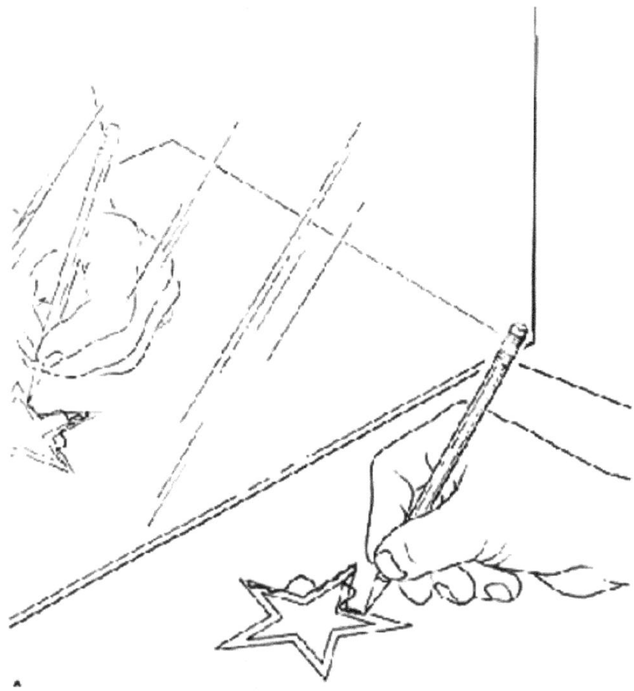

Figure 5.4
Mirror tracing task learned as well by patient H.M. as by neurotypical individuals.

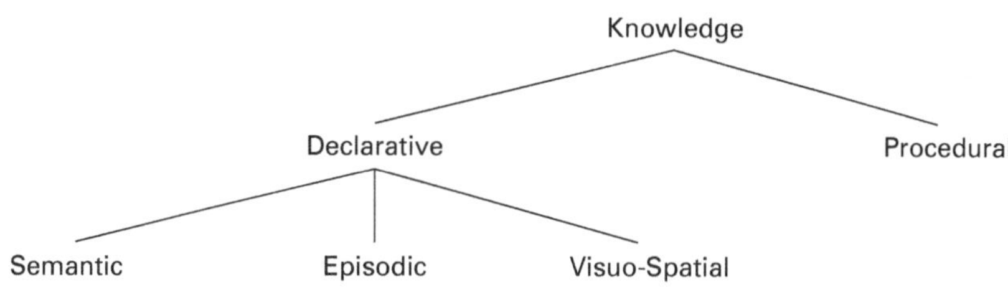

Figure 5.5
Model of knowledge based on work with H.M. and others.

Declarative knowledge consists of three kinds of information: semantic information (the meanings of words and symbols), episodic information (remembering events, including who you have met and when), and visuospatial information (remembering locations). Procedural knowledge underlies action sequencing and control. Those actions can be low level, like moving a pencil through a path with reversed visuomotor mappings, or higher level, like knowing how to play the violin, perform dance routines, or pull tablecloths out from under dishes.[7]

Hippocampus and Striatum

If the hippocampus isn't necessary for procedural learning, which parts of the brain are? You've gotten some information about this topic in this book. In the last chapter, you learned that place cells are found in the hippocampus and that place cells represent learned locations, so they are important in visual-spatial knowledge. If the hippocampus is removed, patients like H.M. should be unable to form new cognitive maps, and indeed that is what happened with H.M. When his family moved, he was effectively unable to learn the new spatial layouts of his new house and neighborhood. When a neuropsychologist came to visit him, he failed to recognize her.

Another brain region is important for procedural learning: the striatum. This area, which has the name it does because of its striped appearance, consists of a number of structures, including the caudate nucleus, putamen, and nucleus accumbens. In healthy adults, the larger the volume of the striatum, the better individuals do on learning procedural skills. This was shown in a study of individual differences in learning a video game. For individuals in the study, the larger the striatum, the better they got on the video game. At the same time, and consistent with the hypothesis that the hippocampus is not directly involved in skill learning, it was found that there was no correlation between hippocampal volume and acquired skill on the video game (Erickson et al., 2010).[8]

Quieter Brains Are More Skillful Brains

If greater brain volume is associated with greater skill, that needn't imply that greater brain activity is associated with greater skill. In fact, the opposite is true. As skill develops, associated parts of the brain tend to become *less* active, not *more* active (Anderson, 2015).

That result may surprise you. How could it be, you might say, that *less* brain activity is seen as skills develop? Shouldn't *more* brain activity occur as skills grow? Aren't more neural helpers beneficial?

To make sense of this finding, think about the state you're in when you're in the earliest stage of learning an activity—for example, when you're just beginning to learn to play the guitar. In the earliest stages, you may find yourself in a kind of blooming, buzzing confusion, to use the famous phrase from William James in his 1890 *Principles of Psychology* when he described the state of knowledge of very young babies (James, 1890). In the first phase of acquiring a skill, you're still trying to sort out the basics. You draw on everything you can think of to help get you started. Your mind is all over the place, so to speak, and so is your brain, figuratively speaking. However, with greater skill, you settle down. Instead of thinking of everything you can do to help you get past first base—"How do I hold up the neck of this guitar? How do I bend my wrist?"—you reach a state of calm, thinking about the music, not the means of making it. Ultimately, you get so deeply absorbed in the music-making that you reach a state called "flow" (Csikszentmihályi, 1990).

Calming down physically has been documented in studies of skilled performers. For example, highly skilled rifle shooters become so quiet before pulling the trigger that they hardly breathe; they squeeze the trigger between heartbeats. This shows that being highly focused reduces the noise that can upset performance. Getting more skilled entails focusing and quieting. As will be seen in a later section, this has implications for dealing with stage fright and choking (having a panic attack) in athletic domains and other high-pressure settings.

Stages of Skill Learning

Do people go through identifiable stages as they learn skills? According to an influential theory, people go through three main stages: (1) the cognitive stage, (2) the associative stage, and (3) the autonomous stage (Fitts, 1964). I'll unpack these in a moment but want to emphasize that the word "stage" shouldn't be taken literally. There aren't discrete jumps from one level to another. No one goes to bed one night unable to play the guitar at all and wakes up the next day being fully able to do so. Over the long term, skill learners progress from can't-play-at-all to play-a-little to now-it's-all-there (all slang terms invented for this text). The transitions are gradual. Stages can be posited for conceptual and didactic (teaching) reasons, giving a general sense of broad statistical features of performance at various points along the way.

In the cognitive stage, people acquire the most basic elements of the activity to be learned, paying a lot of attention to do so. A classic example is learning to drive a vehicle with standard transmission (a manual stick shift) rather than automatic transmission. Even though stick-shift vehicles are harder to find these days, quite a few of my students have used them, so this task isn't as anachronistic as it might seem.

In the earliest stage of learning stick, you have to concentrate on the positions of the stick to bring the vehicle to its various gears: neutral, first gear, second gear, third gear, fourth gear, fifth gear, and reverse. You may resort to verbal codes to remind yourself that "first gear is left forward," "second gear is left backward," and so on. At the same time, you may talk to yourself about pushing in the clutch with your foot, remembering that you need to push the clutch all the way to floor and then release it slowly once the gear shift is in place. During this first, cognitive, stage, you have to devote all your attention to the steps that are needed for your hands and feet to get your car going at all. Dividing your attention in this early period is nearly impossible. Listening to a conversation at the same time is extremely difficult. Engaging in conversation is unimaginable.

In the second stage, you form associations between stimuli and responses. These associations allow you, over time, to reach the *third* stage, where you get to a point of performing with little or no need for conscious control. By the time you reach the third stage, you can shift gears smoothly with little need for attention. You can listen to others, you can talk, and think about other things.

The third stage of learning needn't be the end of the line. In this stage, you can continue to fine-tune your performance. While gear-shifting, you can learn to control the micro-timing of your actions. Sitting in your car on an uphill grade, for example, waiting for a red light to turn green, you can learn to put your car into first gear by very gradually releasing the clutch, modulating the speed of the clutch-release based on the sound of the engine and based on your feel of the car as it vibrates like a purring cat. Becoming finely tuned to the whirring of the engine and feeling the vehicle gliding into gear, you can become a kind of maestro of gear-shifting. In my own case, when I mastered gear-shifting, I enjoyed rocking my car back and forth a bit while waiting on a steep uphill grade for a red light to turn green. Earlier, when I first started out, I avoided steep hills like the plague because my car conked out whenever I tried to get the car going.

There is experimental evidence for the three stages of skill learning. Research has shown that directing attention to details of the way actions are performed helps performers in the early stages while actually *hurting* performance in the late stages. For example, in a study of expert golfers playing miniature golf, the golfers did less well when asked to direct attention to their bodies than when asked to do putt as they normally would. On the other hand, novice golfers were not hurt by the instruction to attend to their bodies, for that is where they were attending anyway (Beilock, Carr, MacMahon, & Starkes, 2002).

These results can be used to help people overcome stage fright and avoid choking under pressure. If you get stage fright or freak out when you need to perform, you may be attending to the *means* of performing rather the *goal* of performing. A leading researcher on this topic, Sian Beilock (Figure 1.11), calls this problem *analysis paralysis*. Her advice (paraphrasing here) is "Don't think! Just do!" She knows whereof she speaks. Beilock is such a level-headed, competent performer that she became the president of a top-flight college, Barnard College (part of Columbia University, in New York City), in her early forties.[9]

Distributed Practice Is Better Than Massed Practice

Many factors affect skill learning. One, obviously, is the amount of practice on a task. Ten thousand hours is the time needed to achieve expertise, as discussed earlier. Another factor is the way practice is scheduled. Distributed practice is generally better than massed practice, at least for long-term learning. You can understand what this means by considering a study concerning the best way to learn touch-typing (i.e., typing that does not requires looking at the keyboard). People were randomly assigned to four training groups (Baddeley & Longman, 1978). One group practiced for one hour a day once a day. A second group practiced one hour a day twice a day. A third group practiced two hours a day once a day. A fourth group practiced two hours a day twice a day. The striking result was that the total number of hours practiced did not predict performance. The group that practiced just one hour a day achieved typing rates rivalling the rates achieved by the groups that practiced two or more hours a day. Bunching up the hours (practicing for two hours at a time, an example of massed practice) wasn't as effective as spreading the hours out (practicing for one hour at a time, an example of distributed practice). This result makes sense from the point of view of fatigue. If you push yourself to the limit while practicing, you'll get tired, which may cost you time and energy. In terms of cognition, getting too much information of one kind may produce interference: material coming late will interfere with what's already there, and material that's already there may interfere with what comes later. Spreading out the learning can help forestall interference and reduce the chance of physical fatigue.

You can carry this lesson forward to your academic schedule. If you take a class on human memory from 9 a.m. to 10 a.m., you won't remember the material very well if the next class you take is Principles of Cognitive Psychology. On the other hand, you'll remember more from the memory class if your next class is swimming.

A related implication is to avoid cramming for exams. Cramming is the paradigmatic example of massed practice. A good night's sleep before an exam will serve you better than cramming all night long, especially if you spread your studying over several days.

Sleep, Interference, and Consolidation

Sleep was just mentioned, and it enters here in an interesting way. One of the first studies of sleep and learning provided a demonstration of memory interference. In the study, published long ago, people learned lists before going to sleep or before starting busy work days (Jenkins & Dallenbach, 1924). The participants were tested several hours after exposure to the lists. It was found that participants remembered more if they slept for several hours than if they spent the same number of hours doing other things. The outcome suggested that the amount of material remembered doesn't simply depend on the amount of time since initial learning. Rather, the nature of what occurs during the retention interval plays an important role. During waking hours, new information tends to interfere with the initially stored material. During sleep, there is less new input and the information can be better remembered.

Does sleep per se have a benefit for learning? Setting aside the fact that it obviously gives you a chance to recharge your batteries physically, sleep may allow for memory consolidation. Memories take time to sink in, crystallize, or consolidate. Fact learning benefits from sleep, and so does skill learning, as demonstrated in a study where people practiced counteracting forces applied by a robot that tugged on their hands as they aimed for targets. The participants did better on the task over the long term if they slept before returning to the task than if they returned to it the same number of hours later without the chance to sleep. Details of the procedure and results helped rule out the uninteresting possibility that subjects in the nonsleep condition were simply more tired (Brashers-Krug, Shadmehr, & Bizzi, 1996). Other studies led to similar results, supporting the view that memory consolidation occurs during, and is specifically aided by, sleeping (Dumay & Gaskell, 2007; Pan & Rickard, 2015).[10]

Blocked Practice, Varied Practice, and Specificity of Learning

Within a practice session, is it better to focus on one specific task and practice it over and over, or is it better to practice a range of tasks that are only somewhat related? The answer depends on your goal. If you want to do as well as possible in the short term, it's best to practice one task repeatedly (blocked practice). By the end of the practice period, you will probably do well on that one highly practiced task. On the other hand, if you want to do well long term on a *range* of tasks, it's better to get varied practice.

Data supporting this conclusion have come from several experiments. In one (Figure 5.6), participants struck barriers as quickly as possible. If the barriers were the same in successive training trials, the participants did very well by the end of the training session. However, later, if they were given new barrier tasks, they did poorly. They had a hard time generalizing what they had learned to novel barrier challenges. Conversely, if the barriers were different during training, the participants did less well on the barrier task at the end of the training period. However, they did better on new barrier tasks that were tested after the initial training stint (Shea & Morgan, 1979).

Does this mean that practicing intensively on one task leaves you unprepared for later performance on *that* task, the task you practiced at first? The question is worth asking because, strictly speaking, a possible conclusion from the result reported above is that specific practice doesn't stick. It turns out that that's not the case. Highly specific practice can be retained, but what's retained is highly specific. This phenomenon is called *specificity of learning*.

A demonstration of specificity of learning came from a study of foul shooting in basketball (Figure 5.7). Basketball players often spend an especially long time practicing shooting from the foul line. A team of researchers wondered how well basketball players would do at shooting baskets at different distances from the basket (Keetch, Schmidt, Lee, & Young, 2005). Would they do especially well if the distance happened to be the distance of the foul line from the basket? As shown in Figure 5.7, the skilled basketball players stood at different distances from the basket along the straight line joining the center of the basket to the center of the foul line. As expected, the greater the distance, the lower the chance of sinking the basket, but of special note, the players did better when shooting from the free-throw line than would be expected from distance alone. With all the practice the players had gotten at that distance, they learned something specific to that skill. Other studies have produced comparable results (Czyż & Moss, 2016; Nabavinik, Abaszadeh, Mehranmanesh, & Rosenbaum, 2017).

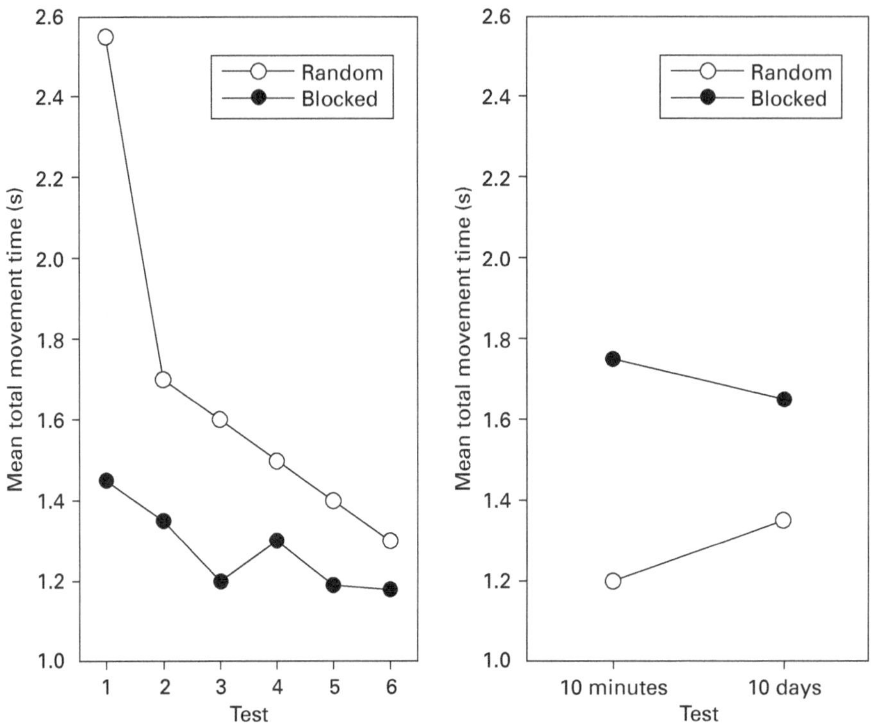

Figure 5.6
Short-term (left panel) and long-term (right panel) effects of blocked versus random (varied) practice.

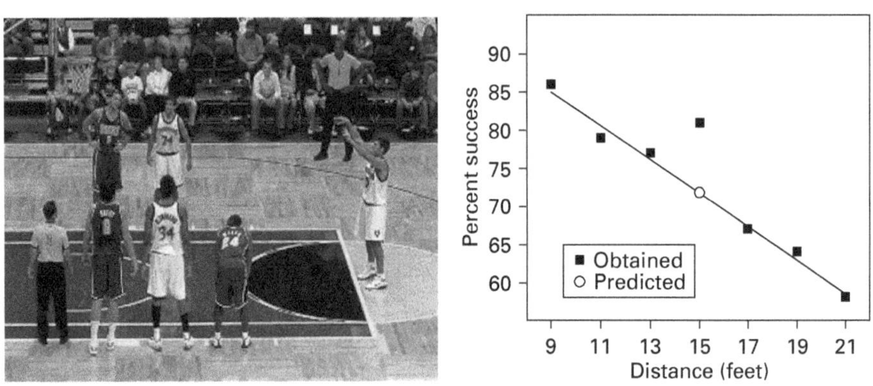

Figure 5.7
Basketball foul shooting (left panel) and the special benefit that comes from making many more shots from the foul line (15 feet from the basket) compared to what would be predicted by simply considering distance from the basket (right panel).

Summing up this section, you can say that it's possible to get super-skilled on narrow activities. At the same time, it's best to prepare for a wide range of demands by getting a wide range of experiences.[11]

Power Law of Learning

Learning a new skill can be exhilarating but also frustrating. It can be exhilarating because, as you progress, you find yourself able to do things you only dreamed of doing at first. But it can be frustrating because you may have to struggle in the beginning to pursue your dream. At first, you may be unable to do even the most basic things. Seeing others engage in the activities you hope to master may leave you wondering whether you'll ever be able to do them yourself. That shouldn't deter you, though. Toddlers fall thousands of times before they learn to walk, but they keep getting up and trying again (Adolph et al., 2012). That said, novices enjoy a special advantage that more experienced performers don't. They improve more from one practice session to the next. Experts only improve by tiny, often imperceptible, amounts as they move from one practice session to another. It may take months for an expert to see as much improvement as a novice sees in a day. This outcome is to be expected from the learning curve shown at the start of this chapter. It was the basis for the suggestion given earlier that college goes on for four to five years because seeing notable improvements afterward might require too many years for the investment to seem worthwhile.

As learning continues, gains get smaller and smaller, as just said. Another way of saying this is that, as skill increases, the rate of improvement decreases, as shown in Figure 5.8. This statement is summarized, mathematically, by the power law of learning. The power law of learning came to light in a study of workers at a cigar factory. The researcher who characterized the relationship, a young man at the time named Ted Crossman, analyzed the number of cigar boxes that the workers produced per work period as a function of how long they worked at the factory (Crossman, 1959; Rosenbaum & Janczyk, 2019). The longer the workers worked there, the more cigar boxes they produced per unit time. Less and less time was needed for each cigar box the more practice the workers had at the job.

Crossman captured this relationship with a single mathematical formula:

$$T = aP^{-b} = a/P^{b},$$

where T was the time needed to produce each cigar box, P was the amount of practice (number of cigar boxes previously filled), and a and b were positive real numbers. Another way of writing $T = aP^{-b}$ is $T = a/P^{b}$, which may look less formidable (less intimidating) because it's a simple fraction with the denominator raised to the b power. The middle expression means the same thing, just expressed in a less familiar way for some. No matter how the equation is written, the larger the value of P (the more practice there has been), the smaller the value of T. The overall height of the curve is scaled by a; the larger the value of a, the higher the curve. And the rate at which T drops with P depends on b; the larger the value of b, the steeper the drop.

The formula above, no matter how it's written, is called a *power function*. It has that name because the independent variable, P, is raised to a power. Depending on the values of P, a, and b, the curve drops less and less steeply as P grows (as practice continues). By adjusting the values of a and b, it has been possible to generate curves that come close to obtained data like those shown in Figure 5.8.

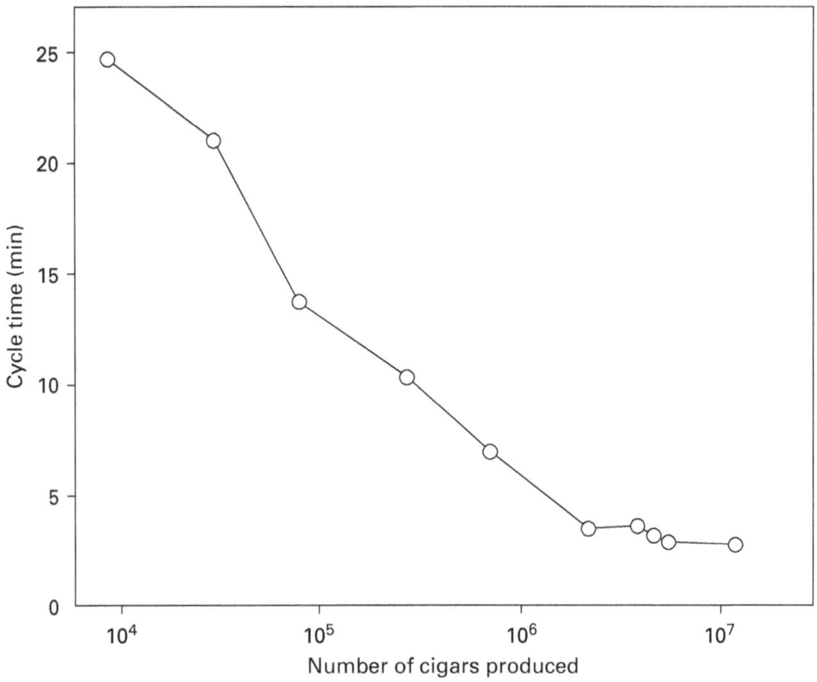

Figure 5.8
Data that led to the power law of learning. The graph shows the time needed to produce a fixed number of cigar boxes as a function of the number of cigars produced. When 10^6 cigars had been produced, roughly one year had passed. When 10^7 cigars had been produced, roughly seven years had passed.

This section is called the Power Law of Learning, but all I've said to this point is that a power function provides a good fit to cigar-rolling times. That's hardly sufficient reason to say that the function should be called a *law*. For a mathematical formula to earn that moniker (that name), it must apply to a very wide range of phenomena. Beyond that, it should say something important.

What is the basis, then, for speaking of the power *law* of learning? First, the power function holds over a remarkably wide range of tasks. As practice continues, performance times shorten at decreasing rates in decision making (pressing buttons in response to light patterns), reading text that has been flipped in various ways, solving arithmetic problems, proving mathematical theorems, and even writing books.[12]

Second, the underlying regularity captured by the power function suggests a deep property of learning. As the denominator P^b gets larger, the ratio $T=a/P^b$ approaches zero, yet T never reaches zero. This feature of the power law suggests that through continued practice, it is always possible to improve, to keep cutting one's time on the task, even if the improvements get harder to see.

What mechanism could yield such an endless potential for improvement? Ted Crossman, the researcher who advanced the power law, suggested that the basis for improvement could be trial-and-error learning. His idea was that learners try different techniques at random, that random changes that happen to result in reductions of task completion times are retained, and that the chance of finding a time-reducing method declines with practice.. If each newly helpful method causes the same or roughly the same time reduction, the result is the power law relation.[13]

Physical Changes and Reflexes

The account just given does not mention physical changes to the body over the course of skill learning. Such changes occur, of course. Chapter 3 showed how the arms of tennis players can change as a result of prolonged use of a tennis racquet. As you grow, you benefit from a larger, typically stronger, body, but you also have to make adjustments for that growth. If you happen to grow to basketball-center height, you need to check that your head clears the door when you enter a bus or airplane. If the way you grow is *out* rather than *up*, say as a result of pregnancy, you have to be similarly mindful, considering the horizontal axis rather than the vertical.

Getting taller or having a protruding belly are changes that occur slowly, but other kinds of bodily changes can occur in short times. If you walk empty-handed and then pick up heavy bags, the signals to your leg and trunk muscles must change for you to keep going. If you pick up a pebble in your shoe, you have to change your gait. Your nervous system must respond to such changes and, of course, it does. How it does so is a core problem in motor neuroscience.

An important step toward understanding how changes in the body affect the expression of skill was provided in the 1980s and 1990s by a developmental psychologist named Esther Thelen (Figure 5.9). She was on the faculty at Indiana University, although here career was all too short. After joining the faculty relatively late in life after

Figure 5.9
Esther Thelen (top left), James Gibson (top right), Lloyd Morgan (bottom left), and Occam (bottom right).

rearing children, she died of cancer at the age of only sixty-three. Thelen demonstrated that changes in the body explain a feature of behavior that, up to that point, had been ascribed to cognitive mechanisms: the appearance, disappearance, and reappearance of the stepping reflex (Thelen, 1995; Thelen & Fisher, 1982).

From birth to about two months of age, babies step. Held under their armpits with their feet touching a solid surface, they move their feet—pitter-patter, pitter-patter—as if taking a hike. Mysteriously, however, the stepping stops and doesn't return until several months later, when full-fledged walking begins.

Why does stepping appear, disappear, and then reappear? Is it the case that all inborn reflexes have this property? If that were so, the temporary disappearance of the stepping reflex wouldn't be unusual and might just be chalked up to some widespread neural change.

It turns out that some inborn reflexes disappear, some inborn reflexes persist, but no other inborn reflexes appear, disappear, and reappear. The stepping reflex is unique in this regard.

One inborn reflex that disappears—so it is in the first category listed above—is the Babinski reflex, named for the French/Polish neurologist, Joseph Babinski. The Babinski reflex is elicited by stroking the sole of a baby's foot. The stimulation causes upward curling of the toes away from the sole, a movement known as dorsiflexion. After two to three years of age, the upward curling of the toes turns into downward toe curling (toward the sole of the foot, also known as plantarflexion). If the direction of toe curling doesn't change, that can be a sign of neurological damage.

Another neonatal response that disappears is the Moro reflex (Figure 5.10). This is a reflex that newborns are tested on as part of their standard initial evaluation. The baby is subjected to forces similar to those encountered in falling. The baby's head is dropped onto a safe receiving surface such as a pillow. The dropping isn't done to

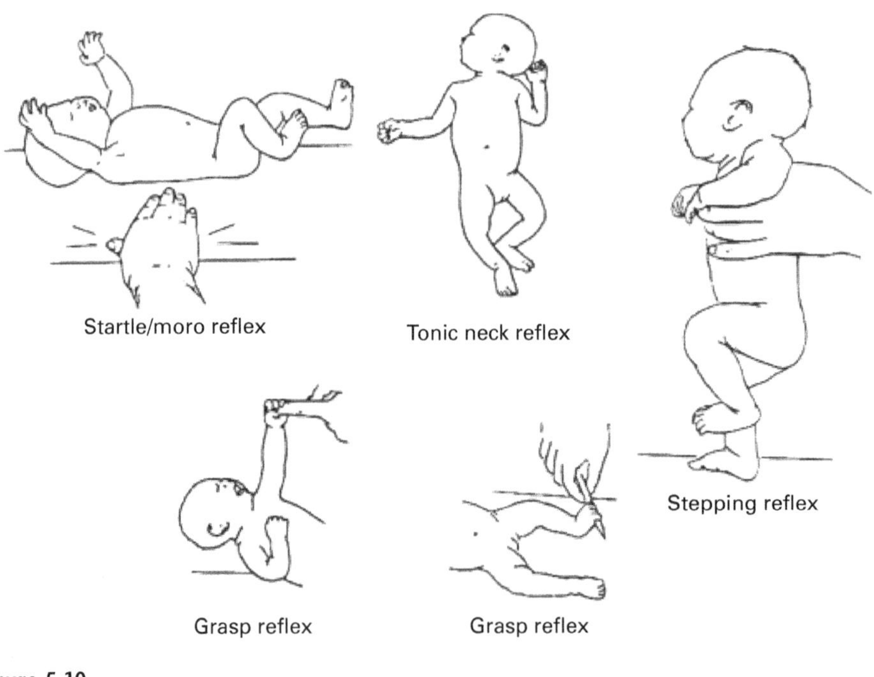

Figure 5.10
Innate reflexes.

torment the baby but to check that the baby extends its arms symmetrically. Some have seen this reaction as analogous to clutching a tree branch or fur of an ancestral mother. It the Moro reflex doesn't appear, that's cause for concern. So too is strong asymmetry or exaggeration of the response. Any of these features may signal brain damage, neurotoxicity, or withdrawal from narcotics. Under normal circumstances, the Moro reflex disappears by three to six months of age.[14]

Whereas the Babinski and Moro reflexes normally disappear after infancy, other reflexes persist. One is the asymmetric tonic reflex, which lasts into adulthood (Figure 5.10). Here, if a baby turns to one side, she bends her elbow and knee on the opposite side. The asymmetric tonic neck reflex is so fundamental to adult behavior that it is seen in adult activities, including fencing and catching flyballs in baseball. The persistence of the asymmetric tonic neck reflex has been taken to show that inborn reflexes provide a substrate for the movements we make as adults, as espoused by Sherrington and discussed in the last chapter (Easton, 1972).

Another inborn reflex that lasts through the lifetime is the startle reflex. Here you jump in response to a sudden loud sound or mechanical perturbation such as an aggressive, unexpected shove from behind. If you startle, you pull your arms in and lower your head. This protective response is present from birth and can be triggered all through life.

The inborn reflexes reviewed above persist, as in the asymmetric tonic neck reflex and startle reflex, or they drop out, as in the Babinski and Moro reflex. None of the reflexes does what the stepping reflex does, which is to appear, disappear, and then reappear. What causes this evanescence?

One hypothesis that was taken seriously was that babies are so busy learning other things that they run out of neural responses for stepping. Sequence learning in particular was viewed as the topic that most deterred stepping, for the stepping reflex disappears when babies showed expanding sensitivity to the sequential structure of repeated light patterns and other stimuli. Esther Thelen was skeptical of this hypothesis for two main reasons. First, babies show the stepping reflex when supine (when lying down) at ages when they don't show the stepping reflex when held up. Second, in this age range, babies kick vigorously when held in water (Figure 5.11). These results showed that the basic neural ability to oscillate the legs didn't disappear. Stepping wasn't masked by cognitive events. It was impeded by physical factors.

What physical factors could they be? Thelen suggested that when the stepping reflex disappears, babies have chubby legs. Raising and lowering their legs while standing on solid surfaces takes more strength than the babies possess. On the other hand, stepping while lying down is easier, for then both feet remain on the ground and merely slide back and forth without the need for greater force for raising and lowering. The same can be said for stepping in water. While floating, less force is needed to raise and lower the legs. So the stepping reflex disappears and then reappears for a simple reason. Babies are simply too weak to move their legs while supporting their own weight in a critical age range, but as they get stronger and their legs thin down, they can move their legs again.

The Ecological Perspective

Now it's possible you're not terribly impressed with that conclusion. In response to the assertion that babies have chubby legs and aren't strong enough to move their legs while standing, you might exclaim, "Oh, anyone could have thought of that!" In retrospect, the explanation does seem obvious, but good explanations often have that property. They seem obvious *after* they've been proposed.

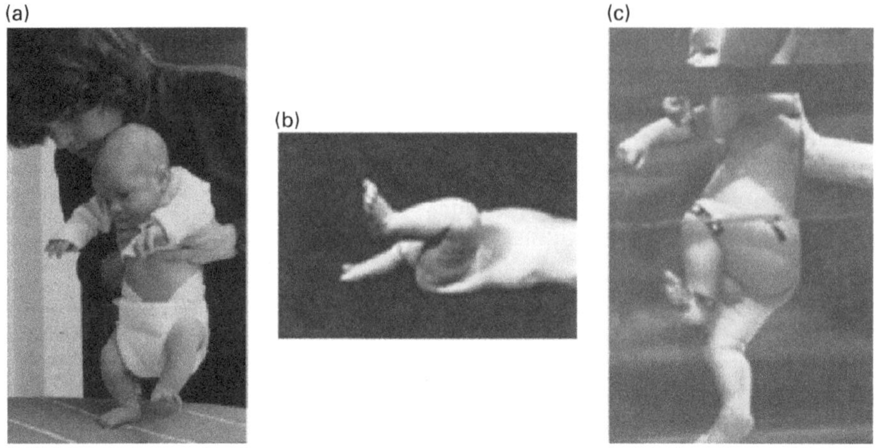

Figure 5.11
Appearance and reappearance of the stepping reflex. (a) A newborn showing the stepping reflex. (b) An older baby who could not step while held up as in panel a but could kick while supine. (c) An older baby who could not step while held up as in panel a but could step while held up in water.

When Esther Thelen offered her analysis, people in her field did not look askance at her explanation (think it was silly or mundane). Instead, they lauded her or her account, for it helped promote a perspective that was becoming popular at the time and has continued to be influential to this day—the *ecological* perspective. That view was popularized by James Gibson (Figure 5.9), whose ideas were discussed earlier in this book. Gibson's ideas were, in turn, traceable to a British psychologist and ethologist (someone who studies behavior under natural conditions) named C. Lloyd Morgan (1852–1936). Morgan (Figure 5.9) spent a great deal of time watching animals behave in their natural habitats. He sought to understand how animals controlled their behavior, and he was especially interested in their learning and decision making.

Morgan wanted to be disciplined in his inferences. He wanted to avoid anthropomorphizing, the tendency to attribute to animals and, for that matter, inanimate objects, human thoughts and feelings. Morgan was one of the first to warn against anthropomorphizing. He expressed this view in a turn-of-the-twentieth-century textbook called *An Introduction to Comparative Psychology*, where he wrote the following passage (https://en.wikipedia.org/wiki/Morgan%27s_Canon#cite_note-2):

> In no case may we interpret an action as the outcome of the exercise of a higher mental faculty if it can be interpreted as the exercise of one which stands lower in the psychological scale.

This passage has come to be called *Morgan's canon*. It can be understood as an appeal to parsimony: given possible explanations of some behavior, prefer the simplest one.

Esther Thelen thought along similar lines. She argued that one should look to the physical makeup of the body to better understand what can and can't be done with it. If it's possible to explain a behavioral phenomenon such as the disappearance of stepping in babies by appealing to the chubbiness of their legs and the strength needed to move them, then that explanation should be preferred over a more complicated account that relies on unobservable mental processes like preoccupations with sequence learning. Thelen wasn't a radical behaviorist who sought to deny mental events. She simply wanted the most principled account she could come up with. In so doing, she relied on observations of body composition and body strength.

As I said above, when Thelen published her findings, she was widely acclaimed, especially by ecological psychologists. These people drew inspiration from James Gibson, who challenged the idea of unconscious inference, introduced in the late nineteenth and early twentieth centuries by Hermann von Helmholtz, who was discussed earlier in this book and will be discussed at some length in Chapter 7. Gibson endorsed the idea that perception is direct, meaning that the way the environment reflects energy specifies the environment's structure. How light rays are reflected from a roughly textured rock is different from the way light rays bounce off a smooth sheet of satin, for example, so the optic array emanating from the objects directly specifies their structure. (Recall the earlier discussion of the diving gannets and the disappearing steps for people walking down stairs. Both relied on this concept.)

Those who follow Gibson's teachings also endorse another concept he advocated, the concept of *affordances*. Affordances are actions that are possible in an environment. A chair affords sitting and also affords standing if one needs to reach a high shelf. Affordances are picked up, Gibson said, by virtue of the structure of the environment, both external and internal (body based).[15]

Thelen understood this idea at a deep level and saw, accordingly, that it is vital to attend to the behavioral capabilities of the body itself. Thelen's view respected Gibson's teachings, which were traceable in way or another to Morgan's canon, as said above, which in turn harkened to Occam (Figure 5.9), whose famous razor cut away unnecessary explanatory frills. The simplest explanation is probably correct, Occam avowed. That perspective has echoed through the ages.[16]

Jolly Jumpers

The final section of this chapter extends these ideas to jolly jumpers. This term is a double entendre (a term that means two things), referring to a device for babies to jump in and the happy state they're in while jumping.

Begin with the observations that between roughly two and seven months of age, babies who try to step find it hard to do so. However, if they never tried to step, they'd remain passive blobs. Babies do, of course, step again. When their legs get thinner and their strength increases, they step, toddle, walk, and run. Along the way, they continually "test the waters," trying out different acts or proto-acts (miniature approximations of full-scale acts) to find out what outcomes arise.

Allowing babies to test the waters can be done by placing them in water, as in Esther Thelen's study, and it can be done by letting babies occupy Jolly Jumpers (Figure 5.12). You may have occupied one of these yourself, or you may have watched a baby in one of these contraptions. A Jolly Jumper is a seat attached with elastic bands to secure points overhead. A wide door jamb above the Jumper makes a good support, as does a crossbeam. The height of the device must be adjusted so the suspended baby's feet can touch the ground.

When a baby is first slung in a Jolly Jumper, nothing happens. The baby just hangs there. In my own experience raising a couple of babies, when I put each of them in the Jolly Jumper, they just hung there, looking at me quizzically, as if to say, "What's this all about?"

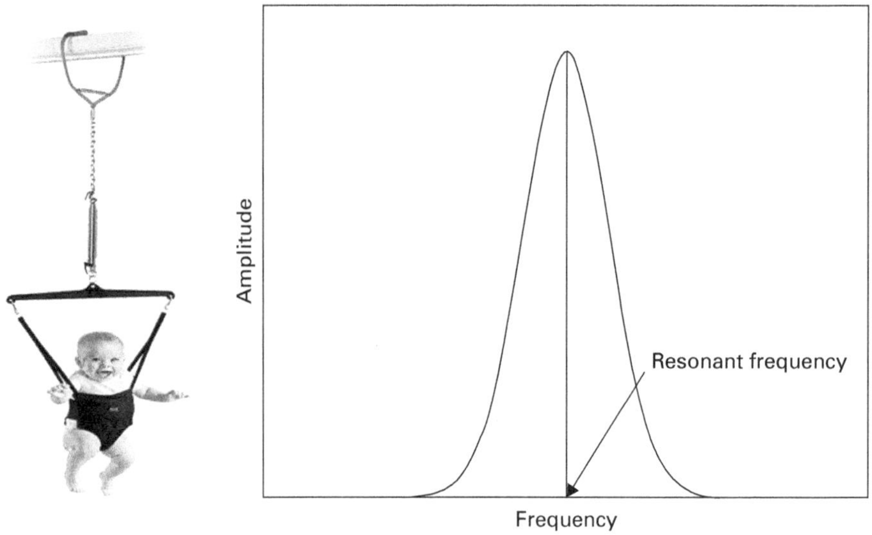

Figure 5.12
A baby in a Jolly Jumper (left panel) and resonant frequency (right panel).

A major challenge for a baby in a Jolly Jumper to stop spinning around, which is what happens if the baby doesn't touch the ground with her foot. If she puts a foot on the ground, she can stop the turning, perhaps to return to staring at the person who put them in the perch. That person may demonstratively bob up and down, jumping repeatedly, causing the infant to think that whatever this strange new seat was for, it certainly makes the grownup seem even sillier than usual.

Over time, things get more interesting. The baby learns to bounce. At first, she bounces barely. If she happens to press a foot on the floor, she manages to rise a bit. This happens over and over until the incidental foot touches become more forceful and purposeful. The touches also become more frequent. If the baby pushes down with her feet each time she lands, she can keep bouncing for a long time, covering larger vertical distances up to some maximum height.

Data bear out this summary (Goldfield, Kay, & Warren, 1993). In successive days of practice after babies get the basic hang of it, stopping their spinning and moving up and down more than side to side, they manage to generate larger and larger jumps with lower and lower variability in jump launch times.

What's happening here? The babies, quite simply, are learning physics. They learn about the principle of *resonance*, the idea that at a certain frequency, you get the biggest bang for the buck (Figure 5.12).[17]

You can understand resonance by recalling what it's like to push a child on a swing. If you push the child each time he or she completes a cycle of swinging—going out and coming back again—you needn't push so hard. The force you need can be less than the force you'd need if you pushed at some other time in the cycle. Timing your pushes to coincide with the natural frequency of the child-swing physical system lets you tap into the system's resonant frequency.

Someone pushing a child on a swing isn't essentially different from a child pushing him or herself off the floor in a Jolly Jumper. The same principle applies. Indeed, the authors of the Jolly Jumper study concluded that the babies they studied were behaving in accord with self-discovery of resonant frequency. Through trial and error, the authors said, the babies learned how and when to launch themselves. The babies were sensitive to feedback from their bodies and developed a capacity to modify their behavior to be as jolly as possible (or at least as "bobby" as they could have been, physically speaking).

The study of Jolly Jumpers is just one of several that point to the facility with which learning can be body based. The study of action, mind, and brain is a perfect domain to analyze this capacity, and the behavioral transitions that occur over time are perfect examples of trial-and-error learning. Through such learning, organisms (such as babies) tend to engage in behaviors that yield rewards. Trial-and-error learning isn't restricted to Skinner boxes; those are just convenient chambers for learning about learning itself. The ultimate Skinner box is the body and the world it occupies.

6 Feedback

The phrase "trial-and-error learning" has come up a lot in this book. In the last chapter, you read about trial-and-error learning as the basis for the power law of learning and as the basis for jolly jumping. In the case of the power law of learning, trial-and-error learning leads to shorter and shorter times to complete tasks such as rolling cigars, reading texts, and writing books. For jolly jumping, trial-and-error learning leads to less and less variable jump times and increased jump heights. Because the outcomes of trial-and-error learning differ in these cases, you can see that trial-and-error learning provides widespread benefits. But how does trial-and-error learning work? What's the basis for its wide application?

Trial-and-Error Learning through Feedback

Consider a cat in a cage who wants to get out but doesn't know how (Figure 6.1). It tries all sorts of maneuvers: scratching, meowing, clawing. After lots of random activity, the cat manages to lower a lever outside the cage door and the door pops open. Out scurries the cat, racing past the man who has been observing this, a Columbia University professor named Edward Thorndike. Thorndike notes the time between when he puts the cat in the cage and the cat manages to escape. Repeating the experiment with the same cat, he finds that the cat gets out more and more quickly the more often it's put in the cage (Thorndike, 1898). The time for the cat to escape decreases at a decreasing rate as practice continues. The drop is large between the first trial and second, smaller between the second trial and third, and so on.

Thorndike made these observations in the nineteenth century and didn't dwell on the best mathematical formula to describe the pattern he observed. In the last chapter, however, you saw that the formula would turn out to be, or would be fairly well approximated by, the power law of learning.[1] Rather than generating a formula, Thorndike formulated a statement that has proven to be one of the most important ideas in the history of psychology and education, the *law of effect* (Gray, 2011, pp. 108–109)[2]:

> Responses that produce a satisfying effect in a particular situation become more likely to occur again in that situation, and responses that produce a discomforting effect become less likely to occur again in that situation.

Take a moment to contemplate the law of effect. The proposition implies that learning works in such a way that actions with positive effects become more probable, whereas actions with negative effects become less probable. The law makes no claims about the inner processes yielding this outcome. For example, it does not posit an inner executive

Figure 6.1
Cat in a box from which it could escape more and more quickly based on trial-and-error learning.

who deliberately tries out different strategies. The law of effect relies on nothing more than trial-and-error learning. Actually, it doesn't even go that far in any explicit sense, for it says nothing about the process by which the observed regularities arise.

How, then, does one get from the law of effect to the observation that there are large improvements at first and smaller improvements later? At least two ingredients are needed: (1) repetition with variability and (2) a diminishing pool of possible means of improvement.

The method of trying over and over again to achieve a goal was embodied in an influential model of behavioral control described several decades after Thorndike's seminal work. The model was described in a book called *Plans and the Structure of Behavior*, by George Miller, Eugene Galanter, and Karl Pribram (1960). You might recognize George Miller's name. He was responsible for the concept of chunking in memory. Perhaps you encountered the chunk concept in connection with the idea that a "magical number" of meaningful elements (chunks) can be retained in working memory. That number, according to Miller, is 7 ± 2. The symbol $\pm$ can be read as "plus or minus"; it's an estimate of the confidence interval that Miller could put around 7. Miller was reasonably confident that he could capture the memory capacity of almost 95 percent of most anyone by saying that they could remember between five and nine meaningful elements—for example, five to nine random digits or five to nine names (Miller, 1956).[3]

In *Plans and the Structure of Behavior*, Miller et al. (1960) suggested how trial-and-error learning might work. They called the procedure TOTE, short for Test-Operate-Test-Exit (Figure 6.2). According to the TOTE model, goal-directed behavior is controlled by trying out an action to see if it works (Test), doing something if it doesn't (Operate), seeing again if it works (Test), doing something again if it doesn't, and quitting (Exit) when success is reached. To avoid going on forever, there can also be a time's-up rule or analogous limit on the number of tries allowed. The most important idea is that with variability in the way acts emerge, success is likely to come along eventually.

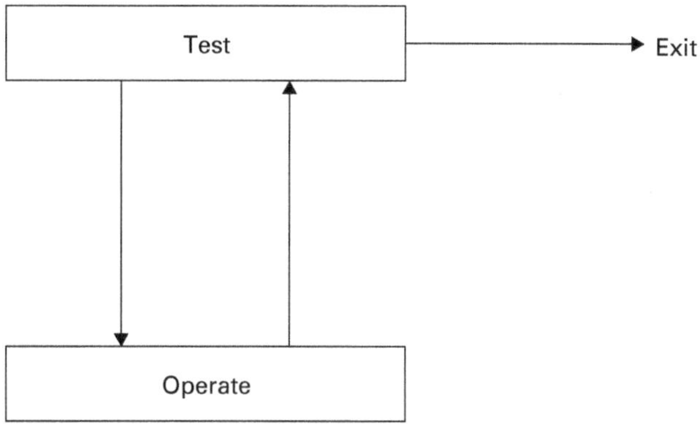

Figure 6.2
The TOTE model.

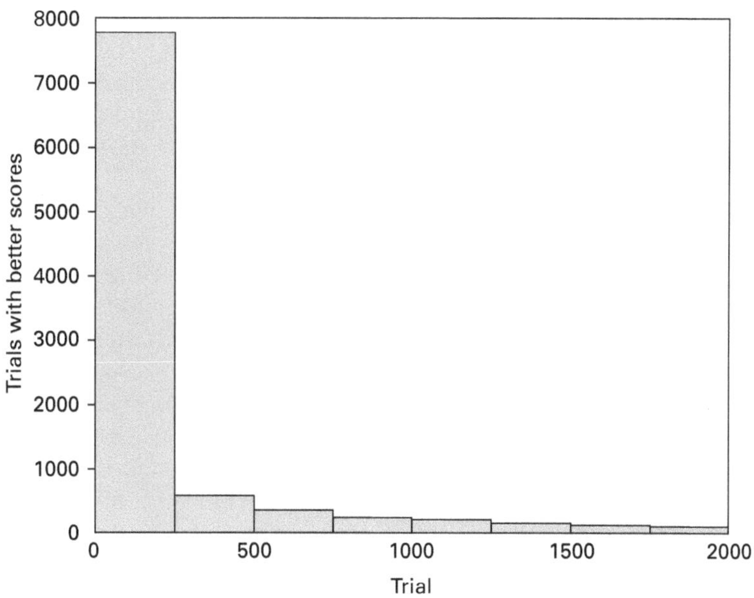

Figure 6.3
Trials with better scores using trial-and-error learning.

Repetition with variability is the first of the two ingredients needed for trial-and-error learning. The second ingredient is a diminishing pool of possible ways to improve. Suppose, as shown in Figure 6.3, that methods that get tried are marked in memory along with their positive or negative effects. Previously used methods with good marks are retried with higher probability than previously used methods with bad marks. When a previously used method is reused, it may emerge a bit differently than it did before because of random statistical noise. Because of the variability, new methods emerge, and they in turn get good marks or bad. Over time, the changes that come along tend to get smaller and smaller, and the improvements that are seen get smaller and smaller as well. Which kinds of improvements come along depend on which outcomes are valued.

If higher speeds are valued, shorter *times* emerge, as in the power law of learning, but if higher *jumps* are valued, as in jolly humping, higher jumps come along. This is how trial-and-error learning can provide benefits of different kinds.[4]

Control Theory

Learning was the focus of the last chapter, and learning is one of the most important sources of information about how feedback is used. Feedback is used in the TOTE model, as just seen, and it makes sense to consider how feedback has been studied more broadly.

The TOTE model instantiated feedback control theory, which centers on the feedback loop (Figure 6.4). The feedback loop has a *reference condition*, otherwise known as a goal state. A typical goal state is getting a golf ball into the next hole on a golf course. For an activity like golf, actions are taken via the *plant*, the physical system producing the output, which is compared to the reference condition. The difference between the reference condition and the feedback is the error. If the difference is zero or close enough to zero to be judged acceptable, the job is done. The ball got into the hole, for example, and in the TOTE model, it's possible to exit. However, if the difference is positive, the output relative to the goal is too large. The ball went too far, say, and corrective action is required. Similarly, if the difference is negative, the output relative to the goal is too small. The ball didn't go far enough and something should be done to reduce the error (to get closer to the hole).

A feedback loop has a cycle time, the time between receipt of feedback and production of the corrective output. The feedback cycle time, or loop time, as it's also called, can be very long or very short.[5] Times between exams and feedback can take a long time—months in the case of bar exams for lawyers, for example. Times between golf shots and sight of the ball landing in or near the hole are much shorter, a few seconds at most. The shortest feedback loop times for human actors are associated with monosynaptic reflex responses, such as the kick of your leg in response to a tendon tap, as discussed in Chapter 3.

An important feature of feedback control systems is whether they are *closed loop* or *open loop*. A closed-loop system is one in which feedback returns and can be used for error correction. An open-loop control system is one in which no feedback returns, or if it does, it can't be used for error correction. The studies of deafferented soldiers and monkeys reviewed in the last chapter illustrated the transition from closed-loop systems (systems with feedback) to open-loop systems (ones without).

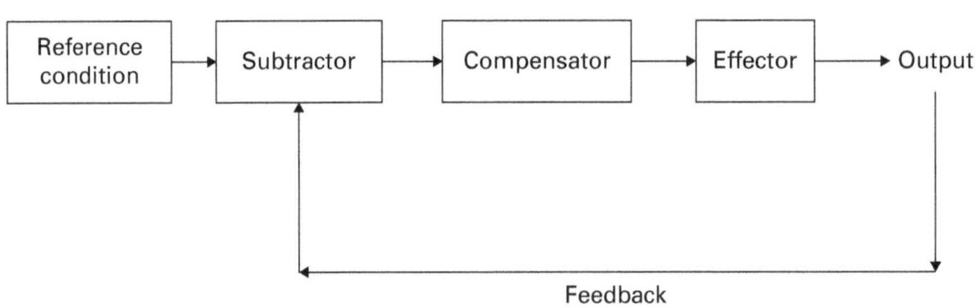

Figure 6.4
Feedback loop.

Open-loop systems don't just exist due to surgery or injury. Some exist by design. If you have an automatic lawn sprinkler that comes on at the same time of day no matter how dry or rainy it is, you have an open-loop lawn-sprinkling system. A baseball thrown from a pitcher's mound is an open-loop system once it's released, for once the ball has left the pitcher's hand, it has no onboard guidance. The ball, once let go, is said to be *ballistic*.

A missile is ballistic if it has no means of control once launched. You may have heard of antiballistic missiles (ABMs), which are missiles designed to knock out rockets traveling with no guidance, typically to wreak havoc by delivering bombs. ABMs are closed-loop systems since they make corrective actions to home in on the ballistically traveling, bomb-carrying rockets. The phrase "going ballistic" means losing control, striking out at others, yelling, screaming, ranting and raving, having a tantrum, and so on. Another phrase where the word "ballistic" is used is ballistic vomiting. In ballistic vomiting, it's heave-ho all the way. Nothing can stop it.

Compensatory Tracking

Closed-loop systems are essential when adjustments must be made. If you alter the pressure on your gas pedal to maintain a roughly constant speed while driving, you engage in *compensatory tracking*. You may rely on your speedometer to check your speed and press harder to speed up or press less hard to slow down. Another driving example is staying in lane. When you use your steering wheel to say in lane, you engage in compensatory tracking for spatial centering rather than speed maintenance.

Compensatory tracking doesn't only rely on mechanical equipment. When you stand still, you make corrections to stay upright. Keeping your balance is harder than you might think. When you were a toddler, it was hard for you to stand. You spent a long time learning how to prevent yourself from falling.[6] Even as an adult, your body sways a bit when you stand, as shown in laboratory experiments where people stand on force plates and have their center of pressure recorded (Figure 6.5). If your center of pressure drifts noticeably based on proprioceptive and visual feedback, you will take corrective action. Loss of proprioceptive feeling (loss of feeling from the feet or legs, for example) makes it hard to stand, especially with eyes closed.

You can demonstrate the importance of vision for standing by doing what was just mentioned: stand still with your eyes closed. If you close your eyes while standing, you'll probably be unable to stand as long as you can with your eyes open. To make the difference more apparent, try standing with eyes closed while standing on one foot. In that condition, you'll probably be noticeably less stable than when you do the same thing while being able to see.

A dramatic demonstration of the importance of visual feedback in the control of standing was developed by David Lee, the ecological psychologist whose work I discussed earlier in connection with diving gannets; see Chapter 2. When Lee was not studying these diving birds, focusing on the way they closed their wings as they approached the sea off the Scottish coast, he wondered what would happen if people stood in a room whose walls moved around them (Lee & Aronson, 1974).

Lee built a device that I refer to, when I teach my Action, Mind, and Brain class, as "David Lee's Swinging Room" (Figure 6.6). The room was a huge box suspended from a ceiling. It was open at the bottom and could slide back and forth. Lee's subjects were asked to stand still in the room, looking straight ahead. While they did so, their body positions were recorded. When the facing wall approached, the subjects swayed back,

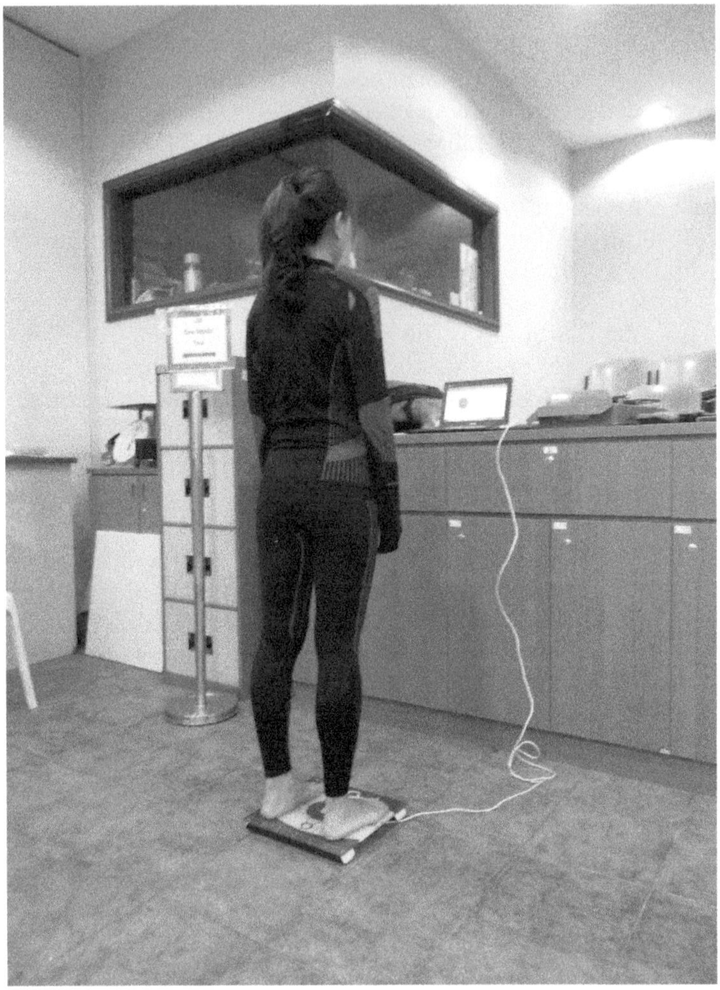

Figure 6.5
Standing on a force platform.

and when the facing wall receded, the subjects swayed forward. When subjects stood on one foot rather than two, their stability was especially affected. Toddlers standing in the room could literally be knocked off their feet by the approaching and receding wall. Even though the subjects only *looked* at the walls, the effect of seeing the walls get closer and farther away was nearly as strong as the effect of being pushed and pulled with actual mechanical forces.

Lee coined a word for the information underlying this effect. He called it *visual kinesthesis*. This term was provocative because, classically, "kinesthesis" refers to the sense of one's own body motion based on signals coming from receptors in muscles, joints, and skin. By adding the modifier "visual" to "kinesthesis," Lee emphasized that the body and environment form a single, coherent whole. This is a core concept for the ecological approach to perception and action, reviewed in the last chapter.

Whereas Lee's swinging room highlights the importance of vision for standing, another setup demonstrates the importance of touch (Figure 6.7). The phenomenon

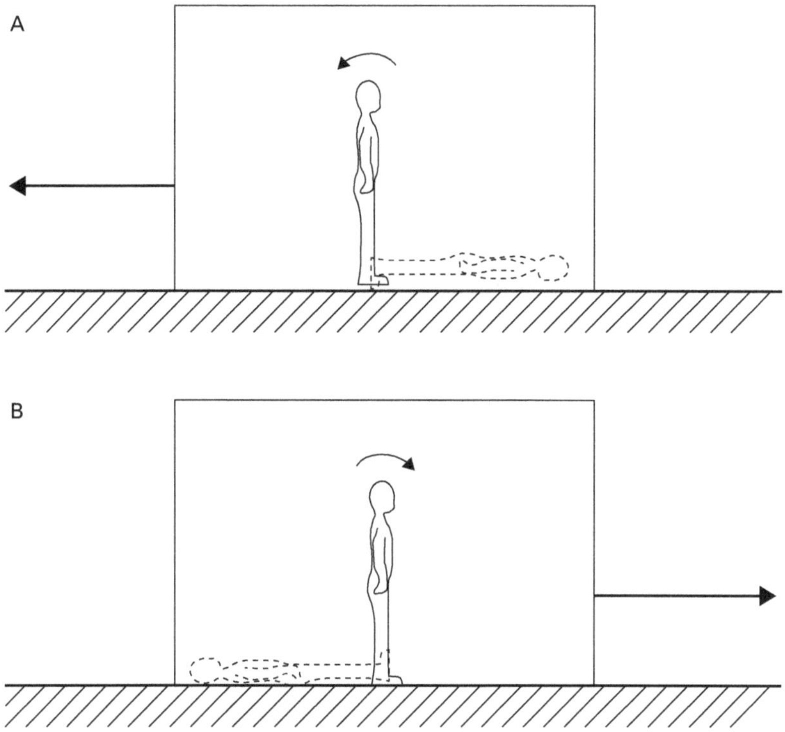

Figure 6.6
David Lee's swinging room along with response to seeing the wall approach (top panel) or the wall recede (bottom panel). The curved arrow above the solid figure shows the direction of sway. The dashed figure lying on the ground shows the outcome prevented by the sway.

involves little more than touching a stable surface while standing on one foot. It turns out that people are better able to stand on one foot when they can touch a stable surface than when they can't (Jeka & Lackner, 1994). Critically, the help comes even if the surface provides no mechanical support. A little finger gently touching a surface works well. This shows that the support provided by the surface is informational, not mechanical. Touching the surface helps you determine, literally, which way is up.

Step Tracking (Saccadic Eye Movements) and Ramp Tracking (Smooth Pursuit Eye Movements)

The last section was concerned with the use of feedback in compensatory tracking. The goal was to keep still. Another kind of tracking has the aim of staying with a moving target. The target can move discontinuously or continuously. If the target moves discontinuously, it jumps from place to place (Figure 6.8). This form of pursuit tracking is called *step* tracking because the graph of the position of the target as a function of time looks like a series of steps. An example is the arcade game whack-a-mole. The mole suddenly appears at one place or another. The player's task is to whack the mole whenever and wherever it suddenly appears.

If a mole leaves its hole and runs nonstop across the surface being watched, the person watching the mole uses ramp tracking rather than step tracking. Pursuing a

Figure 6.7
Merely touching an object aids stability. The author stands heel-to-toe with his little finger lightly touching an unoccupied rocking chair. The chair offers no mechanical support, but touching it promotes steady standing.

continuously moving target is called *ramp* tracking because the graph of the position of the target as a function of time looks like an incline (a ramp) rather than a step (Figure 6.8). Other examples of ramp tracking include keeping your eye on an airplane as it flies across the sky or watching a billiard ball roll across a pool table.

In terms of what your eyes do in these two kinds of tracking, if you step-track with your eyes, you make *saccadic* eye movements. The word "saccade" means jump. You use saccades for a wide range of scanning functions—for reading, for looking at other people's faces, for looking at pictures, and so on. If you sit across from a friend and have them look at your face, you'll see them make saccades as their eyes dart from place to place while they scan your visage. If your friend is neurotypical, his or her eyes will land on your eyes, nose, and mouth, with only occasional looks at your hair, chin, or ears. But if your friend is on the autism spectrum, their eyes may be as likely to land on

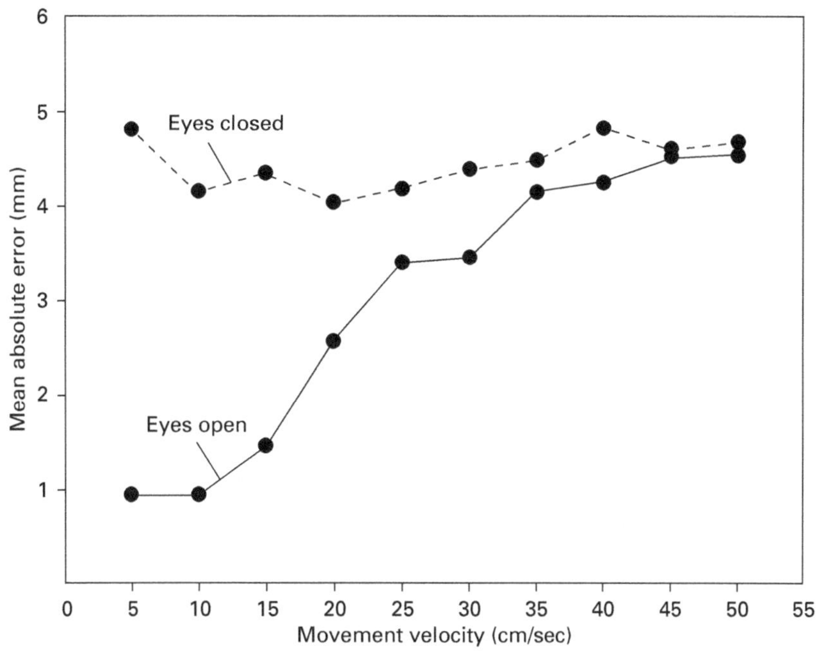

Figure 6.8
Mean absolute aiming error with eyes open or closed.

your hair, chin, or ears as on your eyes, nose, and mouth (Baron-Cohen, Wheelwright, Hill, Raste, & Plumb, 2001).

If you don't step-track with your eyes but instead ramp-track, you make *smooth pursuit* eye movements. These eye movements are used to follow continuously moving objects like the ones mentioned above: the flying plane, the rolling billiard ball, or the mole running continuously across a surface.[7]

Regardless of whether you engage in step tracking or ramp tracking, your performance can reveal the extent to which you're able to predict the motion to be tracked. If a target moves in a simple, repeated fashion, it tends to be easy to track, but if the target follows a more complex nonrepeating pattern, it tends to be harder to track. It takes longer to reach the same level of error for complex patterns as for simple ones.

Tracking reflects learning, as just said, but it can also reflect the extent to which conscious awareness comes into play in learning. This issue was taken up in a seminal study by Richard Pew (1966) at the University of Michigan.[8] Pew had participants move a joystick to keep a cursor on a target that moved smoothly but in a complex repeated path across a screen. Participants got better at staying on the target, although they claimed to have no awareness that the target motion repeated. Pew showed that subjects got better because of pattern learning and not just because of learning incidental features of the task. He did this by altering the pattern from its normal course. When these changes were introduced with no warning and at unpredictable times, subjects made more mistakes than usual. Because the subjects claimed to have no awareness of the underlying pattern, their performance, both verbal and manual, indicated that they were learning implicitly. This was one of the first demonstrations of implicit learning.

Several years after Pew's research appeared, other researchers did experiments similar to his. Nissen and Bullemer asked participants to press buttons beneath lights that turned on in regular but complex, nonobvious patterns (Nissen & Bullemer, 1987). Each light had its own corresponding button. Subjects' reaction times got shorter with practice, and subjects made fewer mistakes as they became more experienced. Their improved performance was due to learning the patterns, as shown by the fact that sudden surreptitious changes in the light sequences cause RTs and errors to increase when the changes occurred. Therefore, Nissen and Bullemer's subjects, like Pew's, learned the patterns implicitly.[9]

Homing in on Targets

When I introduced feedback control theory in an earlier section of this chapter, I referred to golf. Golf comes to mind because golfers take successive shots to reduce errors. They typically cover large distances on the first shot, smaller distances on the second shot, and so on. This kind of homing-in process has been studied in lab tests where, instead of shooting golf balls over vast greens for hours at a time, participants carry out smaller-scale tasks like moving a handheld pointer to a seen target. Some of the most influential scientists in psychology have studied this task. The researchers who have carried out these studies have appreciated that elementary behaviors can shed light on broader principles; this is one of the main ideas in this book.

One of the first studies of this kind was carried out by a colleague of Edward Thorndike at Columbia University, Robert S. Woodworth (Thorndike & Woodworth, 1901). Woodworth made a name for himself based on his doctoral dissertation (Woodworth, 1899).[10] For his thesis, Woodworth conducted experiments that led him to propose the distinction between open-loop and closed-loop control, which was discussed earlier in this chapter. The distinction is widely known in cognitive psychology and neuroscience today, although its origin is little known or remembered. In modern form, the distinction is reflected in the concept of two-stage decision making, with a first pass and an optional second pass. Robert Woodworth was the first to demonstrate this dichotomy scientifically.[11]

How did Woodworth do this? He asked participants to move a handheld stylus back and forth in a slit, trying to reverse course within targets on the left and right. Subjects had to control the placement of the stylus within the target before turning back to the other target. Their performance was paced with a metronome that beat at various rates. The rate stayed the same within each experimental trial but varied over trials. Woodworth asked his subjects to perform with eyes open and with eyes closed. In the eyes-closed condition, they got visual feedback to start with, but then they had to close their eyes while continuing. Woodworth expected subjects to do worse with their eyes closed than with their eyes open. More important, he expected the difference between the eyes-closed and eyes-open condition to appear only when the back-and-forth movement times were long enough for participants to use visual feedback. If the back-and-forth movement times were too short for visual feedback to be used, there would be no difference between performance in the eyes-open and eyes-closed conditions.

Woodworth's data are shown in Figure 6.8. The graph shows mean absolute error (the average distance from each target at either end of the track regardless of its sign) plotted as a function of required movement velocity (the fixed distance between the targets and the rate specified by the metronome). At the lowest velocity (long times between metronome beats), there was a large difference between the accuracy of positioning in the eyes-closed and eyes-open condition, but at the highest velocity, there was no difference between eyes-closed and eyes-open positioning. Between these rates, Woodworth identified the one at which the transition was made from being able to use visual feedback to not being able to do so. That transition point could be viewed as the visual feedback loop time.[12]

Besides estimating the visual feedback loop time, Woodworth also studied the position-versus-time graphs of his subjects' performance. He got these graphs by using the technology of the day, a long strip of paper rolling beneath the slit. The stylus held by the subject left a continuous line on the paper scroll.[13] After the experiment ended, Woodworth turned the paper sidewise and examined the position of the stylus as a function of time. He observed that when the eyes were open and there was enough time for visual feedback to help, participants made corrective movements. If they brought the stylus *short* of the target, they next advanced it to bring it closer, but if they brought the stylus *beyond* the target, they next reversed the path. The series of movements generally continued until the target was reached, provided visual feedback was available and there was enough time to use it. When neither of these conditions was met—when visual feedback was unavailable or when the time was too short to be used—Woodworth observed large moves toward the targets but no follow-up corrections.

Woodworth explained these results by saying that movements to seen targets consist of large ballistic moves followed, if possible, by smaller corrective moves. This proposed sequence of events—ballistic performance followed by corrective performance—may be likened to other forms of cognition and decision making. We often take a first crack at a problem, using rough heuristics to arrive at an answer, and then refine our decision upon further reflection. This idea is best known in connection with Daniel Kahneman's work, referred to in the footnote above. It is noteworthy that Woodworth came up with this idea nearly a century before Kahneman did and surely influenced Kahneman's thinking even if only indirectly.[14]

Speed–Accuracy Trade-offs and Ratio Relations

Normally when you reach for a target, you don't do so at a pace specified by a metronome, as in the method used by Woodworth. You might do so if you were playing kettle drums in an orchestra or hitting piano keys while playing a mazurka, but if you were reaching for a glass of water or extending your hand to press an elevator button, you'd probably move at a speed that you find comfortable. The time would be chosen without conscious deliberation and would be adaptive. What would that adaptive time be? In general, how long does it take to move to a target depending on the aiming requirements?

Consider the following experimental task. You are asked to move a stylus to a target some distance away. In different conditions, the target's width varies. When the target is far away, it can be wide or narrow, and similarly, when the target is close at hand, it can also be wide or narrow. How much time would you need to reach the target in these conditions?

Your intuition might be that you would need more time to reach a target the farther away the target is and that you would also need more time to reach a target the narrower the target is. That intuition squares with observations. As seen in Figure 6.9, movement time (MT) increases with target distance (D) and decreases with target width (W). The way these factors work together was summarized by a person whose theory of skill learning was covered in Chapter 5, Paul Fitts. Fitts advanced a formula pertaining to the variables mentioned above (Fitts, 1954). According to the formula, the time (MT)

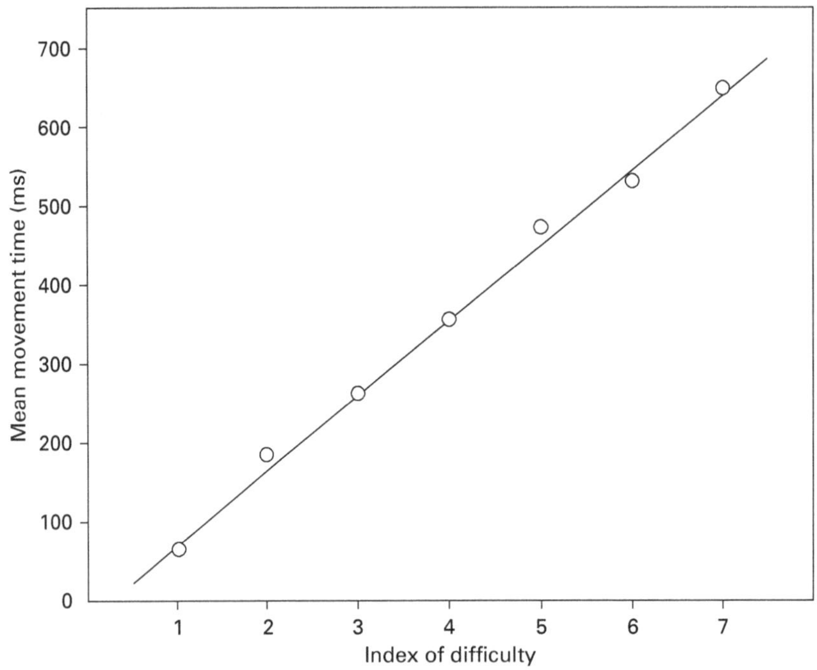

Figure 6.9
Mean movement time (ms) to move to each of a number of targets as a function of index of difficulty, $\log_2(2D/W)$, as per Fitts's law. The points are dummy values whose proximity to the straight-line function mimics what is seen in actual experiments.

to move to a target (and come to a stop within the target) grows with the logarithm of the ratio of the target distance (D) divided by target width (W):

$$MT = a + b \log_2(2D/W).$$

Here, a and b are empirical constants, that is, numbers whose values reflect the efficiency of aiming. Graphically, the values of a and b affect the height and steepness of the straight line that can be fit to data in the manner shown in Figure 6.9.

The formula shown above has been found to apply in such a wide range of situations that it has earned a title rarely used in behavioral and neural sciences. It is considered a "law." One of the few other laws in psychology and neuroscience is the power law of learning, discussed in Chapter 5. The formula that Fitts offered has come to be called Fitts's law. In words, Fitts's law says that aiming time increases by a constant amount with each doubling of aiming difficulty. Aiming difficulty is just $2D/W$, or the distance to the center of target divided by half the width of the target. Fitts's law says that the time to move to a target increases by a constant amount of time with each doubling of $2D/W$. As $2D/W$ goes from 2 to 4 to 8, the corresponding values of $\log_2(2D/W)$ go from 1 to 2 to 3; in other words, $2^1=2$, $2^2=4$, and $2^3=8$. The *ratio* of these successive ratios is always 2 because $2^2/2^1=2$, $2^3/2^2=2$, and so on. The fact that a fixed time is added with each doubling of aiming difficulty suggests a deep principle: the nervous system cares more about *ratios* than arithmetic *differences*.[15]

It turns out that ratios are more important than arithmetic differences all over the map in psychology and neuroscience. The same principle is seen in choice reaction times, where people make unique responses to unique unpredictable stimuli, as covered in the section on Donders in Chapter 2. According to another law of psychology and neuroscience, the Hick–Hyman law of choice reaction times, choice RTs increase by a constant amount with each doubling of the number of stimulus–response alternatives. Essentially, this is the same statement as Fitts's law, just in a different context.[16]

That the same principle arises in so many contexts is probably not an accident. The ratio rule holds even at the neural level (Figure 6.10). Recording from the brain and plotting the number of firing neurons as a function of their firing rate, the distribution becomes symmetric only if the number of firing neurons is plotted as a function of the *logarithm* of the firing rate. Successive values along the firing-rate axis have the same ratio going from one value to the next on the graph, which means that the nervous system, even at the level of individual neural firing patterns, cares more about ratios than arithmetic differences (Buzsáki & Mizuseki, 2014).

The same principle also applies in perception. According to Weber's law, the just-noticeable difference between two quantities depends on the *ratio* of the quantities, not their arithmetic difference. You can feel a .1 ounce added to a 1-ounce weight, but you

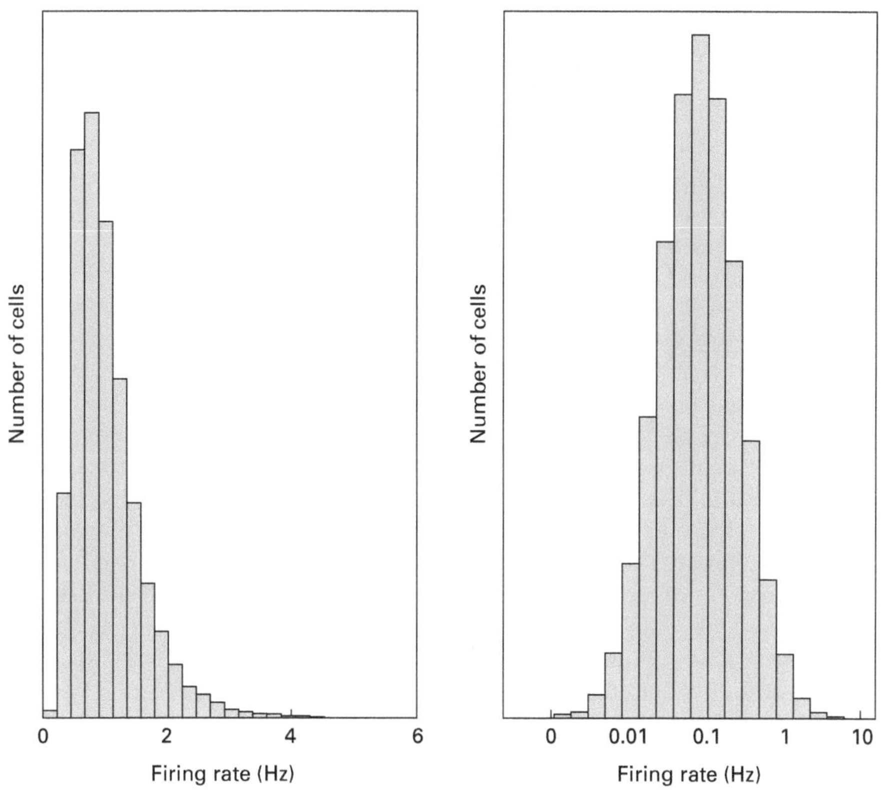

Figure 6.10
Number of neurons with firing rates plotted on a linear scale, where it is skewed (left panel), and on a logarithmic scale, where it is symmetric (right panel). Simulation of data reported by Buzsáki and Mizuseki (2014).

can't feel a .1 ounce added to a 100-pound weight. For the 100-pound weight, you need an extra 10 pounds to feel the increment. Ten pounds added to 100 pounds is the same ratio increment as .1 ounce added to 1 ounce. The ratio is 1/10 in both cases.

These remarks show that aiming reveals fundamental features of the nervous system. But aiming, and Fitts's law in particular, also has important implications for practical fields. Computer designers can't have targets that are so small that it takes "forever" to move cursors to them or so large that too few targets can fit on the screen. Fitts's law helps establish the "Goldilocks" value (the just-right value) for the target sizes used.[17]

Process Models of Aiming

Fitts's law describes a quantitative relation between the variables, D, W, and MT, but it doesn't provide an account of the process by which aiming occurs. How does aiming work such that Fitts's law captures its temporal dynamics? A possible answer was pursued by an investigator who noted that in aiming, participants reduce uncertainty. Consider, by analogy, the case of choice reaction time with four possible lights and four buttons beneath them. A possible strategy is to decide whether a light that lights up is in the left or the right half of the visual field and then to decide whether the light is in the left or right quarter of its respective half of the visual field. Narrowing down the search in this way—like playing twenty questions—can cause the choice reaction time to increase by a fixed amount with each doubling of the number of S–R alternatives, as given by the Hick–Hyman law. For aiming, a similar process can be imagined. Someone aiming for a target can halve the distance to the target, then halve the remaining distance, and so on, until the target is reached. Other proportions can be used; ½ isn't the only possibility.

The investigator to whom I'm referring is the same person who introduced the power law of learning, Ted Crossman. It's notable that so many "lions" of experimental psychology have studied aiming—an homage, one might say, to their appreciation that even very simple acts can reflect broadly important capabilities.

The model that Crossman developed assumed that aiming involves successively homing in on a target, cutting the remaining distance to its center by some proportion, p, of the remaining distance; p could be ½, for instance. With this assumption, it was possible to show that MT grows with D/W in the manner specified by Fitts's law (Crossman & Goodeve, 1963/1983).[18]

As simple as Crossman's aiming model was, it had shortcomings. One concerned the fact that p could never exceed 1. This meant that there could be no target overshoots, but clearly there are. Just as golfers sometimes loft their shots too far, people sometimes reach too far and have to backtrack. Another problem with the model was that discrete submoves were not seen when they should have been if iterative corrections were made until targets were reached. A third problem was that when submoves *were* seen, they didn't always cover a constant proportion of the distance to the target. Even within a series of moves to one target by one person, different p values might appear. Despite these limitations, the general spirit of the iterative corrections model seemed correct. The concept of homing in on a target made sense. The idea led others to pursue a more accurate process model for Fitts's law.

Paving the way for such a model was a finding by Richard A. Schmidt of UCLA, along Howard Zelaznik now of Purdue University, and others (Schmidt, Zelaznik, Hawkins, Frank, & Quinn, 1979). These researchers asked subjects to move a stylus as close as possible to a target point some distance away in some amount of time (Figure 6.11). The researchers varied the distance, D, and the time, T. Critically, after making the move,

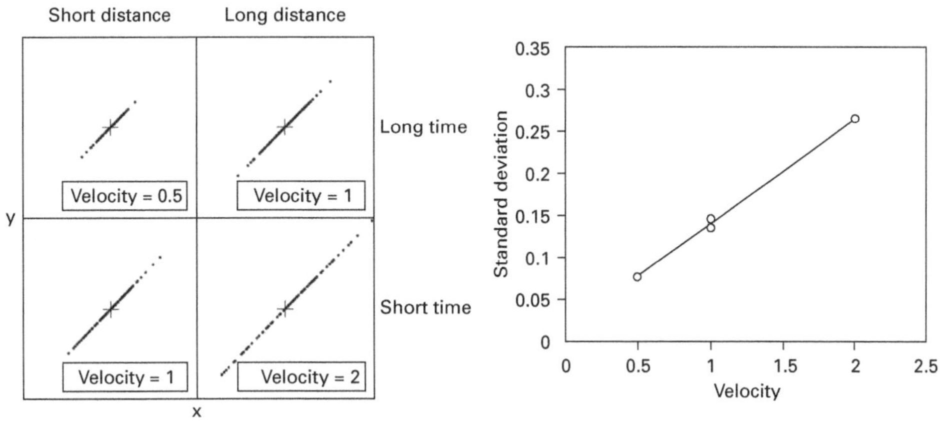

Figure 6.11
Simulated endpoint positions of one-shot movements made along a 45° angle to a target (the plus sign in each of the left panels) when the distance to the target was either short or long and the allowed time was either short or long, resulting in required velocities of .5, 1, or 2 (arbitrary units). The standard deviations along the movement axis are plotted as a function of velocity in the right panel.

the subjects were supposed to refrain from making further corrective moves. Schmidt, Zelaznik, and colleagues were interested in the standard deviation, S, of the endpoint position along the straight line connecting the starting point to the target point. As shown in Figure 6.11, they found that the standard deviation, S, of the position of the movement endpoint was proportional to D divided T. Using the symbol $\propto$ to represent "proportional to," this relation can be written as

$$S \propto D/T.$$

Recall that distance divided by time equals velocity, V, so the expression above can be rewritten

$$S \propto V.$$

In simple terms, the higher the velocity, the greater the spatial variability along the axis of movement.[19]

This speed–accuracy trade-off was used by another team of investigators led by David Meyer at the University of Michigan to lay the groundwork for a process model for Fitts's law.[20] The main idea of the aiming model of Meyer et al. (1982, 1988, 1990) was that reaching to a target as quickly as possible is an optimization problem. According to the model, the performer must find the optimal distances and times that minimize total time. Meyer and colleagues showed that Fitts's law is that optimal solution. Critically and with respect to one of the problems with Crossman's iterative corrections model, Meyer's model allowed that there could be no submoves or several. Also, the proportion of the distance to the target covered per submove didn't have to be fixed, unlike in Crossman's model. More important and more interestingly, the model of Meyer et al. ascribed a great deal of intelligence to individuals making rapid aiming movements to targets. According to Meyer et al., when one aims for targets and tries to get to the targets as quickly as possible, one is actually doing some sophisticated computations, at least implicitly.[21]

Vision for Action

One of the most astonishing discoveries about feedback in action came from an accident. While on a trip, a young woman took a shower in her hotel room. She and her newlywed husband had just checked into the room for the first time. When the young woman decided to take the shower, her husband went out for a few minutes. When he returned, he found her crumpled on the shower floor, unconscious. When she awoke, she was able to hear, understand, and produce speech; her memory was intact; she could move normally; and in all other respects seemed fine—except for one notable problem: she couldn't see. She had gone blind, or so it seemed.

The reason I have written "or so it seemed" is that something unexpected happened in this young woman's case. Although she could not recognize objects, she could reach for objects effectively if the lights were on and her eyes were open. Although she said she could not see what she was reaching for, she could in fact reach for objects essentially perfectly when she could look at the objects she reached for. Here is a dramatization of the testing session where this strange result came to light. It is composed by me with no pretense that these were the exact words spoken at the time.

Neuropsychologist: Please hold this card in your hand. Look out in front of you. I'll show you a slot oriented in various ways. Please turn the card so its orientation matches the slot's.
Patient: But I don't see a slot.
Neuropsychologist: Just do your best.

As expected, the young woman did poorly at orienting the card to match the orientation of the slot. If the slot was vertical, she held the card any which way—upright, sideways, in virtually any orientation you can imagine. If the young woman was shown a horizontally oriented slot, she held the card in a similarly random orientation. This task was not difficult for others, however. Control subjects—young adults who had not suffered carbon monoxide poisoning, which is what happened to this young woman—could do the task essentially perfectly. Their card orientations matched the orientations of the slot, but for the young woman with the brain damage, there was no relation between how she tilted the card and how the slot was oriented. Her behavior matched what she said: "I don't see a slot"

The surprise came when the young woman was asked to reach out and put the card into the slot as if mailing a letter. Her performance was normal. From trial to trial, if the orientation of the slot changed, she did fine, but not because she bumped into the frame and used touch to get the card to fit through. She altered the orientation of the card right from the beginning, from the moment her hand reached out while grasping the card to transport it to the slot. In fact, her hand headed to the correct orientation even before she could see her hand's orientation relative to the slot's.[22]

What was happening here? The researchers who studied her, Mel Goodale of the University of Western Ontario and David Milner, now of the University of Durham, proposed that there are two visual systems, one for consciously recognizing object forms and one for visually regulating actions. Goodale and Milner (1992) offered this proposal based on the behavioral evidence sketched above as well as other behavioral tests, plus scans of the young woman's brain (Figure 6.12). The investigators noted that

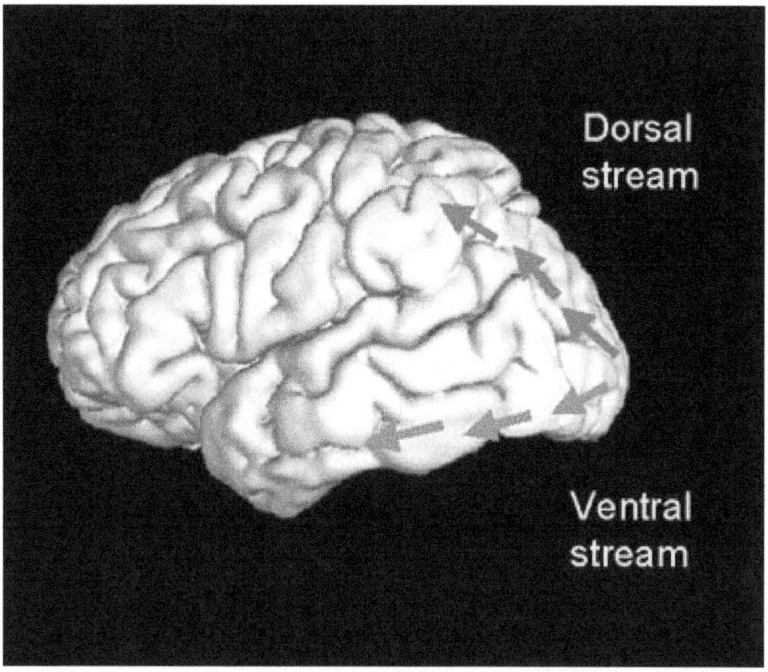

Figure 6.12
The how (dorsal stream) and what (ventral stream) parts of the brain.

damage had been sustained in a part of the brain related to visual processing, the *ventral* stream, which is a nerve tract passing from the eyes to the inferotemporal cortex, which is known to be crucial for the conscious recognition of object forms. Little or no damage was sustained in the *dorsal* stream, a tract that passes signals from the eyes to the posterior parietal cortex, which is known to be important for motion perception. Other scientists had suggested that the ventral system and dorsal system are responsible for object recognition and motion perception, respectively. What was new in Goodale and Milner's suggestion was that the dorsal system is important for visual guidance of self-generated *action*. There are two visual systems, these authors proposed—the *what* system and the *how* system. The *what* system provides information about what's out there. The *how* system provides information about how to act in view of visual input.

How and What in Neurotypical Individuals

It is astonishing to consider how recently Goodale and Milner suggested the dichotomy between the how and what subsystems for seeing. Their book on this proposal was published in 1992. That such a broad claim was made so recently indicates what a young field cognitive neuroscience is.

Given the sweep of Goodale and Milner's claim, others were skeptical about it. That they were is a healthy sign, for scientists are, by nature, skeptical. Perhaps the brain doesn't really have two distinct visual systems, these scientists thought. Maybe this was a freak occurrence. A single mishap in a hotel shower by one young woman may not be enough to seal the deal for all human brains. The case would be strengthened if individuals with no known impairments also showed the dissociation that Goodale and Milner hypothesized.

Who better than Goodale to address this question head on? In other work, Goodale and other colleagues provided evidence that neurotypical individuals do indeed have a *what* visual system and a *how* visual system (Aglioti, DeSouza, & Goodale, 1995). The logic of their studies was ingenious. To appreciate it, look at Figure 6.13, where you can see two disks surrounded by small disks or large ones. If you're like most people, you will see the surrounded disks as having different sizes. The disk surrounded by small disks will probably look larger than the disk surrounded by large disks. This is an illusion, however, as you can tell by putting a ruler up against the two center disks. The two center disks are identical. This illusion was discovered by Hermann Ebbinghaus (1850–1909), one of the first experimental psychologists.[23]

The Ebbinghaus illusion, as it is called, reflects conscious visual recognition, which is the realm of the *what* visual system. Therefore, if there is a visual system for action distinct from a visual system for object recognition, it should be possible to show that the Ebbinghaus illusion disappears during action. A way to test this prediction is to invite people to reach for disks and monitor the spread of their fingers as they do. When people reach for large objects, they spread their fingers farther apart than when they reach for small objects, as shown by Marc Jeannerod, the French neurologist who pioneered the use of cinematic analysis to study reaching and grasping; see Chapter 2. If the Ebbinghaus illusion is a byproduct of a visual system for object recognition, separate and apart from a visual system for action, then action should not fall prey to the Ebbinghaus illusion. This prediction was confirmed. When participants reached for a disk surrounded by small disks, they spread their fingers apart no more (and no less) than when they reached for the same disk surrounded by large disks. Thus, the distinction between a visual *what* subsystem and a visual *how* subsystem applies to normal brains, not just the brain of one person who, sadly, was overcome with carbon monoxide in a shower one day (Aglioti et al., 1995).

When I tell students about Goodale and Milner's proposal, they're often transfixed. It's amazing, they say, that we have two visual systems, one for object recognition and one for action. And it's astounding, some of them also proclaim, that we can see but not know we do. I remind them that we often find ourselves driving for long stretches,

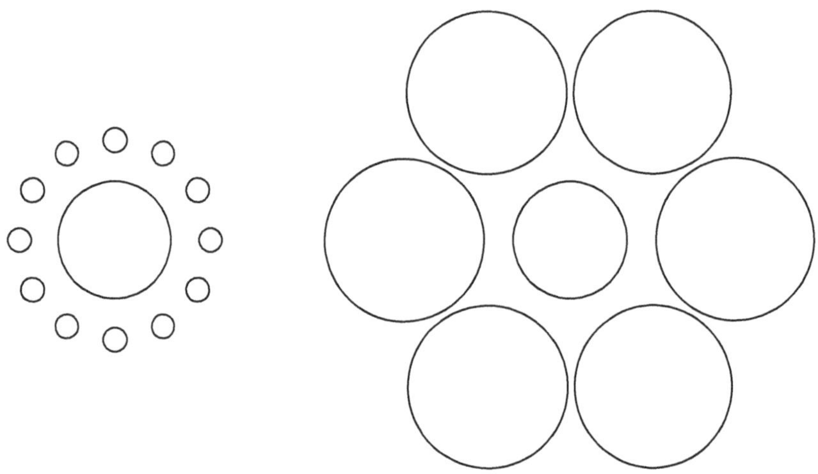

Figure 6.13
The Ebbinghaus illusion. The two inner circles have the same size. Your hand knows this, so to speak, even if your eye does not.

suddenly becoming aware that we have been doing so, having been lost in thought or so deep into conversation or listening to the radio or podcasts that we forget that we're driving. I also tell them about a neurological syndrome called *blindsight*, where people with damage to the primary visual cortex claim to be unable to see, although they perform better than chance alone in two-alternative forced-choice experiments where they have to guess which of two stimuli they were shown. The original stimulus, to be asked about, is presented to the side where its neural projection goes to the damaged side of the brain. The test stimuli are presented to the brain's intact side, which is how the patients can be asked which of the stimuli shown to them was the one shown before; otherwise, they would protest that there is nothing to see and compare to. In this situation, the patients can choose the originally shown stimulus at better-than-chance levels, which means that the stimulus was seen even if it was not consciously perceived—hence the paradoxical term "blindsight" (Weiskrantz, 1986).

The two-visual-system hypothesis is fascinating and raises a host of questions: do the two visual systems actually serve the functions that Goodale and Milner proposed? There has been some debate about this (Glover, 2002). Additional questions concern the nature of the crosstalk between the two systems and whether function can be restored in one system by piggybacking on the other.[24]

Visuomotor Adaptation

One of the most interesting features of action and feedback is how well actions can be adapted to feedback. When you walk on dry land, for example, you take steps that require familiar forces and torques, but when you plunge into water and keep walking, you have to generate much larger forces and torques to propel your legs as far and as fast as they were going before. If you're healthy, you can make this adjustment quickly and easily. Similarly, if you suddenly feel a stone in your shoe, you can instantly change your gait.

Yet another example comes from the domain of visually guided reaching. If you hold a laser pointer and point it at a wall, to get the laser pointer to cover the same distance on the wall when you are varying distances from the wall, you must change the angle covered by your hand. The farther you are from the wall, the smaller the angle. You can adapt in a flash, so to speak, recalibrating the amplitude of your hand movements based on the amplitude of the seen visual motion, typically in a matter of seconds.[25]

Other kinds of recalibration take longer, as in the case of visuomotor rotations. Recall the experiment in which monkeys learned to move their hands to positions rotated 90 degrees away from the targets they were shown (Figure 3.11). At first, the monkeys reached directly for the targets, but with training, they learned to redirect their reaches to the new targets. Recordings from the monkeys' brains revealed that when each target was shown, the neural activity in the motor cortex yielded a population vector that first "pointed to" the initially shown target but then shifted to the target. The authors of that study hypothesized that redirection of the population vector could underlie the mental rotation.

It took a while for the monkeys to learn to adapt in this context, and so it is with people. In one widely cited study, people performed a task similar to the one the monkeys carried out (Mazzoni & Krakauer, 2006). They moved one hand from a central location to a target and then back again without seeing the hand. Instead, what they saw was the effect of the hand movements on a computer display; a cursor moved as a function of how the hand did. The mapping of hand movements to cursor movements was altered so movements of the hand in a given direction caused the cursor to move 45 degrees counterclockwise. What also happened in the human experiment was something that couldn't be done with the monkeys. Some of the human subjects were told explicitly what the transformation would be. They were given instructions along the lines of, "If you see the target at 12 o'clock, move to 2 o'clock."[26] The hypothesis was that people might use explicit knowledge to adapt to visuomotor transformations. If so, they would benefit from explicit instructions.

The participants did in fact benefit from the explicit instructions, at least at first, but over the long term, the verbal explanations provided no extra benefit compared to what was seen with control subjects who simply practiced the task without special instructions. For both groups, after the software rotation was turned off so the cursor moved in the same direction as the hand, there was a comparable aftereffect in both groups as they readapted, which both groups did equally well and quickly. The authors concluded that adapting to visuomotor rotations ultimately engages implicit learning only. Consistent with this conclusion, they noted that participants in the explicit condition were not especially clear in their verbal reports about how they did the task.

Other Forms of Adaptation

The forms of adaptation that can be studied are endless. If you were an astronaut on a planet with more or less gravity than Earth's, you'd have to make adjustments to that unfamiliar force field, and returning to Earth, you'd have to readjust to the *g* you grew up with. If you were shooting arrows at a target and the wind blew steadily to the right, you'd have to compensate by aiming to the left; once the wind died down, you'd have to return to a more straight-ahead strategy. If you were speaking in an auditorium with a powerful echo, you'd have to adjust the timing of your speech; later, returning to an echo-free surround, you'd have to get back to normal speech timing. As these examples show, adaptation occurs in different contexts and with different sensory modalities. It takes time to adapt, and it takes time to readapt to normal conditions. Readaptation generally takes less time than initial adaptation.

Because the possibilities for studying adaptation are endless and the number of studies that have been done is huge, I'll mention just a few of them here. My aim is to give a sense of the range of questions that can be asked. I also want to give you an impression of who has pursued these questions because there is at least one researcher who contributed greatly but did not get adequate recognition owing to her sex.

Consider the fact that you see the world as right-side up rather than upside down. This is unsurprising until you learn that the image of the world on your retina is inverted. Your brain does a kind of backflip to undo the inversion caused by the optics of your eye. Your conscious perception of direction is the opposite of what actually appears on your retina.

In the late 1800s, a professor at the University of California, Berkeley wondered whether perception of up-down is based on learning. Do we have an innate ability to deal with this inversion, or must it be learned, George Stratton asked? To address the question, he wore a special optical system that inverted the image of the world. Never, while the experiment was in progress, did he allow himself to look at the world without the inverter on. His question was whether he would be able to reach for objects and navigate the environment if the image of environment were flipped. He reasoned that if appreciation of the objective orientation of the world was hardwired, he never would adapt; never would he see the artificially flipped world as right-side up. Conversely, if appreciation of the objective orientation of the world was *not* hardwired but was learned or subject to learning, then he *would* be able to learn to see the artificially flipped world as normal, and he would be able to adapt his behavior accordingly.

When Stratton first put on the inverter, he was understandably clumsy. He reached up when he should have reached down, and he reached down when he should have reached up. Over time—after a few days—he adapted, managing to perform as well as he had before. Thus, Stratton learned to adapt. When he later took off the inverting device, he went through a readaptation phase. He had a bit of trouble getting back to normal, but he did so in much less time than he needed for the original adaptation. Stratton was one of the first to demonstrate that readaptation takes less time than initial adaptation (Rock, 1966).

Stratton was a true pioneer. He was one of the first scientists to study adaptation to distorted visual feedback. In so doing, he exhibited considerable curiosity, dedication, and courage, for it may have crossed his mind that he might never be able to readapt to the normally presented world after adapting to the topsy-turvy world he'd confront. He did readapt, however, which surely must have been a relief to him and those in his life who counted on him to put cups *up* in cupboards and debris *down* in dumpsters.

Others who followed Stratton provided more information about the functional mechanisms allowing for adaptation. The work showed that the most important basis for adaptation is picking up on errors following self-initiated actions. Merely looking through prisms that displace the image of the world doesn't enable adaptation. You have to *act* in the world, expecting to see consequences of your actions, and then, getting feedback that you were right or wrong, make corrections. Getting mismatches between predictions and actual outcomes helps you implicitly develop mathematical functions relating desired perceptual outputs to required motor commands, as will be discussed in the next chapter (Held, 1965).

Textbooks that describe adaptation to visuomotor distortions tend to cite several investigators as the key contributors to this area: Herman von Helmholtz, James Gibson, George Stratton, and Richard Held. You have encountered all of their names in this book. Richard Held was cited in a footnote at the end of the last paragraph, and his work will be discussed in more detail in the next chapter. All of these investigators were men, something I say because I want to relate that, while working on this chapter, I learned that an extremely important contribution to this field was made by a woman I had never heard of before. I encountered her name while reading Irvin Rock's 1967 monograph, *The Nature of Perceptual Adaptation* (Rock, 1967). There he mentioned an investigator in a way that was revealing but also a bit disturbing. In fact, the way he mentioned her may explain why I hadn't heard of her before. In the midst of a discussion of Helmholtz, Stratton, and Gibson, Rock included this brief sentence, which I quote here as it appeared, with the parentheses enclosing it. I found the sentence so striking that I set it here in italics on its own line:

(Margaret Wooster had already conducted a careful experiment along these lines in 1923.)

My jaw dropped when I read this. Here, in a parenthetical remark, was an acknowledgment that the pioneering work in this field had in fact been done earlier. Given that I knew a little about this field of study, I was thunderstruck by the fact that the researcher who was named here was someone I had never heard of before. Her name was Margaret Wooster.

Who was this person, I asked myself? I got a hold of her monograph (the writeup of her doctoral dissertation, done at the University of Chicago) and learned that her careful

Figure 6.14
Margaret Wooster.

experiment was, in fact, not *one* experiment but an entire *series* of experiments. Here is the abstract of Wooster's (1923) monograph. Wooster's picture is shown in Figure 6.14.

The purpose of this investigation was to determine experimentally the relative influence of various sensory modes of reaction, as sight, touch, and hearing, in the building up of a new space habit. We determined to use a pair of prismatic glasses which produced an angular deviation of the visual field of about 21 degrees. The glasses used in our experiment consist of two 40 degree optical prisms producing an angular deviation of about 21 degrees. Throughout the experiments we were thus able to work with binocular vision, in which respect our conditions are superior to those of Stratton, who relied on monocular vision. Our investigation, therefore, has to do with the relative efficacy of hearing, sight, and touch, as factors contributing to the formation of the new spatial coordination developing under the conditions of our experiment. The results of the first series of experiments seemed to indicate that knowledge of the direction and extent of the deviation might well be the chief factor in the process of readjustment. No readjustment to the changed visual situation occurred without definite reaching movements of the hand while the eye was fixed on the visual object. In any case it seems that slight changes in the delicate muscular mechanism regulating the eye may profoundly influence the overt reactions of an individual although he may be entirely unaware of the effect on his behavior of the new kinesthetic stimuli. Experimental verification and further investigation of this hypothesis are needed.

So Wooster discovered what others later claimed to have discovered. I don't know whether they were aware of her work beyond the parenthetical one-sentence allusion to it given earlier. I don't mean to imply that I think there was deliberate obfuscation. On the other hand, there clearly wasn't an intense effort to learn more or say more about her work. In any case, Wooster (1923) wrote the following to summarize her project:[27]

Perhaps the most significant result of this investigation is the demonstration that it is possible to secure accurate and extensive quantitative data on the problem of the factors in the development of space perception through the use of relatively simple experimental procedure. (pp. 95–96)

The Rubber-Hand Illusion

As Margaret Wooster said, relatively simple experimental procedures can reveal much about the development of space perception. An extraordinarily simple procedure has indeed revealed a lot about space perception or, more specifically, the perception of the body in space. That method has shown that it is possible to get people to feel their hands where their hands *aren't*.

The setup is shown in Figure 6.15. Videos can be watched on the internet.[28] If you were a subject in this experiment, you'd put your hand under a cover so you couldn't see your hand. Beside your hidden hand would be another hand in full view. That other hand would be made of rubber. Its wrist would be covered so you couldn't see that it's a disembodied limb segment. The experimenter would sit opposite you and he or she would use two brushes to stroke the two hands at once—*your* hand and the *rubber* hand. You wouldn't see the brush applied by the experimenter to your own hand, but you *would* see the brush applied to the rubber hand. Soon, something uncanny would be likely to happen. You might feel your hand where the rubber hand is. The *felt* position of your hand would be drawn to the *seen* position of the rubber hand. If you were asked to point to where you felt the brush strokes to be, you would point with your other hand to the rubber hand, not to your own hidden hand (Botvinik & Cohen, 1998).

This is called the rubber-hand illusion. The illusion is a bona fide scientific phenomenon, not just a party trick or carnival stunt because reliable data have been obtained about it using established methods of psychophysics, and the factors that affect the appearance of the illusion are scientifically plausible. If the two hands aren't stroked synchronously or if the rubber hand and actual hand are too far apart, the

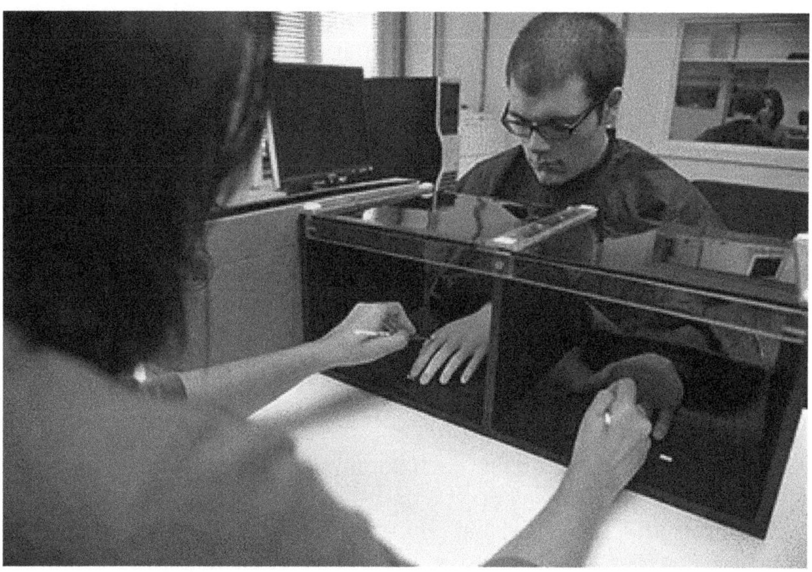

Figure 6.15
Setup used to elicit the rubber-hand illusion. The subject's actual hand is the one under the glass on the right in this photograph. A rubber hand is directly in front of the subject (toward the left in this paragraph) and is being observed by the subject while it and the subject's real hand are being stroked by the experimenter.

illusion doesn't work. In addition, the subjective feeling and observable pointing biases are accompanied by recordable brain changes. With fMRI, it has been shown that the somatosensory cortex becomes activated when the illusion is induced and is less activated if the stimulus conditions are less conducive to the effect. The premotor cortex also becomes activated when pointing is required, provided the timing and spacing of the strokes are conducive to the illusion (Ehrsson, Spence, & Passingham, 2004).

Vision Dominates Touch

How can such shifts of subjective hand position occur? Findings from other laboratories suggested a basis for the effect. First is a result from Irvin Rock, who, along with Charles Harris, published a classic article in *Scientific American* called "Vision and Touch" (Rock & Harris, 1967).[29] The question posed by Rock and Harris was whether touch dominates vision or vision dominates touch. It's unlikely to be one or the other all the time, of course. Presumably, the answer depends on the context. If you're at a party where lots of people are drinking red wine and there's a big red splotch on a sofa, you'll probably touch the fabric to make sure it's dry before sitting down. That's a case where touch dominates vision. Cutaneous (touch) cues determine your final judgment in this case. There are other cases, however, where vision dominates touch, as shown by Rock and Harris.

As a backdrop to their work, Rock and Harris noted that a philosopher in the 1700s named George Berkeley claimed that touch educates vision. For an infant to learn that a two-dimensional image of an object actually corresponds to a three-dimensional object, Berkeley said, the infant must feel the object, for otherwise, the infant would never be able to map two-dimensional visual images onto three-dimensional objects. Berkeley's thinking led to a distinction that is fundamental to psychology and neuroscience today, the distinction between *distal* stimuli and *proximal* stimuli. Distal stimuli are picked up by *exteroceptors*—receptors for signals arising from afar: the eyes, ears, and nose. Proximal stimuli are picked up by *interoceptors*—receptors for stimuli impinging on or in the body: the skin, the tongue, muscles, and tendon organs.

Rock and Harris challenged Berkeley's claim that touch normally dominates touch. They showed that it does not and did so by creating conflicts between vision and touch. In one of their experiments, Rock and Harris had subjects feel a reference cube whose image was smaller than normal, thanks to a minifying glass (the opposite of a magnifying glass). The subjects could not see their hands. A cloth covered the subjects' hand and they felt the cube they were looking at from beneath the cloth. In control conditions, the same subjects just looked at the reference cube without any touch feedback, or they just felt the reference cube without any visual feedback. The critical test came when subjects pointed to one of several test cubes to indicate which cube seemed to be the one that matched the reference cube.

Rock and Harris reasoned that if touch dominates vision, as Berkeley supposed, the test cube that would be chosen would be the one that *felt* the same as the one initially inspected. On the other hand, if vision dominated touch, contrary to Berkeley's supposition, the test cube that would be chosen would be the one that *looked* like the initially inspected cube. The victor was vision. The test cube judged to match the reference cube was the one that *looked* the same. So touch was downplayed relative to vision. As in the rubber-hand illusion, seeing was more important than feeling.

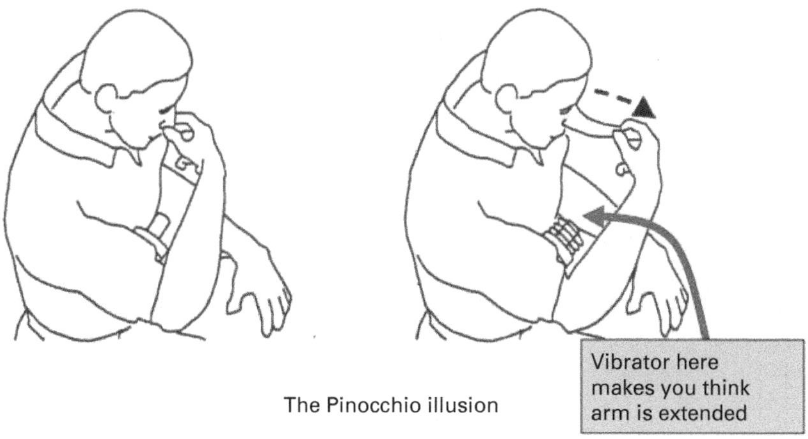

Figure 6.16
The Pinocchio illusion.

The Pinocchio Effect

Another interesting illusion is the Pinocchio effect. Pinocchio, as you may recall, was the fictional character whose nose lengthened each time he lied. It turns out you can feel your own nose grow not by lying but instead by allowing your biceps muscle to be vibrated while you touch your nose (Figure 6.16). Vibrating a muscle stimulates the stretch receptors within it. The signals sent by the stretch receptors via the dorsal afferents to the spinal cord and then to the brain create the illusion of muscle stretch. If vibration is applied to the muscle of one of your arms and you show with your other arm how outstretched the vibrated arm seems to be, you will overextend that other arm, showing that you feel the vibrated arm to be more extended than it really is. This phenomenon shows that muscle stretch receptors contribute to conscious perception of limb position, something that was not understood before the first vibration study was done (Goodwin, McCloskey, & Matthews, 1972).

In a clever extension of this method, James Lackner, whose work with John Jeka was featured in Figure 6.8, thought it would be interesting to see what happens if people touch their noses while undergoing this kind of muscle vibration. Would people feel their noses lengthen? The answer is that they did. Hence, even something as fundamental and familiar as the length of your own nose is subject to feedback-based perturbations (Lackner, 1988).[30]

Changes in Body Perception Based on Sound

The perceived length of limb segments can also be altered with other tricks involving feedback. In another study, subjects touched the floor while blindfolded. Each time they touched the floor, a loudspeaker played a tapping sound. In the control condition, the tap sound was played where the tap was made, but in the experimental condition, the speaker was displaced from the touch site. When the speaker was a bit farther away than the touch site, a remarkable illusion emerged. Subjects felt the tapping arm to be longer than it really was (Tajadura-Jiménez et al., 2012).

How was this demonstrated? Just asking participants to report on the felt length of their arms would have been inadequate, for participants may have just said what they thought the experimenter wanted to hear. The way the subjective lengthening was measured was to have subjects judge the distance between two touches on the *nontouching* arm relative to two touches on the *touching* arm. The result was that the distance on the nontapping arm had to be longer than the distance on the tapping arm for the two distances to feel the same. Thus, the tapping arm felt longer than it actually was as a result of hearing the trick sounds. Confronted with the mismatch between hearing and touch, the touching arm subjectively elongated. Hence, hearing dominated touch, much as vision does.

The research I just described was led by a young woman named Anna Tajadura-Jiménez. She also directed a study that led to another startling result. She and her coworkers fitted shoes with pressure sensors (Figure 6.17). Participants wore the shoes

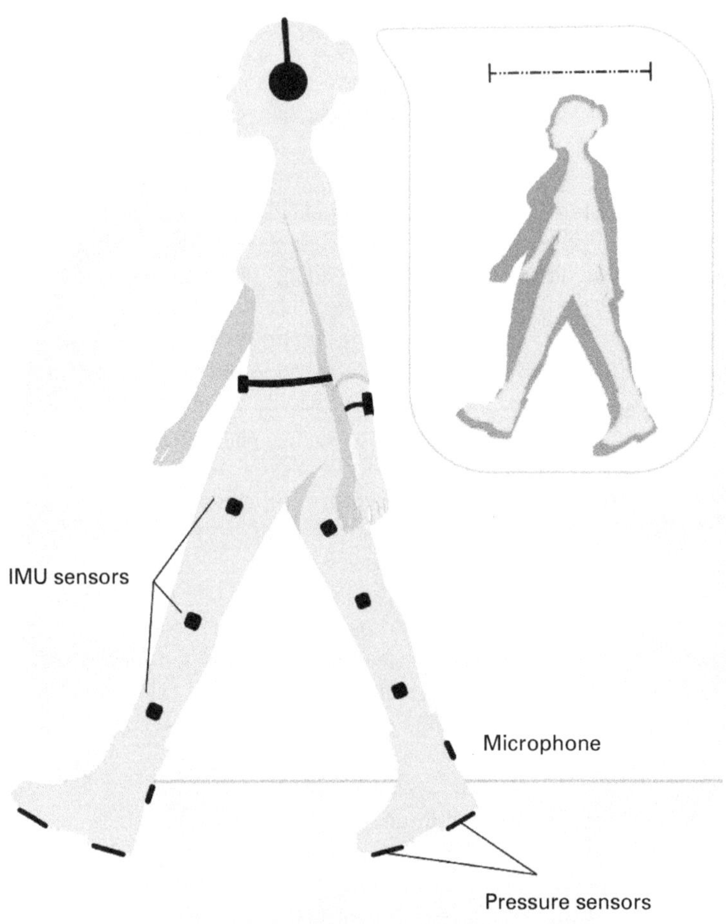

Figure 6.17
Hearing one's own footsteps as high or low sounds based on pressure sensors and inertial measurement unit sensors on the legs. The inset illustrates how the same person can perceive herself to be lighter or heavier when getting high- or low-pitched sounds.

while walking. Each time the subjects took a step, a signal from the pressure sensors triggered a sound heard over headphones. The clever artifice introduced by Tajadura-Jiménez and her colleagues was to have the sounds be high-pitched in some conditions and low-pitched in others. The hypothesis was that high-pitched tones would suggest to the walkers that they were light, whereas low-pitched tones would suggest to the walkers that they were heavy. The prediction was confirmed. Participants reported feeling more positive and energetic when they got high-pitched sounds than when they got low-pitched sounds. When the participants manipulated an avatar to indicate the extent to which its dimensions matched the walkers' own, they thinned down the avatar when the high-pitched tones were presented. In addition, the sounds affected the participants' ongoing gait. High-pitched sounds led to larger upward acceleration of the legs than did low-pitched sounds, and low-pitched sound led to longer heel contact with the ground than high-pitched sounds did (Tajadura-Jiménez et al., 2015).

When I tell students about this study, I walk in front of them making grunts while demonstrating the low-pitched condition and chirps while demonstrating the high-pitched feedback. While making the sounds, I find it hard to avoid a happy stride while chirping and a slouch while grunting. This leads me to mention that Ana Tajadura-Jiménez and her colleagues suggested that the feedback manipulation they developed could be used to encourage people to exercise more. Folks who feel reluctant to take brisk walks might be played high-pitched happy chirps. By helping them feel lighter or happier, they might be more willing to walk more and enjoy the associated health benefits.

7 Feedforward

The subject of the last chapter was *feedback*—information *following* action. The subject of the present chapter is *feedforward*—information *prior* to action concerning the expected perceptual effects. Feedforward representations—also known as internal models—are predictions. The predictions can be conscious or, more often, unconscious. Feedforward is widespread in the animal world and fundamental to successful behavior.

To set the stage for what's to come, I'd like to start with a joke: "Why did the chicken cross the road?" You've heard it many times and know the punchline: "To get to the other side." But why start the chapter with this joke, not a very good one for that matter? The reason is that it serves as a springboard for thinking about feedforward.

Replace the question of *why* the chicken crossed the road with a more psychologically nuanced question: what did the chicken have in mind when it decided that road-crossing was in order? The chicken may have wanted to escape a predator, expecting the predator to be unwilling to cross the road if there were too much traffic. Or the chicken may have seen something on the other side of the road that it found tempting and deemed that item so desirable that the benefit of going for the item seemed to outweigh the potential costs of crossing the road.

We'll never know what the chicken was thinking, of course. The more serious point is that acts are based on mental states. When you reach for an apple, you know you've succeeded once you've gotten hold of the fruit. Seeing your fingers close around the apple, witnessing what you hoped to see and feeling what you hoped to feel, you get information that you achieved your goal.

These remarks raise a question. If you know what to expect, why act at all? Presumably it's because the expectation is just a sketch. If you turn to the right to get a better look at something, you may have *some* idea of what you'll see, but the idea is still vague and incomplete, so you go to the extra trouble of checking it out. The tendency to look at something can be automatic, of course—a so-called orienting response or orienting reflex—but it can be motivated by curiosity, however microscopic.

Another example makes the point more dramatically. Suppose you hear a noise upstairs at home and walk upstairs to get a better look at what's causing the sound. You imagine an open window in the bedroom causing the Venetian blinds to bang against the window, or you suspect that the wind blowing into the bathroom through the open bathroom window is causing the shirt hanging on the shower rod to be hitting the tiles. That particular shirt has oversized metal buttons which could be rattling against the wall. Once you get upstairs, you find out what it was. The action you took was designed to quell your curiosity.

These comments suggest an answer to the question "why act all?" The answer is, "to get more information." Beyond this—and this is what makes the information-attainment hypothesis more interesting, is that in determining which actions to perform, the nervous system may rely on predictions of what may be perceived. In other words, predictions may help determine what should be done. The motor commands that are issued may be the ones that are expected to make the predictions come true (Straka, Simmers, & Chagnaud, 2018; Wolpert & Flanagan, 2001).

Discussions about forward models can get very technical. The reason is that the relations between motor commands and the movements they produce are complex. So too are the relations between movements and their sensory consequences. Because of the technical nature of this material, I will refrain from delving into math-intensive material here. Fortunately, there is plenty of other material about feedforward control that illustrates the main points about it. In addition, I think you won't be disappointed by what's to come because many of the results to be covered here are really surprising. In my opinion, work on this topic has yielded some of the cleverest experiments and most brilliant insights into the nature of mind and brain in any branch of psychology or neuroscience.

Helmholtz, Unconscious Inference, Inflow, and Outflow

You met Hermann von Helmholtz in Chapter 3. Recall that this nineteenth-century German scientist/eye doctor invented the ophthalmoscope and measured the speed of nerve conduction—two remarkable achievements for anyone. But he also noticed something that no one had seen before, even though it was right there before their eyes. Here is what Helmholtz beheld.

If you look at a stationary scene, it doesn't seem to move. However, if you watch a moving object, it does. There's something odd about that, Helmholtz realized. If you look at a stationary scene—a bookcase, say—your eyes dart about, making jumps or saccades (discussed in the last chapter), which cause the image of the bookcase to flit across your retina. On the other hand, if you watch an object move smoothly from one place to another—a frisbee sailing over a lawn, say—the image of the moving object stays at approximately the same place on your retina; nevertheless, you see motion. This raises a paradox. How could it be, Helmholtz asked, that the image that stays still on your retina gives rise to the perception of motion, whereas the image that moves on your retina gives rise to the perception of stability?

Here is the idea Helmholtz came up with. There must be unconscious inference, he realized. Somehow the brain must use some sort of problem solving to help with matters like determining whether visual input is due to objects that are moving or objects that are stationary. The brain has to sort this out, Helmholtz figured, and insofar as it does without our being aware that it does or how, there must be unconscious processes.

Sigmund Freud is usually credited with unearthing the unconscious, but Helmholtz actually preceded him. Helmholtz unveiled the unconscious in a less sexy domain than the one Freud frequented, which is perhaps why Freud's unconscious is better known than Helmholtz's. Postulating an unconscious basis for visual perception is not as sexy as exposing hidden urges, as Freud did. A book called *Fifty Shades of Grey*, a soft porn book that has sold millions of copies, tends to sell better than a book about the visual perception of achromatic (colorless) stimuli.

What answers did Helmholtz give to the question of how we know whether objects are seen to move or stay still? He realized that it's essential to factor in action states, as seen in Figure 7.1. What is taken into account here is whether the retinal image moves or stays still and whether the eyes move or stay still. If the image of the world passes

		Retinal image moves?	
		Yes	No
Eyes move?	Yes	*Object seen as still*	*Object seen as moving*
	No	*Object seen as moving*	*Object seen as still*

Figure 7.1
Determining whether an object is seen as still or moving depends on whether the retinal image moves or not and on whether the eyes move or not.

across the retina and the eyes move (the top left cell of the table), then, according to Helmholtz, the retinal image shift can be attributed to eye movements. But if the image of the world does *not* pass across the retina and the eyes move (the top right cell), the retinal image shift can be ascribed to external motion. The bottom left cell is the case of the eyes not moving but the image doing so. Here the object is seen to move. Finally, the bottom right cell is when neither the eyes nor the image moves. In that situation, the object is seen to stand steady.

The four cases just listed indicate that information is needed about the positions of the eyes to see or not see external motion. This was Helmholtz's great insight. He realized that there are two kinds of perceptual inputs. One is due to external stimulation alone—what has since been called *exafference*. The other kind is due to one's own motion—what has since been called *reafference* (von Holst & Mittelstaedt, 1950). Helmholtz realized that there are two possible ways of distinguishing exafference from reafference. One, as seen in Figure 7.2, would come to be called *inflow*. The other would come to be called *outflow*.[1]

Inflow is feedback from receptors in and around the eyes. It takes two forms: retinal inflow and extra-retinal inflow. Retinal inflow consists of neural signals from the eye's photoreceptors, especially the rods in the periphery, which are mainly responsive to motion, and the cones in the fovea or center, which are mainly responsive to shape and color. Extra-retinal inflow comes from eye-muscle stretch receptors and from cutaneous (touch) receptors in the eyelids.

Outflow is the set of signals related to expected visual changes and/or commands sent to the muscles. Helmholtz understood that if outflow were used to interpret retinal images, the brain would compare retinal information with internal signals related to expected visual changes. Outflow-based visual processing would begin *before* commands are sent to the eyes, whereas *inflow* processing would occur *after* the eyes move.[2]

Outflow theory makes predictions that inflow theory does not. One prediction is that if the eyes are moved by forces other than normal oculomotor commands, illusory

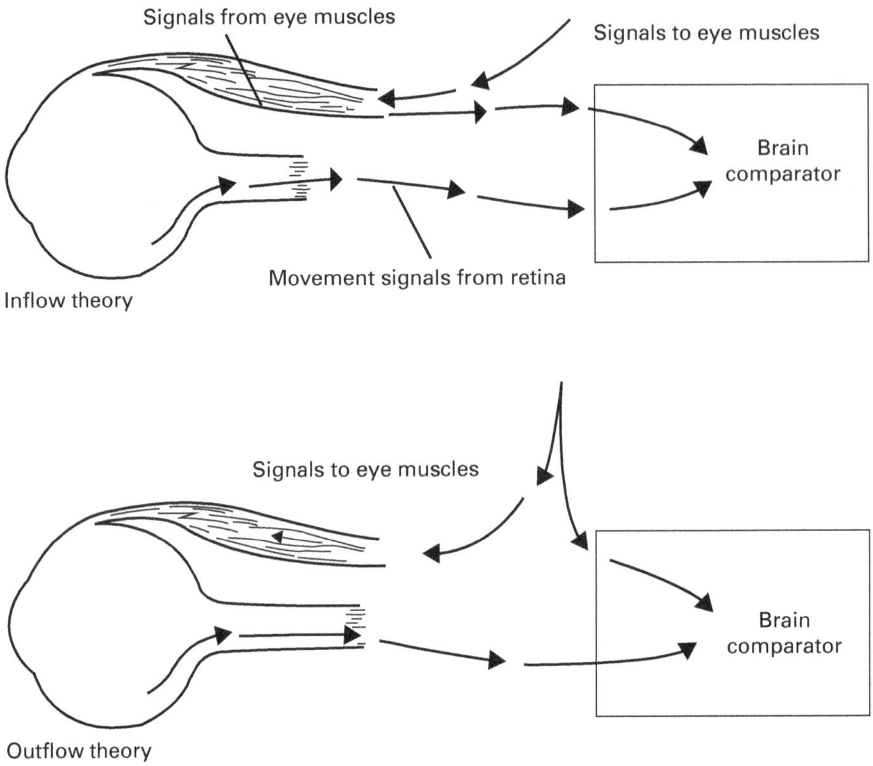

Figure 7.2
Inflow theory (top panel) and outflow theory (bottom panel).

visual changes should occur. You can test this prediction by pressing on one of your eyes with your finger. Be gentle. Don't press hard. Close one eye and press on the lid of the open eye. Look straight ahead at a stationary object—a bookcase, say, or a piano. Press gently and let go, repeating this a few times. If you're like most people, you will see the stationary scene bounce back and forth.

How is this possible? Helmholtz reasoned that in this situation, the eye muscles stretch and the eyeball slides beneath the lid. Therefore, if inflow helps interpret retinal changes, the retinal changes caused by jiggling the eye should be canceled; the world should seem stationary. On the other hand, even with all that inflow, a critical piece of information is missing. Commands are not sent to move the eye muscles, so the visual center of the brain is not apprised of oculomotor (eye movement) commands. The resulting unconscious inference is that a visual jump must have come from outside.[3]

What if, instead, commands are sent to the muscles but the retinal image does not shift? Now the visual system should interpret retinal nondisplacement as external motion; see the top right cell in Figure 7.2. This is just what happens. If the eye muscles are paralyzed with a nerve block such as curare, a poison that, among other uses, is applied to arrowheads to immobilize prey in the Amazon and elsewhere, the patient or the intrepid volunteer undergoing eye-muscle paralysis should have an unsettling visual experience. Each time he or she tries to move the eyes, the world should appear to swing wildly from place to place with each intended eye displacement. Indeed, this is what happens (Kornmuller, 1930; https://www.scientificamerican.com/article/stability-of-the-visual-world/).

Another demonstration of outflow avoids resorting to muscle paralysis. Were you to participate in this demonstration, you would allow yourself to be exposed to a brief bright light such as a camera flash. If you then closed your eyes or moved your eyes in an otherwise dark room after the flash, you'd see the spot of light move with your eyes.[4] Here your brain interprets nondisplacement of the afterimage on the retina as being due to motion of the external object. The relevant source of eye-movement information could, logically, be outflow or inflow. However, the outflow account is preferable because it explains this phenomenon as well as the others reviewed above. If the same mechanism can be used to explain afterimage motion as well as illusory motion during oculomotor paralysis, simplicity recommends that mechanism.

Predictive Changes in the Parietal Cortex

Is there a way to check the outflow model neurophysiologically? Figure 7.3 shows a schematic diagram of a study that provided confirmation (Duhamel, Colby, & Goldberg, 1992). Here, recordings were made in the parietal cortex of monkeys as they looked at a screen. The monkeys could move their eyes when and where they pleased. By recording the monkeys' eye movements as well as the neural activity of selected cells, the researchers could determine, both before and after the saccades, what the cells' visual receptive fields were. The visual receptive field of a neuron is the seen region to which the neuron responds. If you look straight ahead and a visually responsive neuron in your brain has a receptive field centered at 2 o'clock relative to your gaze line, the neuron will fire when an insect, say, happens to appear in that spot.

The researchers who conducted this study reasoned that if the receptive field of the parietal-cortex neurons were defined solely in terms of retinal information, the receptive fields for those neurons would be the same before and after saccades. If the receptive field was at 2 o'clock relative to the line of gaze *before* the saccade, it would also be at 2 o'clock relative to the line of gaze *after* the saccade. This turned out to be true, but something remarkable happened around the time of the saccade. Before saccades were initiated, the visual receptive field changed to where the eye *would be after* the saccades. So, referring to the image in Figure 7.3, if the original receptive field was centered on the *cloud* during the initial fixation, it switched to the sun *before* the saccade was made. The sun would be at the center of the receptive field *after* the saccade was completed. Therefore, the brain anticipated where the eye *would* point, presetting the visual receptive field accordingly. The information defining the visual receptive field wasn't just retinal, therefore. It was also extra-retinal. Because the change happened *before* the saccade, it reflected outflow rather than inflow.

When I tell students about this experiment, I encourage them to stop and think about how amazing it is from a technical standpoint and in terms of the discovery it yielded. To obtain this result, the scientists had to record eye movements as well as neural discharge patterns while presenting visual stimuli to their monkey subjects. Through these combined techniques, the researchers obtained direct evidence for prediction in the brain.

Finding that a neuron's receptive fields anticipated future positions would have surely amazed behaviorists from the first half of the twentieth century who saw the brain as an impenetrable block box. That they felt this way was understandable, for in the early to mid-twentieth century, the sort of experiment described above would have

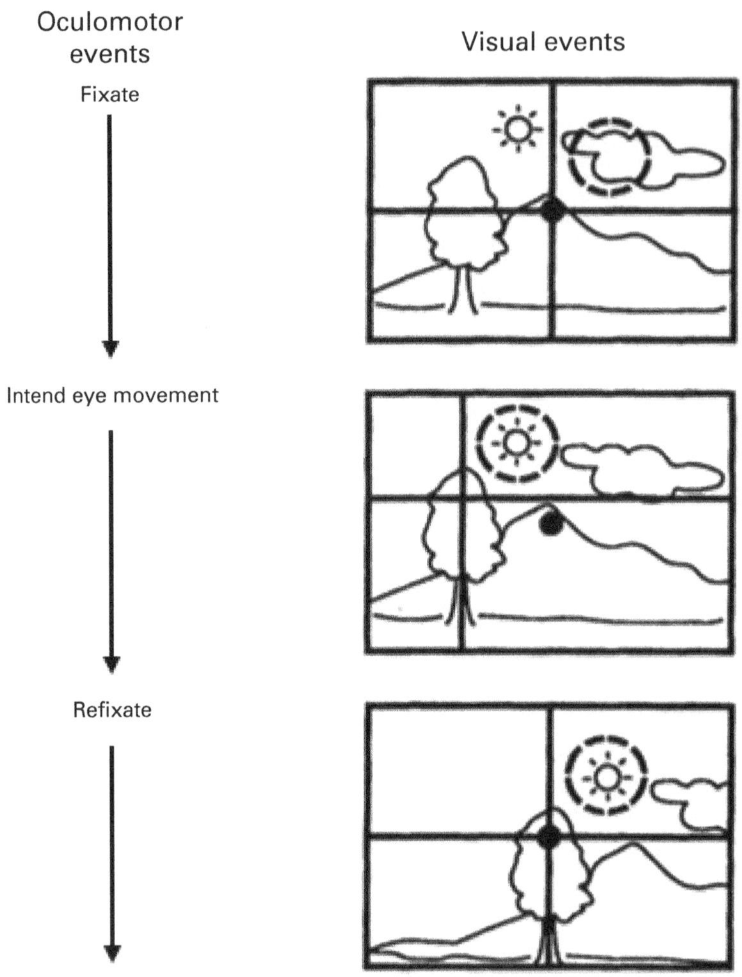

Figure 7.3
Receptive field of a neuron in the parietal cortex of a monkey before and after a saccadic eye movement.

been unthinkable. By the end of the twentieth century, however, when this study was done, researchers could open the black box, so to speak, and check for neural indicators of prediction. That evidence having been obtained, it showed that any radical behaviorist denying prediction could be overruled. Karl Lashley, who had challenged the radical behaviorists by saying that plans (predictions) are real, was right after all. And of course, Helmholtz, who had foreseen all this in the nineteenth century in broad terms if not in detail, was not just right but prescient.

Saccadic Suppression

The experiment I just told you about was highly advanced from a technical standpoint. An earlier observation relied on a less technically specialized method but allowed for a conclusion that was similar. In 1900, a phenomenon was described that you can replicate for yourself with no special equipment at all (Dodge, 1900). Look into a mirror and try to see your own eye jumps. You won't be able to. You can try hard to see your saccades, but you won't see them. This is not because your eyes move too quickly to be seen. You can see your own saccades in a video recording of yourself, and others can see your saccades if they watch you produce those jumping eye movements while you look at yourself in a mirror. If you trade roles, so your friend is the one looking in the mirror and you are the watcher, you will be able to see their saccades, although they won't be able to see those saccades when they search for them in the mirror they're starting at.[5]

What's going on here? A not-very-interesting possibility, although one that needs to be taken seriously, is that when you make saccades, your eyes move so quickly that the image projected on your retina is smeared to the point where you can't see anything. If that were the source of the effect, sensitivity to visual inputs wouldn't be expected to change *before* saccades.

It turns out that visual sensitivity does drop before saccades happen. This was demonstrated in experiments where participants indicated whether they could see a light flash at different times relative to their own saccades (Matin, 1974; Volkmann, Schick, & Riggs, 1968). As seen in Figure 7.4, visual sensitivity dropped even *before* saccades began, which meant that the reduced sensitivity was outflow based. The reason you can't see your own saccadic eye moves is that your visual system shuts down when your eyes jump from one place to another. This phenomenon is called *saccadic suppression*.[6]

One of the most amazing features of saccadic suppression is how often it occurs. It's worth doing some math here. If you make three saccades a second on average and

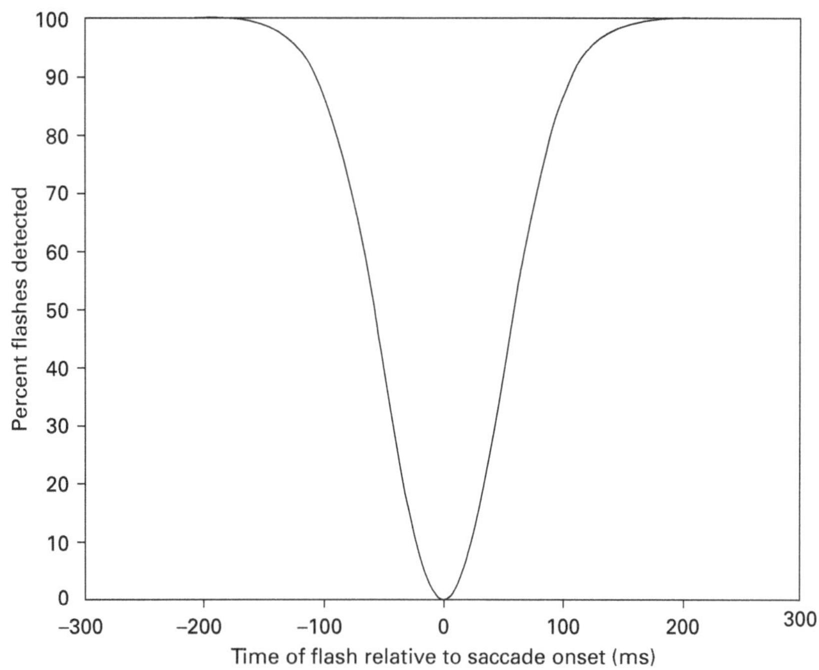

Figure 7.4
Change in visual sensitivity before, during, and after saccadic eye movements. Simulation of data reported by Volkmann, Schick, and Riggs (1968).

the average time to make a saccade is a tenth of a second, then your visual system shuts down for 30 percent of your waking time. If you're awake 16 hours a day, that means you undergo saccadic suppression for 4.8 hours a day. You don't feel yourself plunging into darkness, of course. You don't "see darkness." Instead, you remain oblivious to the darkening—as oblivious, in fact, as someone who was born completely blind and cannot imagine what it means to be in the dark versus in the light.

Blink Suppression

There is another time when you go blind momentarily. That is when you blink your eyes. When you close your eyelids, which you do to moisten your cornea and protect your eyes from approaching objects, your eyes remain covered for about a quarter of a second. If you blink once every 4 seconds on average, your eyes are covered for about one-sixteenth of the time, which means that if you're awake 16 hours a day, your eyelids are closed for about 1 hour a day. Add that time to the 4.8 hours of being saccadically suppressed, and you're functionally blind for 5.8 hours a day. You've probably been unaware of this until this now. I remember being astounded by this fact when I first learned about it.[7]

Why don't you realize you're deprived of visual input when you blink? Do feedforward processes cause blink suppression? If you hypothesize that the basis for obliviousness to darkening during blinking is the same as the basis for obliviousness to darkening during saccades (feedforward suppression of visual input in both cases), then an interesting challenge awaits you as a researcher. How can you show that there is reduced sensitivity to light during blinking?

Demonstrating that people can't see when their eyelids are closed won't get you a publication in *Science* magazine (one of the top scientific journals in the world). What *will* get you a publication in *Science* is a clever insight about how to test visual sensitivity when blinks occur.

Such an insight was had by Frances Volkmann and two of her colleagues, who published a paper in *Science* demonstrating blink suppression (Volkmann, Riggs, & Moore, 1980). These investigators made use of the fact that a bright light in your mouth can be visually perceived as a diffuse red glow. Shining a bright light in your mouth with a pen flashlight, for example, passes light through your sinus cavity back to your retina.

Volkmann and her coworkers thought it would be advisable to get precise control of the illumination and timing of intraoral lights, so they took advantage of a new invention at the time, the optic fiber. With an optic fiber—the kind of synthetic tubing that sends email, stock orders, and other signals over vast stretches—Volkmann and her coworkers delivered brief bursts of light of varying intensity at random times relative to participants' eye blinks. Because the eye blinks occurred haphazardly, Volkmann and her colleagues had to look back at the data to tell when the lights were presented relative to when blinks occurred, how bright the lights were, and whether the lights were detected. They reasoned that if a light had to be brighter to be detected at blink times, then that would show that there was centrally triggered blink-related visual suppression.

Volkmann and her colleagues found that sensitivity to light did in fact decrease during blinks, much as sensitivity to light decreases during saccades, as shown in Figure 7.4. And as for saccadic suppression, Volkmann and her colleagues found that sensitivity began to decline *before* blinks were made. They concluded that there is centrally produced blink suppression, much as there is centrally produced saccadic suppression. Because both forms of suppression anticipate their corresponding motor events (blinks and saccades, respectively), both kinds of suppression can be said to reflect the operation of feedforward modulation of sensory input.

Calls across the Brain and the Zen of Suppression

When I teach about saccadic suppression and blink suppression, I often play-act communication between parts of the brain, playing the role of a vocal oculomotor brain center. I yell with my hands cupped around my mouth like a megaphone, "Hey, visual system, we're sending signals to the eyes to make a saccade. Pay no attention to the visual swoosh that will come your way. It's all due to the commands we're sending to the eyes to make them saccade." When I teach about blink suppression, I change the referred-to actions to blinks. My aim is to bring feedforward signaling to life.

I also stress to my students in the classroom and now to you, dear reader, that as a result of this inner signaling, the resulting perceptual experience is not perception of darkening. Instead, it is no experience at all. Just as people who are congenitally blind can't know what it means to see or not see, when you blink and when you saccade, you neither see nor don't see. You are, as I tell the students, in a brief Zen state.

Why mention Zen states? The reference isn't based on religious views. I don't mean it mystically. Instead, the reference comes from something I read once in *Zen and the Art of Motorcycle Repair*, an influential counterculture work that swept the United States and other parts of the world in the 1970s (Pirsig, 1974). The passage that struck me, which relates to saccadic and blink suppression, pertains to the fact that computer logic gates are generally said to be in either of two states: 1 (on) or 0 (off). All of computer science is based on this idea. It's why we speak of "bits" (short for binary digits, 0 or 1), "bytes" (8-bit strings), and the digital age (nothing but the digits 0 and 1). However, the author of *Zen and the Art of Motorcycle Repair* noted that there is actually a third state for computer logic gates, which is neither on nor off. Computer gates are in that other state when the electricity is shut off. Being in that other state isn't mystical or ethereal. It's simply a fact of the world, the most basic fact. When a computer is turned off—when no electricity passes through it—its logic gates are neither on nor off. It only makes sense to speak of those states when the machine is activated.

When you move your eyes saccadically and blink your eyes, your visual system momentarily enters a Zen state like this. It may be reassuring to learn that you enter a mini-Zen state for several hours each day when you move your eyes saccadically or blink. You're not aware that you enter that Zen state, but that makes it "especially Zen."

Tickle Suppression

During blink and saccadic suppression, visual sensitivity drops to the point where seeing is essentially impossible. Something similar occurs for tickling. When you tickle someone, the person laughs, provided the tickling is welcome (Figure 7.5). If someone tickles you, you laugh, too, provided the tickler tickles your fancy—if you like them enough for the interaction to be welcome. There's one person who can't make you laugh through tickling, however. That person is you.

It turns out that people can't tickle themselves. When I teach about this, I ask whether any pair of students who are emotionally close to each other might like to come to the front of the classroom and tickle each other. I only offer the invitation when it's obvious that there's at least one such pair. In my many years of teaching, I've had the pleasure of teaching quite a few sweet-looking pairs (both straight and

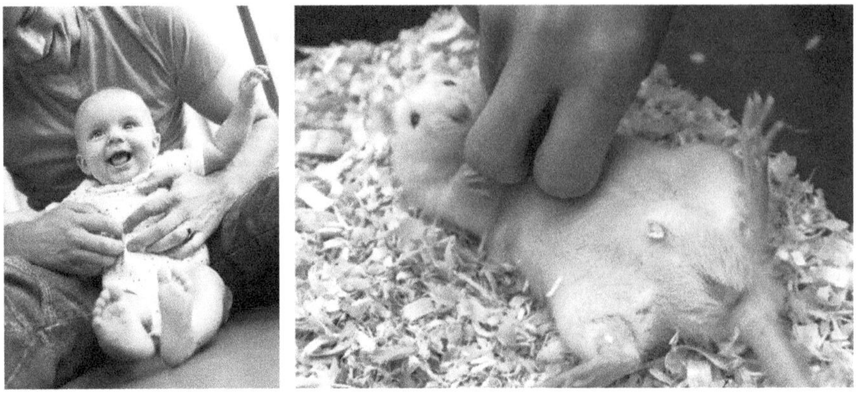

Figure 7.5
Tickling is enjoyed by people and rats.

gay) who sit next to each other for the entire academic quarter holding hands. When couples agree to come up to the front of the room and tickle each other in public, it's fun for all. Everyone laughs, not just the tickler and ticklee. After the laughter subsides, I invite each of the ticklers to try tickling themselves. After a few quizzical looks, the students try it. In short order, it becomes clear that self-tickling just won't work.

Why is this? Why can't people tickle themselves? You can probably figure out the answer based on the previous discussion. Internal signals related to the expected sensory consequences of tickles are sent to somatosensory centers. "Disregard the tickles you're about to receive," those motor-planning centers call out to the somatosensory centers in suitable neural code. When the tickle-related signals arrive, the response is a muted "blah" rather than a raucous "ha!"

When you tickle yourself, you feel the pressure of your fingers on your skin, so in this respect, tickle suppression doesn't amount to complete touch suppression, unlike saccadic suppression and blink suppression, where there's momentary blindness. In tickling, touch isn't deactivated, perhaps because there's little chance of getting useful information during saccades or blinks given their brevity, whereas there *is* a good chance of getting useful information during the relatively long time that tickling persists. Tickling not only lasts long enough to provide useful information; it's also pleasurable. In addition, its positive affect seems to generalize over mammalian species. Rats (Figure 7.5) chase after human hands that tickle them and chirp with seeming delight as their bellies are prodded in the same playful way that humans like (Ishiyama & Brecht, 2016).

When you tickle yourself, you don't lose your sense of touch, as said above. Instead, you lose your sense of humor, albeit temporarily. The temporary lack of mirth (humor) might be due to the fact that the element of surprise is gone. Either that, or the social aspect is missing. Regardless of the exact reason, the bottom-line point is the same. Feedforward signals arise during preparation for actions. Those signals concern the perceptual consequences of the act to be performed and lead to sensory suppression or, in the case of tickling, suppression of tickling's emotional effect.

Reduced ticklishness in self-administered tickling has been documented in laboratory tests (Blakemore, Wolpert, & Frith, 1998; Lametti, 2010). A robot was used to tickle subjects based on inputs from the subjects themselves. The subjects made hand movements in one location. The hand moves were picked up and echoed by a robot in the form of tickle-type motions on the same subjects' skin (with the subjects' permission). When the robot's tickles exactly matched the subjects' own movements, the subjects didn't find the robot tickles very amusing; the tickle ratings they gave were low. But when the robot's moves differed from the subject's, the rated tickliness increased. So it was the unpredictability of the robot moves that made its tickles more ticklish. Being unable to predict tickle inputs heightens their effectiveness. If you tickle yourself, you can predict what tickles will come and the effect is less amusing.[8]

Schizophrenia

Might there be people who find it amusing to tickle themselves, who laugh uproariously when they rub their own sides and bellies? If there were, they would be folks in whom a breakdown could be said to exist between intentions and actions. Those people, if they exist, might not form feedforward representations for tickles, or if they did, those internal models (yet another term for feedforward representations) might not be compared with action-based feedback in the typical way.

If such people exist, you might expect to see other signs of their problem. For example, you might see them standing on street corners having animated discussions with others when no one is nearby. After you check to see if they're wearing earbuds or holding a cellphone and see that they're not, you might dismiss them as crazy.

It's easy to write people off if they gesticulate wildly and swear at no one in particular; you're wise to keep your distance. But beyond avoiding such people and categorizing them as psychotic, it might be more helpful in the long run, for the advancement of psychiatry, to form more detailed, enlightened hypotheses based on growing understandings of the brain. For example, you might hypothesize that the individual is speaking to no one because the voice they're hearing isn't one they recognize as their own. That state of affairs is very different from the normal or typical case, where, when one speaks, one understands that the voice being heard comes from within. Hearing your own voice and knowing it has come from you frees you from having to wander where it came from, who said it, what they want, and, if you feel compelled to reply, how you should answer.

Long ago, it was suggested that hearing voices is indeed due to a failure to monitor one's own self-generated speech (Bick & Kinsbourne, 1987). To the extent that hearing voices may be a symptom of schizophrenia, recognizing the problem as such may suggest interventions.[9]

Returning to self-tickling, individuals with schizophrenic features do not, as far as I know, laugh uproariously if they tickle themselves, so that possibility can be laid to rest. On the other hand, people with schizophrenic characteristics do perform differently than neurotypical individuals in a task designed to probe sensitivity to self-produced touch (Shergill, Bays, Frith, & Wolpert, 2003). The task was developed by Daniel Wolpert of the University of Cambridge after noticing that his sons punched each other harder and harder while engaging in a game where the rule was to hit the puncher just as hard as the puncher had just hit the "punchee." Wolpert noticed that his sons, like any other pair of rambunctious siblings or pals, hit each other harder and harder in successive punches, although they claimed to have no awareness that they were doing so. Consistent with insights from Helmholtz, von Holst, and others, Wolpert developed an understanding of what was happening here. Each time a boy punched his brother, the force the brother thought he was delivering was less than the force he actually gave because feedforward signals dampened the feedback the puncher felt upon delivering the blow.

To test this hypothesis, Wolpert and colleagues set up an experiment (Figure 7.6) in which one person pressed a lever to apply force to another person. Then that person, the recipient of the first touch, pressed the lever to apply the same amount of force to the other participant. Back and forth this went, with the two participants taking turns applying forces to one another with the instruction that they keep the force levels constant. Sure enough, the subjects applied escalating forces to one another. By

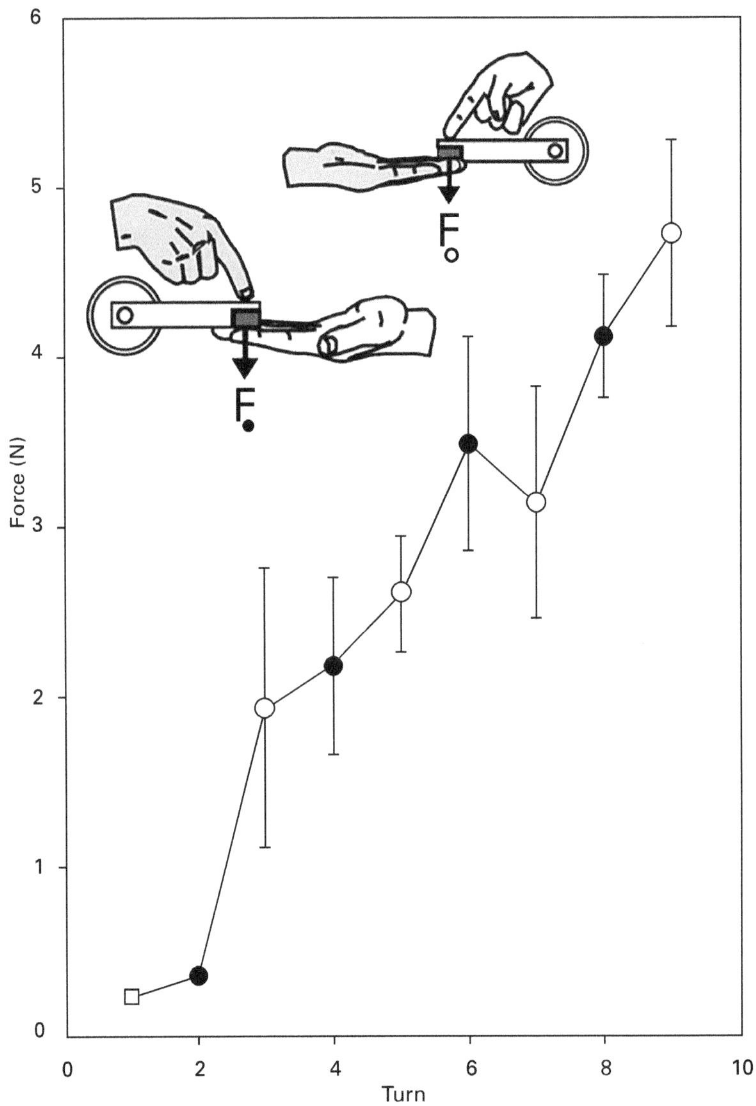

Figure 7.6
Forces recorded when two people took turns applying subjectively equal forces to one another. The gray hand and dark data points are for one person. The white hand and white data points are for the other person. The colors are for illustration purposes only.

hypothesis, the force that was applied was subjectively weakened as a result of feedforward attenuation (internal neural weakening). To compensate for this attenuation, the applied forces kept on growing.

There were some participants who showed a smaller illusory effect, however. These participants were ones whose responses on questionnaires tended to be associated with schizophrenia—for example, "Do you ever feel as if you have been chosen by God in some way?" This outcome was consistent with the hypothesis that schizophrenic delusions might be related to problems with feedforward signaling (Teufel, Kingdon, Ingram, Wolpert, & Fletcher, 2010).

Still more evidence for the hypothesis came from the late Marc Jeannerod and colleagues. Recall from Chapter 2 that Jeannerod was the French neurologist who pioneered the use of motion-picture recordings in action science. He used motion-picture recordings of people's hands as they reached for objects to characterize the coordination of their fingers during prehension (Figure 2.8). In later work, Jeannerod and coworkers had people move their hands while watching their hands on a video monitor (Franck et al., 2001). The hand motions that were shown on the monitor were delayed by various times and angular offsets. The subjects were asked to indicate whether the hand motions they saw matched their own. Virtually all subjects gave "yes" answers when the delay was zero and the angular offset was zero, which was the correct response because what the subjects saw on the video was exactly what they did at the time. But as the delays and angular offsets increased, more subjects said "no" to the question. The most important result vis-à-vis schizophrenia was that more extreme departures had to be interposed between the self-generated movements and the observed movements for the schizophrenics to recognize the mismatch. Whereas it took only a small delay or small angular offset for neurotypical subjects to declare "That's not my hand!" (or its equivalent in French), the delay or offset needed for that response in the people diagnosed as schizophrenic was greater. This outcome fits with the hypothesis that schizophrenics may be less sensitive than neurotypical individuals to the correspondence between what they do and what they see. Schizophrenia may therefore be, in part, a feedforward control problem.

Learning by Doing

What is the role of experience in the formation of feedforward representations? One possibility is that one is born knowing what to expect when carrying out actions. Another possibility is that one *learns* those perception–action relations. You may remember from the last chapter that George Stratton learned to adapt to a visually inverted world. That work helped show that it is possible to learn perception-action relations. Margaret Wooster (Figure 6.14) conducted experiments that supported and extended Stratton's conclusion. She showed that adaptation requires active exploration. That principle was supported in later work with kittens (Held, 1965). Kittens were ideal subjects because their developmental histories could be varied in ways that could not be done with humans. Furthermore, the actions they were needed to test the hypothesis under investigations were ones that kittens can do.

Pairs of kittens were given the chance to circle around in an enclosure surrounded by vertical stripes (Figure 7.7). The kittens were physically yoked to one another so the visual experience of one was replicated in the other. One kitten was the "driver." The other was the "passenger." The passenger kitten sat in a gondola moved by its mate. The question was whether both kittens would learn to see equally well. One possibility

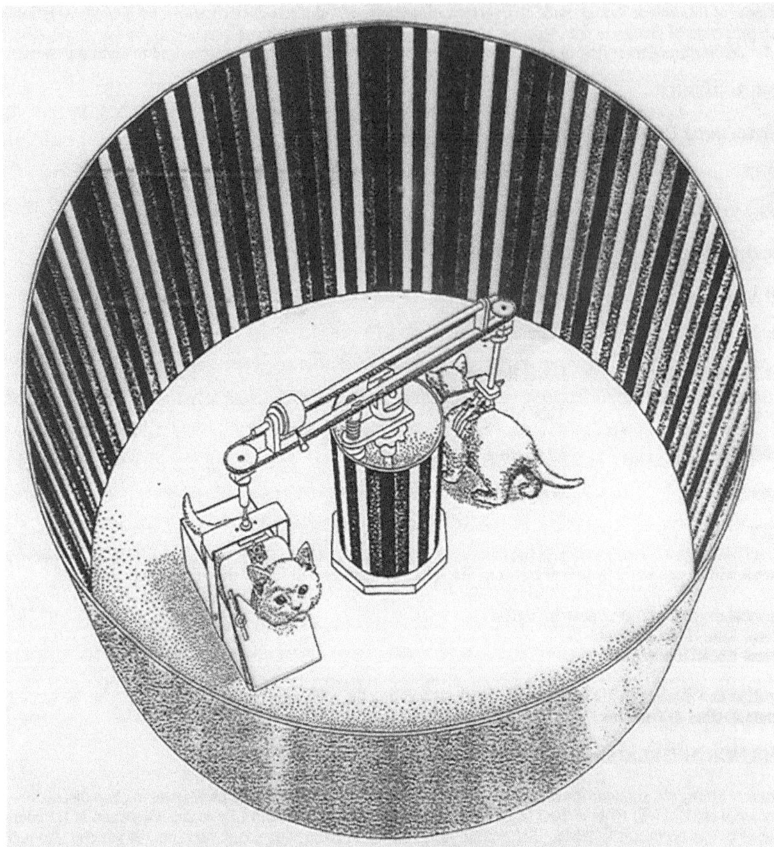

Figure 7.7
Kitten gondola experiment. The kitten in the gondola was moved by the kitten that could walk. The roles of the two kittens stayed the same whenever they were in this lit environment. Otherwise, the kittens remained in the dark and could move about freely.

was that they would. The alternative of interest was that the kitten that walked would learn to see better than the kitten that was moved passively. The basis for this prediction was that the kitten that walked could associate its actions with ensuing visual perceptions; the other kitten could not.

Before telling you the result, I need to say a bit more about the experiment to lay the groundwork for the logic. A critical feature of the two kittens' upbringing was that they were housed in a dark home cage at all times except when they occupied the striped, always lit, enclosure. In the dark home cage, the kittens could romp and play as they wished. Because the kitten had equal chances to exercise, it was implausible to think that by the time of the final test (to be divulged below), one kitten was strong and agile and the other was weak and clumsy. Both kittens were strong and agile. Another critical feature of the kittens' upbringing was that the kittens' roles were fixed. The kitten in the gondola was always the rider and the kitten on the ground was always the "chauffeur." Which kitten was assigned to which role was decided at random ahead of time.

After several opportunities to walk and ride through the lit environment, the kittens were tested for their visual sensitivity. This was done outside the striped enclosure.

Each kitten was lowered to a lit surface. The research question was whether each kitten would extend its paw as a normal kitten would.

The result was that the kitten that was able to move freely extended its paw when lowered to the table. The other kitten did not. In another test, when an object was brought toward each kitten to see whether it would pull away as a normal kitten would, the result was similar. The kitten that had been able to move freely pulled back (the normal response), but the kitten that had been driven around did nothing. It failed to appreciate that it could be struck in the face (which it never was) and remained impassive.

What interpretation was given for these outcomes? The researcher who did the study invoked feedforward control. He noted that the kitten that could move freely learned to associate visual inputs with the actions that produced them. The kitten that was moved passively could not learn those associations. So action-related vision had to be learned through active exploration.

When I tell students about the kitten experiment, they often ask how the gondola kitten fared later on. Did the gondola kitten remain functionally blind for the rest of its life, the students ask. I assure them that after the main test, both kittens were allowed to live normal lives, and both kittens grew up to be equally "cool cats." The kitten that was functionally blind after removal from the gondola regained normal visual sensitivity.

Students also ask whether the kitten result applies to humans. I tell them it does, and I describe experiments in which human subjects participated in experiments that were similar in overall logic to the gondola experiment. For example, people made marks that appeared at unexpected locations as viewed through a displacing lens. The only subjects who could adapt well to the lens displacement were ones who could actively produce marks and observe where the marks appeared. Other subjects who merely looked at the marks or had their hands passively moved adapted less well.

The main conclusion is that the best way to learn is to learn by doing. To learn to act effectively, you are best served by associating actions with feedback the actions produce. The feedback comes to be anticipated before the actions are produced. What begins as feedback becomes feedforward as well.

Expectations in House Flies

Based on the preceding discussion, you might expect feedforward to be pervasive among animal species because, as you have read, feedforward holds in monkeys, kittens, and humans. Does it even appear in insects? If the answer were yes, that would suggest that feedforward representations emerged eons ago in the course of evolution, when insects and monkeys, kittens, and humans shared a common ancestor.

I will now tell you about a demonstration of feedforward control in house flies (von Holst & Mittelstaedt, 1950). When I teach about this study in the classroom, I tell the students that the experiment I am about to report is my favorite experiment in all of behavioral and neural science. I tell them I feel this way about it because the method was incredibly clever and the inference was truly profound. The experiment was conducted by Erich von Holst, a physiologist and director of a Max Planck Institute in Germany.[10] Von Holst and his associate, Horst Mittelstaedt, were interested in *reafference*, which they defined as stimuli caused by one's own movements. The alternative to reafference is *exafference*, stimuli caused by external sources. The question posed by von Holst and Mittelstaedt was whether insects can distinguish reafference from exafference. If the answer were yes, that outcome could suggest that even flies have feedforward representations.

The subject of interest in the experiment was the common housefly (Figure 7.8). This unassuming insect walks around, minding its own business, so to speak. While doing so, it gets optical changes on its light detectors, those hundreds of tiny receptors (ommatidia) on the fly's face. Can the fly appreciate that the optical changes picked up by those receptors are due to its own motions rather than to motions in the outer world? Could it be that the fly actually expects visual changes when it walks one way or another?

To address this question, von Holst and Mittelstaedt placed a fly in a striped cylinder (Figure 7.8), a favorite enclosure for this kind of research, as in the kitten gondola experiment described above. The fly, finding itself in this world, engaged in ordinary fly behavior, circling a bit, standing still for a while, circling a bit more, standing still a while longer, and so on. All of this was unremarkable.

Next, von Holst and Mittelstaedt rotated the cylinder around the fly to see what would happen. The fly turned with the cylinder, keeping itself pointed in the same direction with respect to the stripes before it. The fly acted as if it wanted to minimize visual slippage. Remember, though, that when the same fly had moved on its own when the surrounding tube was *stationary*, it didn't seem to mind that the visual world passed before its light receptors.

What enabled the fly to respond in these two ways? One possibility was inflow. Feeling its feet turn, the fly may have determined that it was moving, in which case the

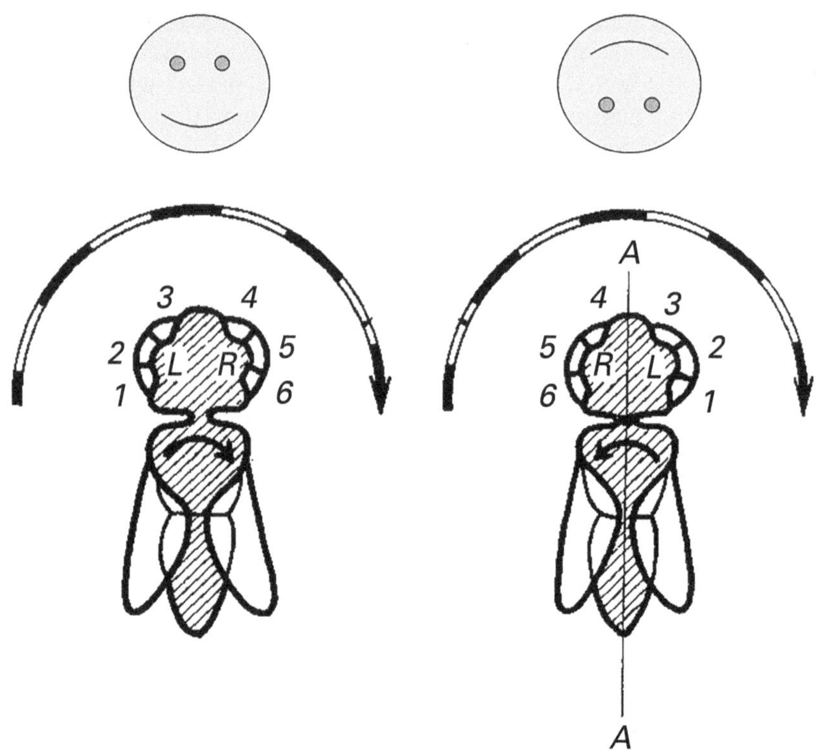

Figure 7.8
A fly with its head oriented normally (left panel) and with its head rotated 180° about the long axis of its body, line A (right panel). The fly faces black and white stripes of a drum like the one used for the kittens shown in figure 7.7. The smiley faces indicate how the fly's head is oriented relative to its thorax.

image shift was due to the fly's motion, not the world's. Continuing with the inflow possibility, when the fly *didn't* move, *not* feeling its feet turn, the fly may have known it was not moving, so the visual slippage was due to the world's motion, not the fly's.

The alternative to inflow would be outflow. A feedforward representation might be formed in the fly's brain forecasting the visual changes that would accompany its forthcoming movements. That's a pretty wild hypothesis when it comes to a housefly, for it amounts to saying that the fly can *think* in a manner of speaking, that it has *ideas*, as it were. Could this be demonstrated?

The ingenious solution that von Host and Mittelstaedt came up was to carefully twist the fly's head 180° and then glue the fly's head to its thorax. With its left eye now on its right and with its right eye now on its left, the fly was in a state similar to a person wearing left–right reversing prisms (similar to the inverting prism worn by George Stratton).

When the fly was returned to the base of the tube with the surrounding vertical stripes and the researchers turned the tube clockwise, the fly turned counterclockwise. Likewise, if the researchers turned the tube counterclockwise, the fly turned clockwise. This pair of outcomes showed that the fly could see and move and sought to keep itself visually aligned with its outer environment, as it had done before. But now it did so in the way that made sense given the left–right reversal of its eyes.

The most dramatic outcome came when the tube stood still and the fly could turn on its own. In this situation, the fly turned one way (clockwise, say) and then immediately turned the other way (counterclockwise, say), and then immediately turned clockwise again and then counterclockwise, and so on, going back and forth over and over again, only stopping when the lights were turned off. The fact that the fly stopped when the lights went off showed that its behavior was visually driven and wasn't just an uncontrolled series of spasms.

Why did the fly go back and forth over and over again when it moved spontaneously in a static lit surround? von Holst and Mittelstaedt concluded, in effect, that the fly *expected* to see visual changes related to its forthcoming movements. If the fly was about to turn clockwise, it in effect *expected* a counterclockwise visual shift. And if the fly was about to turn counterclockwise, it in effect *expected* a clockwise visual shift.

Think about what a remarkable conclusion that is! For a fly to expect, even in a subtle neurally registered way, one kind of visual shift or another is quite an achievement. The fly's expectation is hardly "great" in the sense of a high-level hope and aspiration. Still, it is a true prediction, a feedforward representation, however modest.

I remember first learning about this experiment and thinking it exceeded all my expectations about what could possibly be learned through behavioral research. To this day, I am amazed by this experiment. When I tell students about it, more than one student has told me it has given them second thoughts about killing flies, for the experiment suggests that flies may be more sentient than any of us, or at least most of us (or at least I), ever imagined.[11]

Ideomotor Theory

The fly experiment described above does not provide evidence of learning, but the kitten experiment does, and so does human work carried out in the same lab where the kitten work was done. Whether flies can learn to adapt to perceptual-motor changes brought on by having their heads turned is, as far as I know, an open question. On the other hand, quite a bit of work has been done on learned feedforward representations in humans. Much of this work has been done in connection with *ideomotor* theory. According to ideomotor theory, before a voluntary act is performed, there is a subtle image of what could be achieved motorically. The image is a sketch to be filled in with information obtained through action. The image is incomplete but instigates action. According to *ideomotor* theory, having an incomplete image that can be filled in through action serves as the intentional scaffolding for the action.

Ideomotor theory has its roots in the thinking of two nineteenth-century sleuths Hermann Lotze, a German philosopher,[12] and William James, an American philosopher widely regarded as the founder of American psychology. In his *Principles of Psychology*, a book in which James enunciated some of the most seminal ideas in all of psychology, he wrote the following (James, 1890, p. 500):

> If I will to write "Peter" rather than "Paul," it is the thought of certain digital sensations, of certain alphabetic sounds, of certain appearances on the paper, and of no others, which immediately precedes the motion of my pen.

What James was saying here was that before writing a word such as "Peter," he had a thought, barely conscious, which anticipated what he would see, feel, and hear. Why write "Peter," James asked, except to see the word formed and, in seeing it formed, feel it being written and to hear it said, at least subvocally? Without a reference condition for the form the word would take, how else could one tell that the word had been written correctly?

James's argument was based on logic and introspection. Subsequently, his thinking was supported with data. Enter Anthony Greenwald, an experimental psychologist at the University of Washington, Seattle, who conducted an experiment on seeing or hearing letters and saying or writing those same letters (Greenwald, 1970).

The conditions and results of Greenwald's study are summarized in Figure 7.9. The column headers indicate how participants found out what letter to produce. The row headers indicate how participants produced them. The italicized entries in the table summarize the results. Subjects were quick to say letters they heard and to write letters they saw but were slow to say letters they saw and to write letters they heard.

		Stimulus modality	
		See	Hear
Response modality	Say	*Slow*	*Fast*
	Write	*Fast*	*Slow*

Figure 7.9
Summary of response times in the four stimulus–response conditions used by Greenwald (1970).

Greenwald's interpretation of this pattern of results was the one William James had given earlier: before preparing to *say* a letter, people form an auditory representation of how the letter will sound, and before preparing to *write* a letter, people form a visual representation of how the letter will look. By hypothesis, then—and this was the novel experimental hypothesis that Greenwald introduced—the stimulus–response translation process prior to *speaking* should be aided if the stimulus were auditory rather than visual because preparing to speak entails hearing (or "almost hearing") the forthcoming sound. Similarly, the stimulus–response translation process prior to *writing* should be aided if the stimulus were visual rather than auditory because preparing to speak entails seeing (or "almost seeing") the forthcoming sight. The data supported these predictions.

Many experiments have been conducted with the line of reasoning that Greenwald established. Ideomotor theory has proven to be remarkably on-target whenever it has been tested (Hommel, Müsseler, Aschersleben, & Prinz, 2001; Prinz & Hommel, 2002). I'll describe one such experiment that makes the further point that even when the perceptual consequences of actions are incidental, they are incorporated into the actions' feedforward representations (Elsner & Hommel, 2001).

As shown in Figure 7.10, during training, participants learned to press a left key when they saw an X or a right key when they saw an O. Each time the left key was pressed, it was followed by a high tone, and each time the right key was pressed, it was followed by a low tone. After training, the X was shown along with the high tone (the tone that previously sounded after the X-triggered response), or the O was shown along with the low tone (the tone that previously sounded after the O-triggered response). On some trials after training, the opposite mapping was used. In those cases, the tone that sounded with the X or O was the one that had gone with the *other* response. The question was, what would happen when the tone that was played along with the visual stimulus was the one that had previously sounded after the other response?

As seen in Figure 7.10, the time to press either key was short if the tone that sounded when the X or O was presented was the tone that originally followed the keypress. When the *other* tone was used, the reaction times increased. Evidently, participants

Training phase

X → Left hand → High Tone

O → Right hand → Low Tone

Testing phase

X + High Tone → Left hand (Short RT)

X + Low Tone → Left hand (Long RT)

O + High Tone → Right hand (Long RT)

O + Low Tone → Right hand (Short RT)

Figure 7.10
Design and main result of the study conducted by Elsner and Hommel (2001).

learned to associate the perceptual consequences of their actions with the actions themselves. Representations of the actions' perceptual consequences were anticipated as the actions were prepared following appearance of the associated visual cue. These feedforward representations—in this case, anticipations of the sounds to be heard—were formed even though the perceptual consequences were incidental to the actions.

It's easy to see why the brain would have this capability. It's beneficial to pick up on the consequences of your actions. You may not have expected acting in a certain way would bring about a particular consequence, but if that consequence arises reliably from what you do, it's useful to be able to call up that action when its consequence is needed.

This is the basis for instrumental conditioning, á là B. F. Skinner. It's how Skinner explained superstition (Skinner, 1948). If a pigeon happens to do something that affects the environment in a fortuitous way, the pigeon will tend to do that behavior more and more and will come to refine its actions heightening the chance that the fortuitous outcome recurs. If the consequence is unpleasant, the behavior will be less likely.

Embodiment

The last section provided evidence that preparing to act entails expecting perceptual consequences of the acts to be performed. What about the other way around? Is there evidence that perceiving entails preparing to act? Because this book has endorsed the view that action and perception form a cycle, with neither being the perennial cause of the other, it would be reasonable to expect evidence for the second proposition, that perceiving entails preparing to act. If you play the violin, for example (the example I often come back to because of my hobby), you will not only form a forward representation of how the notes will sound as you get ready to play, but also, while seeing or hearing someone else play those notes, you might feel the penchant (the desire) to play the notes yourself. Similarly, if you play basketball, you will not only form a forward representation of how the ball will travel before you toss it but also, while watching someone else prepare to take a shot, implicitly prepare to take the shot yourself. Having those tendencies would be consistent with *embodiment*, the view that perception (and cognition more broadly) is represented in terms of what the body can do.

Personal experience supports the embodiment view. You may have felt the urge to imitate others. Someone near you may have yawned, and that may have caused you to yawn as well. Or you may have stood beside others at a red light while waiting for the light to change and then felt yourself dragged as if by an invisible force when the others started crossing the street. Imitation is a powerful force.

Studies of the brain shed light on the neural basis of such effects. When people listen to speech, areas of the brain responsible for speech production become active.[13] Similarly, when people watch others reach for objects, areas of the brain responsible for manual control become active as well. Neurons in the brain have been found to be specially tuned to observation of others doing things the observers can do themselves. These *mirror neurons*, as they're called, were discovered accidentally when a researcher absentmindedly nibbled on some food in view of a macaque monkey whose brain activity was being recorded. The neurons whose activity was being monitored turned on at the sight of the nibbling neuroscientist.

Follow-up studies showed that the neurons weren't activated by low-level visual properties of the scene or by low-level smells or sounds of the food being eaten—or, for that matter, by emotions related to hunger or feeding. Rather, the neurons were tuned to the combination of perceiving actions and being able to do those actions oneself. If the monkey could not do what it saw someone doing, the neurons didn't fire. The neurons were sensitive to the mirroring of the monkey's own capacity for action, so they came to be called *mirror neurons*.

The discovery of mirror neurons was a landmark event in neuroscience, not least because it suggested a provocative hypothesis. If mirror neurons exist, and if those neurons mediate perception and action, then their absence or malfunction in some individuals might explain the challenges those individuals face. The challenged individuals might be people on the autism spectrum who have difficulty reading others' emotions. Perhaps those individuals lack mirror neurons, or if the individuals have such neurons, the neurons may not communicate properly with other neural subsystems. This hypothesis has been widely discussed, and although it remains controversial, it has the virtue of tying a specific neural function to a specific psychological problem, which is better, all else being equal, than saying a psychological problem is just a systemwide malady, baffling in its complexity.

Controversy exists about mirror neurons. The debate centers on whether individual neurons actually do what mirror neurons are said to do—code the intersection of actions and perceptions. The alternative hypothesis is that larger networks of neurons serve this function but individual neurons do not. It's fair to ask whether the controversy matters for present purposes. I don't really think it does because the intent of this book is to lay out functional properties of action, mind, and brain without attempting to decipher the nitty-gritty of neural circuitry. Whether larger networks converge on single neurons isn't the pressing question here. What *is* relevant is the extent to which perception affects action and vice versa. The fact that there has been considerable suggestive evidence for mirror neurons suggests that each side affects the other to a considerable extent.

Effects of Possibility for Action on Perception

Is perception affected by the possibility of action? Consider people looking up at a hill preparing to judge the hill's steepness. In one condition, they're told that they will need to scale the hill while wearing a heavy backpack. In another condition, they're told they'll need to scale the same hill while wearing a light backpack. Will the subjects say the hill looks steeper in one condition than the other? If so, in which condition?

The answer, as you might guess, is that the hill is judged steeper in the heavy-backpack condition. When students at the University of Virginia were tested by Dennis Proffitt and colleagues, that's what was found. The subjects adjusted a level with their hands to match the perceived steepness of the hill. They set the angle to a more acute angle (a steeper slope) when they expected to scale the hill with a heavy backpack than when they expected to scale the hill with a light backpack (Proffitt, Bhalla, Gossweiler, & Midgett, 1995).

Does that mean the participants actually *saw* the hill as steeper? Perhaps they simply picked up on demand characteristics (what they thought the experimenter wanted) and responded in kind (Durgin, Klein, Spiegel, Strawser, & Williams, 2012). A great deal of research has been done on whether that was the case or whether the results and others like them reflect actual changes in perception. I am persuaded that they do, although I won't go through all the arguments. Instead, I'll just give a brief list of some other findings that fit with the idea that perception of the external environment is affected by what actions can be done in it.

1. Right-handed individuals tend to say that an object some distance away to the left is a bit farther away than the same object the same objective distance away on the right. Although the subjects' arms are the same length, they see the object on the right as more reachable than the object on the left (Linkenauger, Witt, Stefanucci, Bakdash, & Proffitt, 2009).
2. Holding a tool that can be used to reach otherwise unreachable objects makes the objects look closer (Witt, Proffitt, & Epstein, 2005).
3. Holding an object increases the chance of seeing the object in a scene. For instance, if you hold a gun and are given a brief glimpse of a scene with an object that vaguely looks like a gun, you will be more likely to see the object as a gun than as some other object (Witt & Brockmole, 2012).
4. When golfers judge the size of golf holes, they see the holes as larger on days when they golf better (Witt, Linkenauger, Bakdash, & Proffitt, 2008).
5. Football place kickers judge the width of goalposts to be larger on days when they do well than on days when they do less well (Witt & Dorsch, 2009).

All of these results fit with the hypothesis that perception is tuned to action possibilities. Thus, perception is not disembodied. It is not an abstract, purely symbolic description of the outer world removed from what can be done bodily.[14]

The embodiment view has been extended to other domains as well, including language comprehension. An influential (although somewhat controversial) study followed logic similar to Greenwald's. Participants heard sentences (Figure 7.11) and were asked to indicate whether the sentences were sensible. In some conditions, they were told to *push* a lever to make positive decisions or to *pull* the same lever to make negative decisions. In other conditions, the mapping of lever directions to sensibility judgments was reversed. Participants were quick to *push* the lever but not to *pull* the lever to indicate that the sentence "Close the drawer" was sensible, but they were quick to *pull* the lever but not *push* the lever to indicate that the sentence "Open the drawer" was sensible. The interaction suggests that comprehending sentences involves bringing to mind body-relevant

		Sentence to judge	
		"Close the drawer"	"Open the drawer"
Positive response	Push	Fast	Slow
	Pull	Slow	Fast

Figure 7.11
Design and main result of the study conducted by Glenberg and Kaschak (2002) on the time to confirm the sensibility of sentences either by pushing or pulling a lever. Just two of the sentences are listed here. The entries within the cells refer to the speed of responding.

information, including action tendencies. If language comprehension were disembodied, this result would be unexpected (Fischer & Zwaan, 2008; Glenberg & Kaschak, 2002).

The embodiment perspective has also informed education. Increasingly, it has come to be recognized that concepts can be taught more effectively by combining kinesthesis (movement and body feelings) with teaching otherwise abstract concepts. Using gestures to explain mathematical concepts helps, and gesturing helps facilitate problem solving (Beilock, 2015; Beilock & Goldin-Meadow, 2010; Goldin-Meadow & Wagner, 2005; Thomas & Lleras, 2009). Using self-produced movements also helps memorization (Engelkamp, 1998). If you need to remember where you park your car in a multilevel parking garage, all of whose tiers look alike, hold your hand up to your face while leaving the lot and hold up the number of fingers for that floor. Chances are you'll find your car more easily if do this than if you simply rely on unexpressed memorization.

The embodiment perspective is traceable to philosophers and linguists as much as to experimental psychologists. It has revolutionized the understanding of mind and brain (Clark, 1997; Lakoff & Johnson, 1999; Merleau-Ponty, 1945; Michel, 2020). For those who have placed the study of action front and center in cognitive science, the perspective is welcome but hardly surprising, for if acting is what you believe your brain controls, then it's to be expected that perceiving and understanding are body based.

This chapter has shown that you can go far with the feedforward perspective. With it, you can understand how prediction has roots in action, for even in the most elementary kinds of action like blinking or making saccadic eye movements, and even for creatures like houseflies, predictions are made before actions are carried out. The predictions that are made concern the perceptual consequences of actions that will be performed, playing a key role in choosing those actions. The fact that schizophrenia may be better understood by considering feedforward control mechanisms hints at the new insights that may be possible by having action researchers and clinicians join forces.

8 The Degrees-of-Freedom Problem

If you had taken my class on Action, Mind, and Brain, you would have probably been asked at some point to touch your nose. The request would have come as part of a classroom demonstration aimed at showing that virtually everyone in the class would complete the act in essentially the same way. That would be remarkable because nothing in the request required such consistency. You probably would not have touched your nose with the back of your hand, nor would you have snaked your hand over your head. In all likelihood, you also wouldn't have gotten up and hopped to the other side of the room and pressed your nose against the wall. Those bizarre methods of nose-touching would have been possible, but you wouldn't have used any of them. No student I've taught ever has.

There are an infinite number of ways that you can touch your nose. The exact angle of your hand near your nose can vary, the spatial path of your hand on the way to the final contact can vary, and for each of the infinitely many spatial paths your hand can follow, the timing of the movements can vary as well. With infinitely many ways of completing this task, how is one of them chosen? By what inner process does the unique solution emerge?

If I asked you this question in person, you might be a bit impatient. You might say, "I just did what's most efficient." Secretly, you might wonder how I could be so clueless.

If you said you do what's easiest or most efficient, I'd answer, "Of course, but what does 'most efficient' mean? What makes a movement efficient? What makes your movement or mine *most* efficient?" The answer, it turns out, isn't obvious.

The Meaning of the Term "Degrees-of-Freedom Problem"

The overarching question in this chapter is the one just implied: how are actions chosen when an infinite number of actions can achieve a goal? That state of affairs—having infinitely many possible actions—isn't the exception; it's the rule. The reason is rolled up in the *degrees-of-freedom* problem, which is the title of this chapter.

The degrees of freedom of a system are the number of independent dimensions on which the system can vary. A point in 3-space, such as the tip of your nose, has 3 degrees of freedom: x (left–right), y (in–out), and z (up–down). But your hand has more degrees of freedom than that; it has 6 degrees of freedom: the x, y, and z values for some reference point on the hand (making 3 degrees of freedom), plus 3 more degrees of freedom corresponding to the hand's orientation, namely, its *pitch, roll,* and *yaw*. (The actual degrees of freedom are even more numerous because the hand isn't a rigid three-dimensional object, but for now, the oversimplification will suffice.)

You can understand what the terms "pitch," "roll," and "yaw" mean by considering an airplane (Figure 8.1). The *pitch* of the plane corresponds to its angle about the

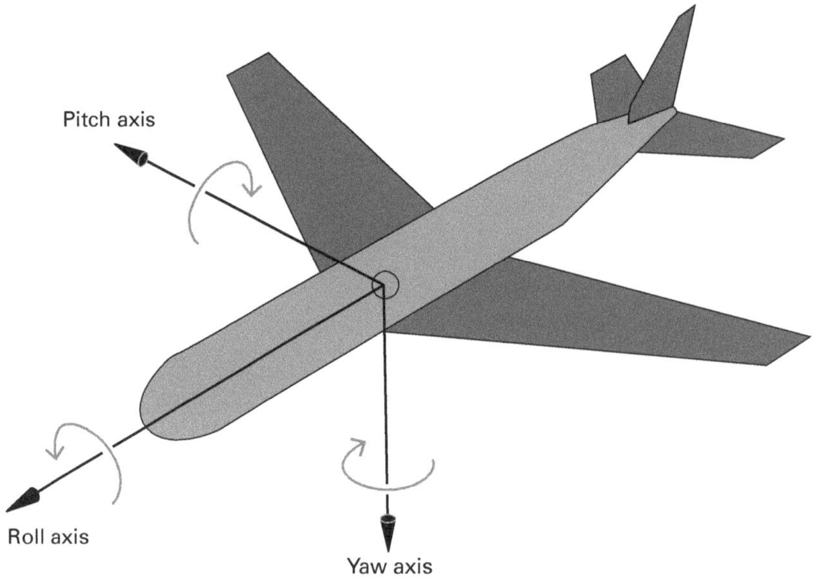

Figure 8.1
An airplane's orientation.

axis running from the left wing to the right wing. It's the degree to which the plane points up or down. Pitch is upward as the plane leaves the runway and down as it heads back down. The *roll* of the plane corresponds to the plane's rotation about its long axis from stem to stern (from front to back). If you're on a plane sitting above the left wing and find yourself considerably higher than someone sitting above the right wing, you can tell that the plane has rolled up for you and down for the other passenger. The *yaw* of the plane depends on its rotation about the plane's vertical axis. A plane heading north has a different yaw than a plane heading east.

Your hand (treated as a rigid object), likewise, has 6 degrees of freedom corresponding to its orientation and position, but your hand's position also depends on the limb to which it's attached (Figure 8.2). Your wrist can turn about two axes: up and down and side-to-side. Your wrist's position depends on your elbow, which can turn about two axes: it can twist and it can bend. And your elbow's position depends on your upper arm, which can turn about three axes at the shoulder: twisting, rising and falling, and swinging in and out. If you count the number of axes just referred to, you get 7 degrees of freedom. Even more degrees of freedom come into play considering the path taken by your hand to its final landing site and the timing of the hand along any given path. What you have, then, is a plethora (an overabundance) of action alternatives, and that gets you to the problem at hand here, the degrees-of-freedom problem: how does the nervous system get from many possible actions to one?

Origin and Scope of the Degrees-of-Freedom Problem

As you begin to think about the degrees-of-freedom problem, you might want to know who first identified it. Credit is usually given to Nikolai Bernstein, a Russian scientist who lived from 1896 to 1966. Bernstein, whose picture is shown in Figure 8.3, was

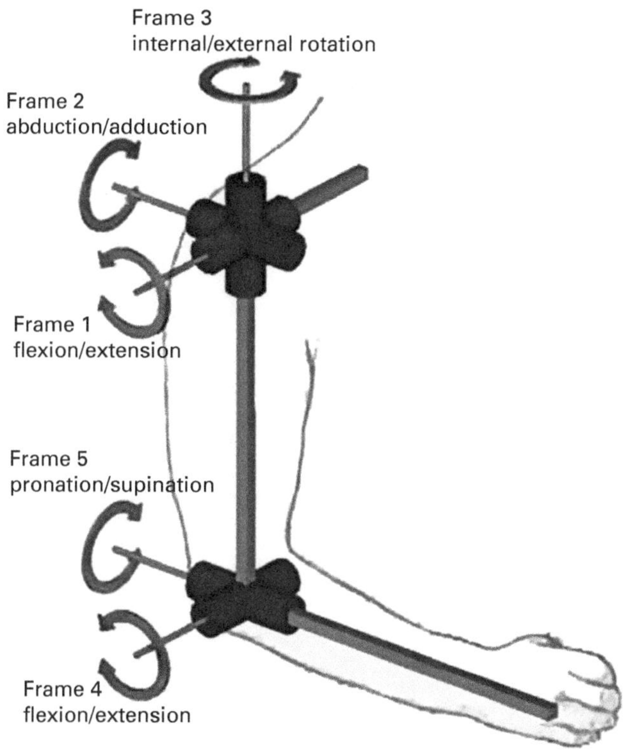

Figure 8.2
The 5 degrees of freedom of the human elbow and shoulder joints. The wrist has 2 more degrees of freedom. It can bend up and down, and it can swivel from side to side.

the first to notice that even the simplest of acts is implicitly the result of narrowing down a vast number of possible acts to one (Bernstein, 1967). No one had noticed this before, so like Helmholtz, who was the first to see that a problem attends the difference between reafference (internally caused perceptual changes) and exafference (externally caused perceptual changes), Bernstein was the first to detect a problem that had gone unnoticed before, making him, in my view, one of the true vanguards of this field.[1]

The degrees-of-freedom problem doesn't only arise in connection with the generation of physical actions. It also arises in vision. Given a two-dimensional (2D) retinal image, an infinite number of three-dimensional (3D) arrays could produce it. The problem for the visual system is to determine which 3D array was the most likely source of the 2D display. In vision, there is a one-to-many mapping of visually registered external sources to single images. In motor control, there is a many-to-one mapping of possible acts to single acts.

It's important to be clear that the degrees-of-freedom problem isn't a problem for you to worry about. Instead, it's a problem—an interesting challenge—for scientists trying to understand the processes by which perception and action work. In fact, the degrees-of-freedom problem is, if anything, an *opportunity* for you.

Why? Why is the degrees-of-freedom problem a good thing for you, a source of liberation, you might say, rather than a source of angst? Think back to touching your nose. If you were unlucky enough to be a kidnap victim with your hands and feet tied and, on top

Figure 8.3
Nikolai Bernstein.

of all those woes, you suddenly developed a terrible nose itch, you might need to carry out one of the extreme actions I described earlier: You might hop to the wall and press your nose against the wall, raising and lowering your torso until you got rid of the itch. You might at that point declare, "Thank goodness for those extra degrees of freedom!"

The idea, of course, is that extra degrees of freedom let you do things in novel ways. This lets you tap into a kind of inventiveness you may not know you have, for you constantly generate new solutions to physical challenges no matter how mundane the acts may seem to be. As a kidnap victim with your hands and feet tied, you could pick up a pen with your mouth to write "Help! I'm being held hostage!" on a sheet of cardboard which you could clutch between your teeth, carry to a window to the outside world, hopping there with your feet bound together, prop it in the window, hop back to the place you were told to remain, and hope for the best.

Such inventiveness may underlie another kind of creativity that is more commonly celebrated in cognitive science: the production of language. You constantly produce new sentences, saying things you never have and that no one else may have ever said either. An example is the present sentence, which I am writing while walking on my desk treadmill, which has pictures of my daughters above it. If anyone ever produced that last sentence, I'd be surprised.

Just as you constantly create new sentences, you constantly create new actions. If you couldn't create new actions, you probably couldn't create new sentences either, because nonverbal action generation comes earlier in phylogenetic and ontogenetic development than language. This point is tremendously important vis-à-vis René Descartes, who was discussed in Chapter 1. Recall that Descartes thought that only human beings have free will, which they receive, per Descartes, from God (Figure 8.4).[2]

Descartes based this idea on the fact that people can use language in endlessly creative ways. He proposed *res cogito* (mind) for humans but not for animals because, Descartes thought, only humans display free will. Animals, however, have *res extensa* (body), as humans do, because both kinds of creatures can carry out physical actions; this was why Descartes granted *res extensa* to both. Animals, however, are endlessly creative, too, at least in the way they act physically. Descartes apparently failed to notice this.

If endless innovation is the criterion for free will, then it makes sense to ascribe free will to animals no less than to humans. Viewing language as the sine qua non of intelligence is understandable; we humans should be proud of it. But it is a mistake to see language as the only vehicle for intelligence. By appreciating the capacity for endless innovation in action, made possible by the excess degrees of freedom that exist for virtually every task, it can be said that animals, no less than people, have free will.

It's worth considering a specific example of action generativity that has been widely discussed in the research literature. The example (Figure 8.5) concerns writing achieved with the right hand, left hand, foot, and mouth. Despite these different ways of writing, the writer—it was the same person writing in all these ways—managed to produce the same style throughout. Getting the same output through different means illustrates the capacity for *motor equivalence*.

Many other examples of motor equivalence can be identified. Nose-touching might be one such task, although there are infinitely many more. Conceivably, the diversity of

Figure 8.4
The Creation of Adam, as depicted by Michelangelo in the Sistine Chapel.

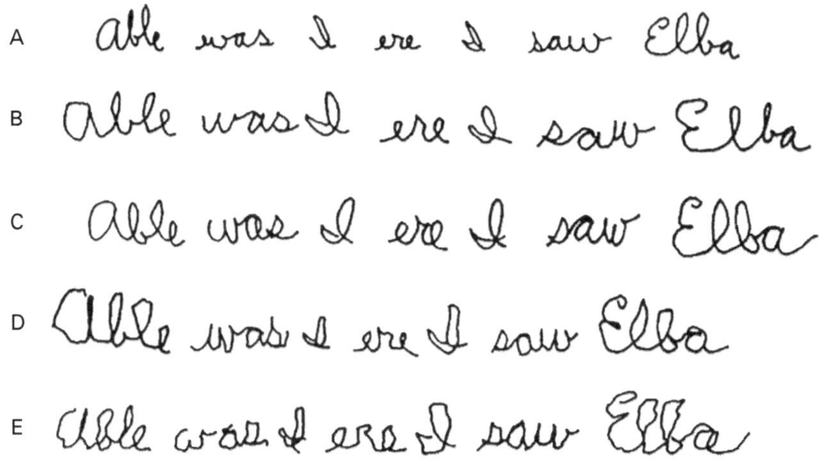

Figure 8.5
The same style of writing achieved in different ways by the same person: with the right, dominant, hand (A); with the right arm but with the right wrist immobilized (B); with the left, nondominant, hand (C); with the pen held with the teeth (D); and with the pen on the foot (E).

methods that one comes up with indexes one's creativity. That is a new hypothesis, as far as I know, and might be worth testing.

Consideration of acts illustrating motor equivalence can give rise to hilarity, on one hand, and inspiration, on the other. YouTube has a hilarious example of musicians playing their instruments in all sorts of crazy ways.[3] An inspiring example comes from a man born with no arms. He plays the drums with his feet and has other talents as well, the most important of which is wisdom. If you tend to complain about "little things," I recommend this video.[4]

Coupling or Synergies

How is the degrees-of-freedom problem solved? One method that Bernstein proposed was reliance on coupling. The idea was that interactions between effectors can limit the control they require. If different parts of the body work together through coupling, then each body part needn't be controlled on its own.

Do different parts of the body work together synergetically (another word for coupling)? Figure 8.6 shows the result of a relevant study by Erich von Holst (1939), the scientist responsible for the fly experiment described in the last chapter (Figure 7.8). The graphs in Figure 8.6 show coupling between limb segments. The top graphs come from a fish. The bottom graphs come from a person.

When the fish's right and left pectoral fins were still, the dorsal fin moved back and forth at a regular rhythm, as shown by the fact that dots superimposed on the dorsal fin graph at equal times occupied the same position in the back-and-forth dorsal-fin cycle. Once the pectoral fins began to oscillate, however, the dorsal-fin points jumped around; the equal-time dots now occupied different positions in the cycle. If the dorsal fin and pectoral fins had moved independently, the points would have remained on the same flat line. Coupling caused the dots to move around.

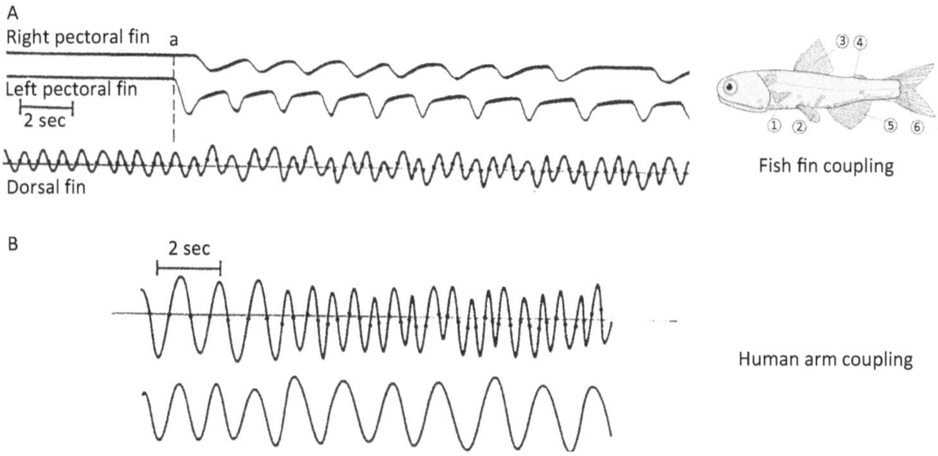

Figure 8.6
Coupling in the fins of a fish and the arms of a person. (A) The oscillation of a fish's dorsal fin changes when the right and left pectoral fins begin to oscillate. In the picture of the fish, the dorsal fin is labeled 3 and the pectoral fin is labeled 2. (B) In a person, when the right arm (upper curve) is supposed to oscillate at increasing frequencies, the left arm is affected. In both panels, the dots superimposed on the curves occupy equal time intervals. Variations in the dot positions along the vertical dimension indicate that the limb occupies different positions at the same time in the cycle.

The bottom graphs in Figure 8.6 come from a person trying to move their arms independently. When the frequency of the person's right-arm waving increased, the left arm came under the right arm's spell. Similar to the fish, the equal-time dots did not stay put but wandered, reflecting intermanual coupling.

These results show that people, like fish, have dependencies between extremities. Because fish and people both show limb coupling, the outcome suggests that such synergies are deep-seated. In all likelihood, they arose before fish and people split off in evolution, eons ago.

Coupling within Limbs and between Lips

Can coupling occur *within* a limb or does it only occur *between* limbs? It can occur within limbs. Try flexing your wrist while flexing your elbow. Then try extending your wrist while extending your elbow. Perform these two actions over and over again, doing the flex–flex action followed by the extend–extend action, and repeat this. You should find this easy; your arm motions will be probably be well coordinated as you move back and forth.

Now try extending your wrist while flexing your elbow, and then try flexing your wrist while extending your elbow. Do this over and over again, going back and forth once more. This task will probably be much harder. If you're like people who carried out these exercises while having their arm motions recorded in a laboratory, you'll be less coordinated in the second case than in the first (Kots & Syrovegnin, 1966). Moving the joints in different ways is harder than moving them in the same way. Only if the elbow and wrist were coupled would it matter whether they moved the same way.

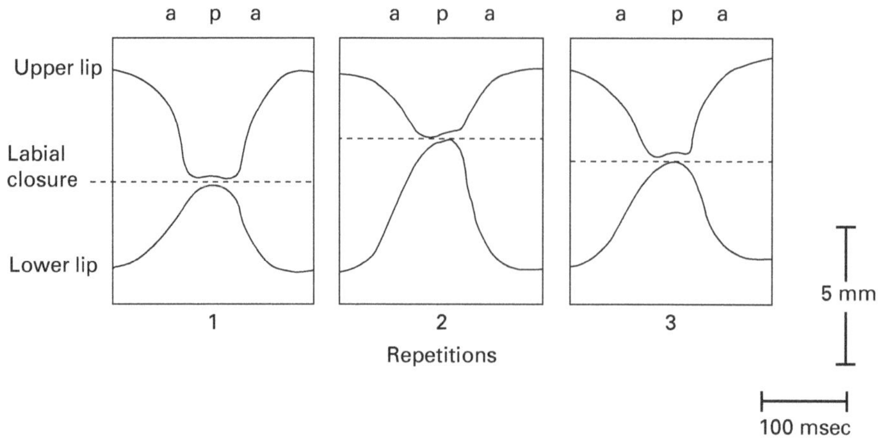

Figure 8.7
Coupling of the upper and lower lip while saying the nonsense word "apa."

Another example of coupling is shown in Figure 8.7. This example comes from speech. The graphs show the positions of the upper and lower lips when someone says "apa" repeatedly. There is variability in how far the upper lip goes down and in how far the lower lip goes up, but the positions are correlated. If the upper lip happens to go down less far than usual, the lower lip goes up farther, and if the upper lip happens to go down farther than usual, the lower lip goes up less far. In this context, the lips are coupled; they work as a team. However, if the sound being produced is "afa," no such coupling occurs. Making an "f" sound does not depend on whether the lips meet, but making the "p" sound does. So coupling between the upper and low lips depends on the momentary speech requirements (Abbs, 1986).

How can coupling be turned on or off? A simple neural network, shown in Figure 8.8, provides a possible answer (Rosenbaum, 2010). If the two lips must meet to produce a desired sound, a hypothetical bilabial closure unit is activated. The unit activates a controller for the upper lip and a controller for the lower lip. The bilabial closure unit also activates interneurons that, when activated by the upper and lower lip controllers, inhibit the opponent controllers. This has the effect of dampening the activity of the opponent lip controller depending on how active the driving lip controller is. The upshot is a pattern of results like those shown in Figure 8.8. If the bilabial closure unit isn't activated, the interneurons aren't activated and the coupling of the lips is turned off.

Because this neural network is simple, it should apply to other behaviors as well. Functional tuning of coupling should be observed elsewhere in behavior, including in behaviors that evolved before speech. That expectation is borne out in other studies. Functionally modulated responding is seen in cat reflexes. As seen in Figure 8.9, stimulation of a reflex pathway during standing causes cats to extend the foot, but stimulation of the reflex pathway during walking causes cats to flex the foot (Forssberg, Grillner, & Rossignol, 1975). A neural network like the one shown in Figure 8.7 can explain such changes.

Coupling and Dynamical Systems

Many other examples of coupling have been documented. One that has received quite a bit of attention is something you can try yourself. Point your two index fingers straight

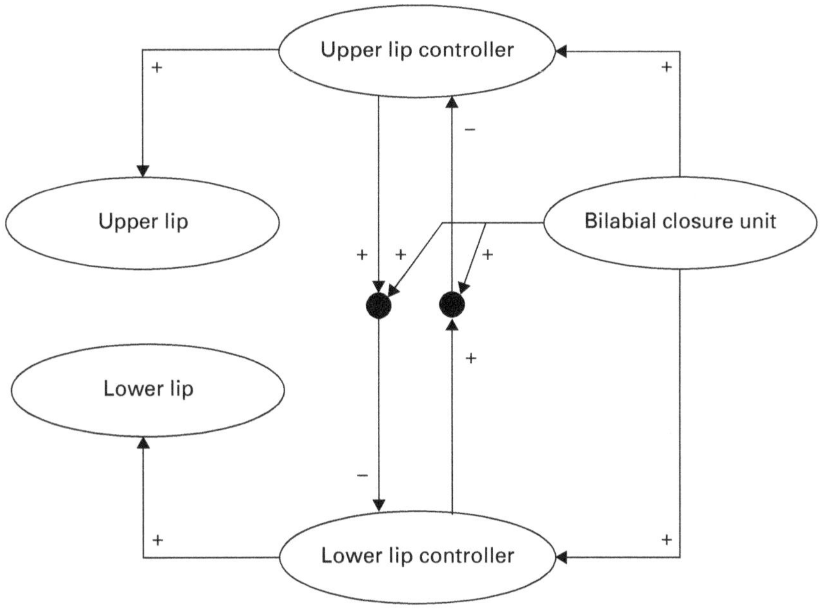

Figure 8.8
A neural network that can explain coupling of the upper and lower lip when such coupling is required (when the bilabial closure unit turns on). Plus signs denote excitatory connections. Minus signs denote inhibitory connections.

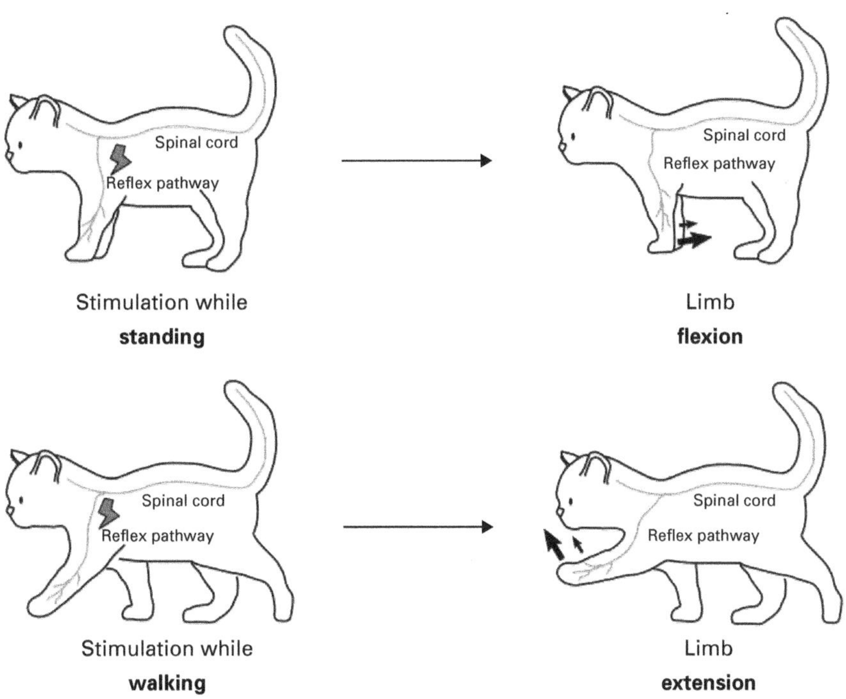

Figure 8.9
Modulation of leg reflexes in the cat based on whether the cat is standing or walking when stimulation is applied to the reflex pathway. Stimulation applied during standing promotes limb flexion. Stimulation applied during walking promotes limb extension.

ahead of you, as in the top left panel of Figure 8.10. Swing your fingers back and forth, pointing them inward, then outward, then inward again, and so on. You should be able to maintain this symmetric mode of performance quite well, even as you increase the tempo (Kelso, 1984).

Now try the same thing a bit differently, in the manner shown in the top right panel of Figure 8.10. Start as you did before, with your index fingers straight ahead, but instead of moving your fingers symmetrically (pointing them the same way *anatomically*), move them asymmetrically (pointing them the same way *spatially*). In that case, when your left index finger points to the left, your right index finger should also point to the left, and when your left index finger points to the right, your right index finger should also point to the right. If you move your index fingers back and forth between these two positions, you should be able to do so relatively easily if the tempo is slow, but as the tempo builds up, your fingers may undergo a "catastrophe." Your fingers may switch from a mode of pointing in the same *spatial* directions (both to the left or both to the right) to a mode of pointing in the same *anatomical* directions (both inward or both outward).

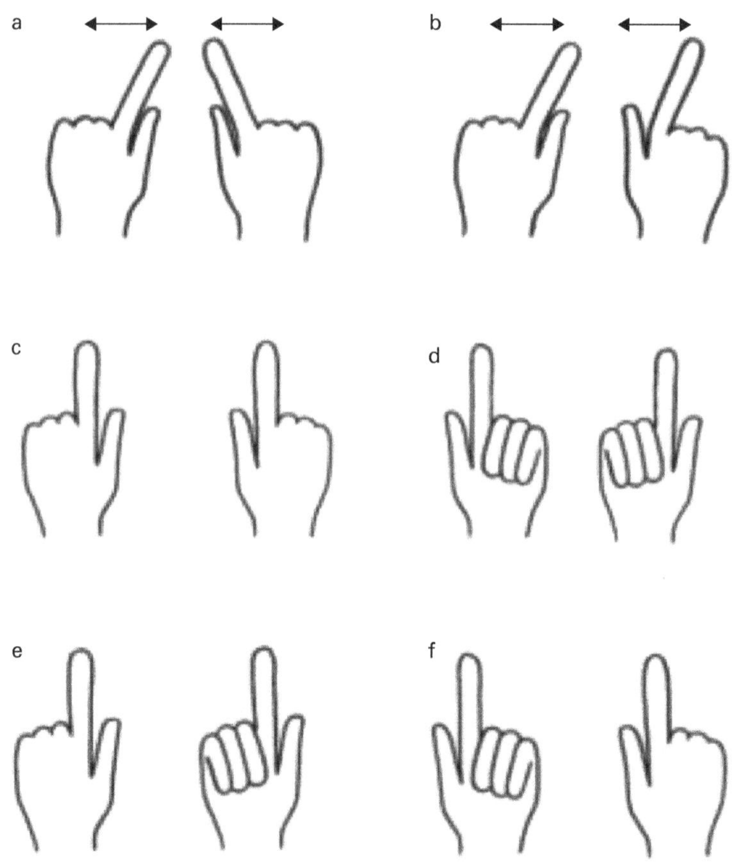

Figure 8.10
Back-and-forth movements of the left and right index fingers performed in the two ways shown in panels a and b: (a) Symmetrical movements and (b) parallel movements. These two ways of performing were carried out with the hands positioned in four different postures, as shown in panels c through f: (c) both palms down, (d) both palms up, (e) left palm down and right palm up, and (f) left palm up and right palm down.

Why did I use the word "catastrophe" in the last paragraph? Isn't that term overblown? This isn't a hurricane or stock market crash we're talking about. It's just a couple of fingers wiggling back and forth!

The reason I used the word "catastrophe" is that researchers who have studied this two-finger pointing behavior have argued that it provides an example of a type of system that can, in fact, give rise to large-scale catastrophes. Both kinds of systems—large-scale ones like weather patterns or the stock market and small-scale ones like pairs of fingers moving back and forth—can be modeled the same way, as *dynamical systems*.[5]

A dynamical system is one whose state at each moment depends on the state of the system at the immediately preceding moment. The changes that occur from one moment to the next may be unpredictable and, within the confines of the system being modeled, unremarkable. On the other hand, the changes can also be wild and downright stupefying. That's when catastrophes happen. Catastrophes that arise in the weather and the stock market can be modeled in a way that is not very different from small-scale catastrophes like the one discussed above concerning two-finger wiggling, which is what makes two-finger wiggling more interesting than it might be otherwise.

What makes dynamical systems so intriguing is that their underlying mathematics are often extremely simple. The equations that characterize successive changes may look like the stuff of elementary math classes, yet astonishingly complex changes may emerge from them anyway. An example is

$$x_{n+1} = rx_n(1 - x_n).$$

This incredibly simple equation can give rise to unpredictable times between drips of a faucet, heartbeats, and a vast number of other phenomena depending on the value of *r*.[6]

Coupling, Marr's Three Levels of Explanation, and Cognition

If you study dynamical systems, your primary aim is likely to be coming up with equations that, when allowed to run in time (typically in a computer simulation), will yield patterns that closely resemble what actually occurs. Typical applications are forecasting weather, earthquakes, and the spread of disease. Modeling these systems has permitted new practical insights, such as the discovery that perfectly regular heartbeats may be less healthy than somewhat irregular heartbeats. Another domain where dynamical system modeling has proven useful is weather forecasting. Predicting the path of a hurricane is a well-known example. The swirls seen in satellite images, along with other relevant data, have enabled dynamical-systems modelers to make increasingly precise predictions about when and where hurricanes will make landfall, and so on.

Despite these advances, some have been dissatisfied with purely mathematical descriptions of body movement. The question these investigators have been interested in is not just which dynamical-system regime (which differential equation with which set of parameters) applies to a given class of behavior but what neural or psychological mechanisms account for the behaviors that are observed. A colleague of mine once put it this way: "If I have trouble moving, I'll go to a neurologist, not a dynamical-systems modeler."

My colleague's complaint may miss the point that an explanation isn't the same as a treatment. We may not have a full explanation of why aspirin works, but the fact that it does can inspire us to develop better explanations of why it does. In addition, there are different kinds of explanation. As the brilliant cognitive scientist David Marr observed in his 1982 book, *Vision*, published after his untimely death at the age of thirty-five from leukemia, there are three levels of explanation for cognitive/behavioral data (Marr, 1982).

The most abstract is the *computational* level. Here explanations are given in terms of abstract mathematical descriptions. This is the level at which dynamical-systems modelers do most of their work.

A less abstract level is the *algorithmic* level. Here the explanatory elements are functional processes as well as representations on which they operate. This is where most cognitive scientists, including cognitive psychologists like me, do their work. Roughly speaking, the algorithmic level is concerned with software.

The most concrete level (a term not meant to disparage) is the *implementation* level. This is the material level, where the work concerns neurons, muscles, bones, and other physical components comprising the physical building blocks of behavior. This is where physiologists and anatomists do their work.

At issue here apropos the study of action is whether the computational level is the best level for analyzing how actions emerge. Is the computational level sufficient for describing the nonlinear changes that occur in two-finger wiggling, for example? One answer from some parties is that we have no choice but to pursue the implementation level. To really understand the system, they say, we must dig down into its physical makeup (Churchland & Seijnowski, 2016). The problem with this view is that if it is pursued to the exclusion of other approaches, it is easy to lose sight of the broad functional characteristics of the system under investigation.

Is there a place for the algorithmic level as well as the computational and implementation levels? The algorithmic level sits between the abstract computational level and the material implementation level. The algorithmic basis of two-hand finger wiggling became the focus of a study published by Mechsner, Kerzel, Knoblich, and Prinz (2001). In this article, published in *Nature*, one of the preeminent journals in science, a team of cognitive psychologists reported a replication of the two-hand finger-wiggling task, along with variants of it in which the hands were oriented in different ways. The tasks that were studied included the ones already mentioned, where the palms of the hands faced down and the index fingers moved either symmetrically (Figure 8.10a) or asymmetrically (Figure 8.10b). The authors observed that in these conditions, movements that looked symmetric to the performers used symmetric muscle movements, whereas movements that looked asymmetric to the performers used asymmetric muscle movements. The symmetry or asymmetry of muscle movements could have accounted for the data patterns observed before, they contended. By flipping one of the hands, as shown in panels (e) and (f) of Figure 8.10, the researchers could dissociate muscle symmetry and visual symmetry. Now when the muscle movements were symmetric, the seen movements were asymmetric, and when the muscle movements were asymmetric, the seen movements were symmetric. Because the original study had not included a case where both palms were up, that condition was also included in the 2001 report, as shown in panel (d) of Figure 8.10.

The result was clear. Symmetric performance was better than asymmetric performance when both palms were down, replicating the original study. Symmetric performance was also better when both palms were up, which was a new demonstration. Most important, when one palm was up and the other was down, visually symmetric movements were still better than visually asymmetric movements, even though the muscles were asymmetric in the visually symmetric case. Therefore, visual symmetry, not muscle symmetry, explained the data.

The researchers who did this study were advocates of ideomotor theory, which was discussed in the previous chapter. From the perspective of ideomotor theory, having a clear perceptual (visual) representation would have been vital for planning and controlling two-hand finger wiggling. By contrast, muscle activity would have been less important. Finding that the visual representation mattered so much fit with ideomotor theory. Those interested in only an abstract mathematical characterization of performance might never have asked whether hand orientations would matter. Hand orientations *do* matter, it turns out, so the mathematical model alone was insufficient or needed to be beefed up to handle the importance of visual symmetry.[7]

Mechanical Coupling

The last section provided a glimpse into the effects of coupling. The argument was that coupling may help reduce the degrees of freedom to be controlled. This has been an influential idea. So too has been the idea that the degrees-of-freedom problem may be reduced through reliance on physical mechanics. The idea is that exploiting physical properties of the body and the environment of which it is a part may cut down the degrees of freedom to be controlled. This idea was sanctioned by Nikolai Bernstein.

Exploiting physical mechanics is not distinct from exploiting coupling, for coupling can have a physical basis. This was first observed by the seventeenth-century Dutch physicist Christian Huygens (pronounced "Howchens," with the "ch" produced as a soft guttural sound). In working with pendulum clocks, he noticed that two pendulum clocks close to one another would exhibit what he called "an odd kind of sympathy."[8] If the two clocks' pendulums were initially set in motion at different points in their back-and-forth swing cycles, they would tend to synchronize; they would *entrain*.

You can see entrainment by conducting an experiment like the one depicted in Figure 8.11, or you can watch a video of the experiment on the internet.[9] Identical mechanical metronomes are set to run at the same frequency by putting their bobs at

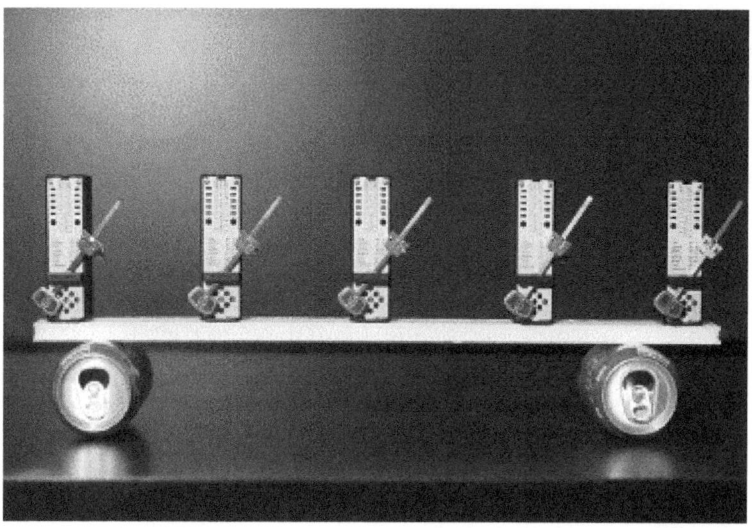

Figure 8.11
Metronomes that have synchronized after starting out desynchronized. The metronomes sit on a light wooden board atop two empty soda cans that roll slightly back and forth as the metronome arms swing back and forth.

the same points. The pendulums are set in motion so they are highly desynchronized, but in short order, the metronomes come together and run in lockstep. This is possible because the metronomes transmit energy to one another via the pliable board on which they stand, which in turn lies on a couple of empty soda cans that can roll back and forth. The metronomes are weakly linked and finally synchronize, clicking together as if under the baton of a conductor, although no conductor is needed.

Christian Huygens in the seventeenth century didn't have soda cans or the internet, but he was able to see and analyze such coupling by realizing that multiple, physically connected pendulums can act as a single physical system. Two pendulums of lengths L1 = 2 and L2 = 3 and masses M1 = 4 and M2 = 5 will tend to oscillate as a single virtual pendulum with an equivalent mass of 9 and an equivalent length of 2.65. It's easy to see how the equivalent mass is computed; it's just the sum of M1 and M2. Calculation of the equivalent length is a bit less obvious; the details aren't important here. The important point is that the multiple-pendulum system tends to act in a manner analogous to a single-pendulum system.

If pendulums hanging from a beam or standing on a board atop soda cans exhibit coupling, do people do the same? Researchers at the University of Connecticut, led by a distinguished investigator named Michael Turvey, showed that they do (Turvey, 1990). Turvey and his colleagues asked participants simply to swing two pendulums with their two hands. The participants were asked to swing the pendulums back and forth in a way that felt comfortable. The researchers varied the lengths and masses of the pendulums. It was found that the way people swung the pendulums followed Huygens's predictions. Although the people holding the pendulums were unaware of Huygens's equations, they behaved in a way that indicated their implicit sensitivity to its underlying dynamics.

Walking versus Running

The last section crossed the bridge from considering coupling to considering physical mechanics. The overarching theme was that both factors play a role in limiting behavioral options. In this section, we turn to another way in which physical mechanics resolves choices. The question concerns walking and running: When you walk at low speeds and build up speed, why do you shift from walking to running? At what speed do you make that switch, and what accounts for that particular transition point?

Answers to these questions were provided by someone who served simultaneously as professor of biomechanics at the University of Leeds (England) and director of the London Zoo: R. McMahon Alexander. Alexander (1984) understood that running could be viewed as a series of leaps, while walking could be viewed as a series of falls, so he pointed out that it would be counterproductive to try to walk (and so to fall) more quickly than gravity allows. What makes more sense, he said, is to switch from walking to running at the speed where falling would have to occur faster than gravity allows, roughly 3 meters per second. That, indeed, is the rate at which people switch from walking to running (Figure 8.12).

Alexander further tested this prediction by measuring oxygen consumption during locomotion at different speeds, not just in humans but in a variety of species, including elephants. Figure 8.12 shows setups used for this purpose. The data confirm that at a critical speed close to 3 m per second, it becomes more efficient to run than to walk. (The most efficient means of human-powered transportation is cycling.)

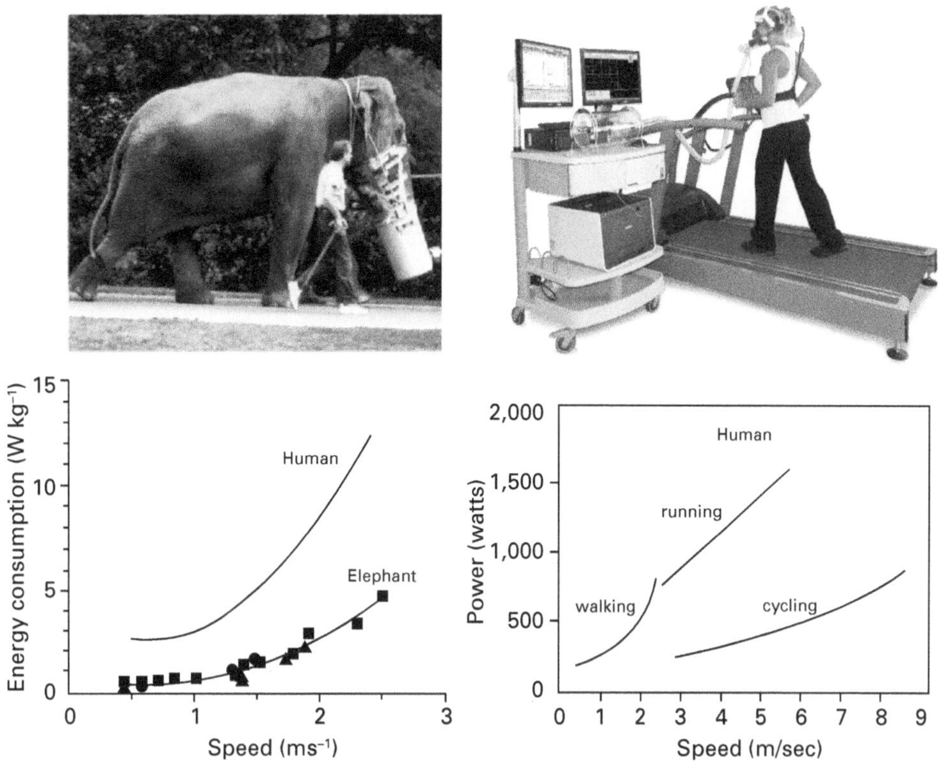

Figure 8.12
Measurement of oxygen consumption in an elephant (top left) and a person (top right). Energy consumption as a function of speed in the two species (bottom left) and power as a function of speed for humans when walking, running, or bicycling (bottom right). The switch from walking to running occurs at the speed where walking power exceeds running power.

Alexander's analysis is another example of the value of Morgan's canon: [10]

In no case may we interpret an action as the outcome of the exercise of a higher mental faculty if it can be interpreted as the exercise of one which stands lower in the psychological scale.

You may recognize this quote from Chapter 5, the chapter about learning. There the message was that physics or biomechanics alone could explain the disappearance and reappearance of the stepping reflex; there was no need to explain that phenomenon in terms of high-level thinking. The same sort of simplicity was achieved here in accounting for the speed at which we switch from walking to running.

Preflexes

Relying on physical mechanics has impacted robotics. In the early days of robotics, robots were mammoth devices. One robot developed in the 1980s was so huge that a human operator sat on top of it and tried to steer it with levers. The behemoth was so hard to control that it had to be warehoused. The difficulty of the human driver's control problem was compounded by the fact that the device didn't exploit physical mechanics. It was subject to physics, of course, but the design didn't capitalize on what physical mechanics could offer the machine for free—a means of simplifying control.

What do I mean? How can physical mechanics be used a simplify control? Consider a softball pitcher who swings her arm backward over her shoulder to release the ball so it barrels down the path toward home plate. The ball's flight is ballistic (see the last chapter), meaning that its passage can't be controlled after it's released. That doesn't mean an enormous amount of control must be used before the release, however. The details of the arm swing needn't be specified in detail. Rather, the way the pitcher swings her arm and opens her fingers can rely on a combination of neural commands and—the main point here—the arm's momentum as it swings around prior to release of the ball. By exploiting physics, the pitcher can reduce the amount of detailed control that needs to be provided. The pitcher just needs to launch the ball at the right time, place, and speed.

Animals, it turns out, use similar tricks. Kangaroos rely on the tendons of their legs to rebound when they hit the ground while hopping. The time for kangaroos to rebound after hitting the ground is literally zero. This is because the kangaroos' tendons have elastic properties that cause the roos' legs to bounce back with no delay whatsoever. The response is entirely mechanical, just as is a ball's bounce when it impacts a surface. There is no pause between when the ball hits the surface and it bounces back. The reason is that the ball and the floor are part of a single physical system in which, for every action, there is an opposite and equal reaction. The universe doesn't need to think about what it should do with a ball that strikes a surface within it. The ball and surface are part of the same universe where, if there weren't instantaneous reactions to balls landing on floors, the laws of thermodynamics would no longer hold.

A clever name has been given to such mechanically based responses in animals: *preflexes* (Loeb, 1995). A preflex is a response that literally takes no time at all, which means that it takes less time than a reflex. Reflexes are impressively quick, on the order of 30 to 60 ms, but preflexes have latencies of 0 ms.

A number of preflexes have been found by scientists studying gait and posture. An example is the helmeted guineafowl, a bird whose response to stepping on uneven ground is entirely "preflexive." This pheasant-like native of Africa has musculoskeletal features that allow it to rebound instantaneously when it steps into an indentation on the ground (Daley & Biewener, 2006).

We humans have preflexes, too. When we crouch down briefly on the way to jumping up, a significant part of the power for the upward jump comes from the elastic recoil in our leg muscles and tendons following the downward crouch. Power from the leg muscles and tendons kicks in (pun intended) in literally no time (Bobbert, Huijing, & van Ingen Schenau, 1986).

Equilibrium Point Control in Monkeys

Preflexes don't just occur in walking and jumping. They also occur in reaching. An example is shown in Figure 8.13. Here a monkey is unable to see, hear, or feel its arm while pointing toward a target light in a surrounding arc. The monkey can't feel its arm because the nerves entering the dorsal side of its spinal cord have been severed. (Recall from Chapter 3 that axons entering the dorsal side of the spinal cord carry sensory signals, whereas axons exiting the ventral side of the spinal cord carry motor signals, as per the Bell–Magendie law). After recovery from surgery, the monkey, who

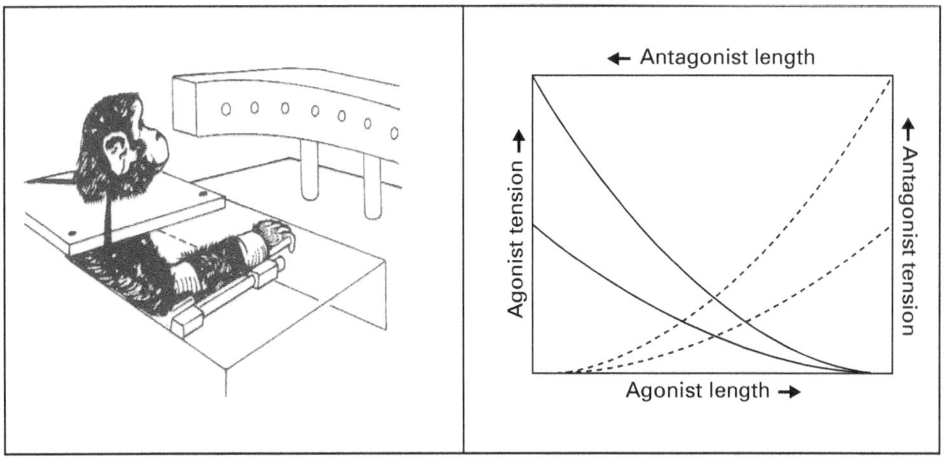

Figure 8.13
A monkey orienting its arm toward a light (left panel). The task could be done before and, more interestingly, after deafferentation. The right panel shows hypothesized length–tension curves for muscle agonists (lengths increasing from left to right) and muscle antagonists (lengths increasing from right to left). The muscles are at equilibrium where the agonist tensions (dashed lines) and antagonist tensions (solid lines) intersect.

had practiced the pointing task before, was returned to the apparatus and was exposed to each of a series of lights. The monkey was supposed to point to each light—that is, to direct its paw in the direction of the light—to get a juice reward. It was found that the monkey could point accurately even though it could not see or hear where it pointed. Remarkably, the monkey could point accurately even though it could not feel its arm (as tested physiologically). Most remarkably of all, even if the monkey's arm was jostled to one side or the other with a torque motor, it could *still* point to the target!

At first glance, this result seems impossible. How could the monkey possibly correct for mechanical perturbations of its arm if it couldn't feel the perturbation? The investigator who led this study, Emilio Bizzi of MIT, came up with a surprisingly simple answer (Polit & Bizzi, 1978). Think of a limb as a hinged rod with springs or rubber bands on either side of it (Figure 8.13, right panel). These elastic bands keep the hinged rod at a position where the forces acting on the rod are equal and opposite. If the rod is pulled to one side and then released, it instantly heads to the position where the opposing forces balance out. This happens entirely automatically with no special feedback controller to cope with the perturbation. By virtue of physical mechanics alone, the rod gets to its equilibrium position. For all intents and purposes, the reaction is a preflex.

Can the nervous system actually regulate muscles this way? Suppose the muscles serving antagonistic functions at a joint, such as extending or flexing elbow, are like two rubber bands (right side of Figure 8.13). Consider how those "muscle-rubber-bands" might work together. The right side of Figure 8.13 shows tension as a function of muscle length for one muscle (from left to right with solid lines) and tension as a function of muscle length for the opposing muscle (from right to left with dashed lines). The points where the solid and dashed lines cross are the equilibrium points—the points where the opposing forces are equal and opposite. Changing the slopes of the lines can yield different intersection points (different equilibrium positions). If the intersection points lie above the same point along the abscissa (the *x* axis), the joint will stay in the same place, although the joint's stiffness will change. If the equilibrium point *descends* without moving sideways, the joint will become less stiff, although the joint position will remain the same. Conversely, if the equilibrium point *ascends* without moving sideways, the joint will stiffen with the joint position again remaining the same. Finally, if the intersection point moves to one side or the other, above a *different* point along the abscissa, the limb will move to a new position.[11]

Does the nervous system use this method? The experiment sketched above, which involved deafferenting monkeys, was taken by Bizzi and his colleagues to suggest that it does. According to Bizzi and his coworkers, the nervous system changes muscle stiffnesses to define new equilibrium positions and, in so doing, exploits mechanics.

Equilibrium Point Control in Frogs

If the method described above is used by monkeys, it might be used by other animals too. It might be seen across the evolutionary spectrum. In fact, equilibrium point control holds for frogs (Figure 8.14). In the relevant experiments, also done in Bizzi's lab, frogs' legs were held in position on a horizontal surface so that forces could be recorded from the legs—a more controlled situation than one in which the legs could leap to and fro (Bizzi, Mussa-Ivaldi, & Giszter, 1991). Brief electrical pulses were delivered to the frogs' spinal cords, which were surgically separated from the brain to enable study of the spinal cord alone. Forces generated by the foot were recorded, and it was found that stimulating the frog spinal cord triggered forces that would have brought the leg to a specific final position if the leg had been free to move. Where the leg would go depended on where in the spinal cord the stimulation was applied. Stimulation of the spinal cord at two different places simultaneously (a condition not depicted in Figure 8.14) caused the frog's leg to generate the average of the forces produced from single-site stimulations. So vector summation worked here, much as it did in Georgopoulos's studies of monkey motor cortex (discussed in Chapter 3).

Equifinality in Frogs and Primates

An important feature of the frog data, not yet mentioned, was that the same final position was reached no matter where the frog's leg happened to be at the time of stimulation. This outcome is shown in Figure 8.14, in which the frog's leg was held at different positions before applying electrical stimulation to any given place in the spinal cord. Although the leg occupied different positions, the leg generated forces consistent with the view that the leg "was trying" to get to the same place when a given spinal spot was stimulated. The tendency to reach the same final position regardless of the starting position is called *equifinality*.

If equifinality holds in frogs, does it hold in other animals such as mammals? The answer is yes. An example is shown in Figure 8.15. Here a monkey adopted the same posture when its brain was electrically stimulated at a given brain site even though the monkey happened to be in different postures at the time (Graziano, Taylor, & Moore, 2002). The stimulation was applied in the monkey's motor cortex or premotor cortex and lasted about .5 seconds, which is about ten times longer than the electrical stimulation applied by Fritsch and Hitzig to the motor cortex of dogs and also about ten times longer than the electrical stimulation applied by Penfield to the motor cortex of human patients. The brief stimuli applied by Fritsch and Hitzig and by Penfield triggered muscle

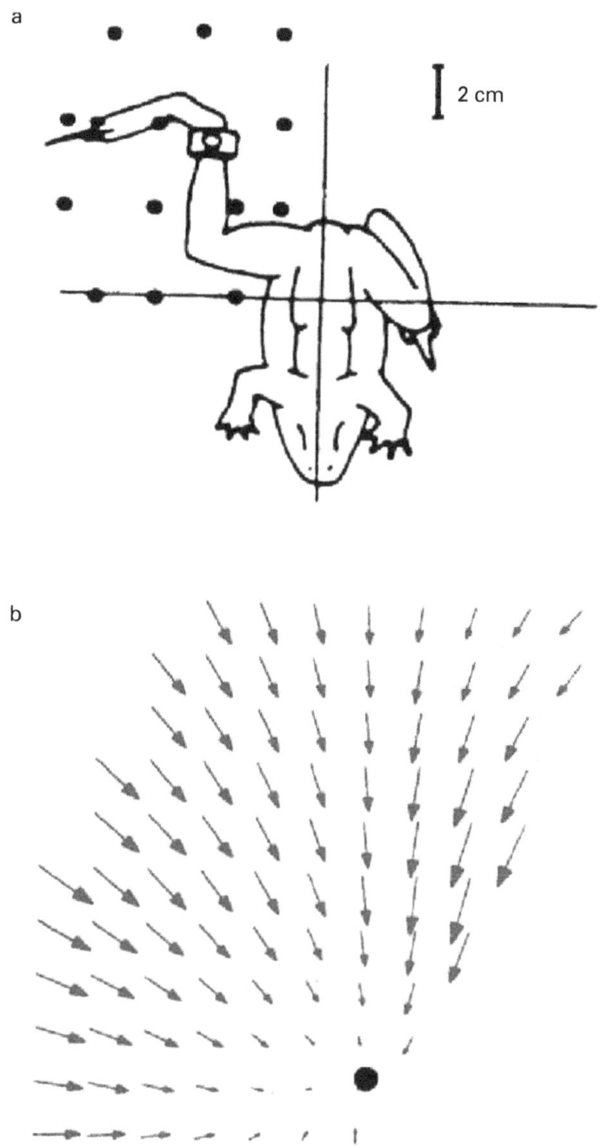

Figure 8.14
Force fields in the frog. The top panel (a) shows a spinalized frog with a force transducer attached to its ankle. The dots show positions occupied by the ankle in different trials. The crosshairs indicate where microstimulation was applied to the spinal cord when the ankle was positioned at each dot and elsewhere. The bottom panel (b) shows the force field obtained when one spinal site was stimulated while the leg was in different positions. The force field is directed to a single place, illustrating equifinality. Force fields for other spinal sites converged on other locations.

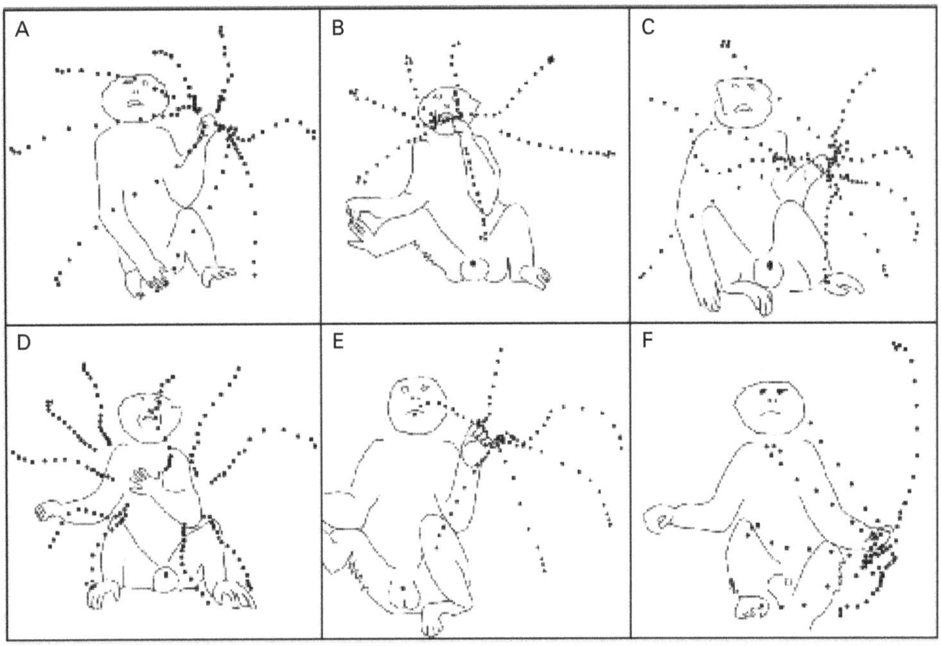

Figure 8.15
A monkey displaying equifinality for distinct body positions in each of panels A to F. The dotted lines show handpaths to the same goal position reached from different hand starting positions at the time of stimulation.

twitches. In the experiments summarized in Figure 8.15, the longer-lasting electrical stimulation led to movements to the same final body postures no matter which position the monkey was in at the time. The final body posture that was moved to depended on where the stimulation was applied within the brain. So the same outcome was obtained in monkeys with brain stimulation as in frogs with spinal cord stimulation.

Equilibrium Point Control in Humans

Because humans evolved later than frogs and monkeys, it is natural to ask whether the equilibrium point hypothesis applies to humans as well. An experiment by a young researcher at the time yielded a positive answer. Anatol Feldman, working in Moscow in the 1960s, asked people to push with one hand in the horizontal plane against a block (Asatryan & Feldman, 1965). At any time, the block could be suddenly removed. Feldman told his participants not to intervene if this happened. "Go with the flow," he said to them, in so many words (in Russian). When a participant complied with the instruction, his or her hand "flew" to a new position where it came to rest. The harder the person pressed on the block, the farther the person's hand traveled once the block was removed. Feldman took this to mean that the block simply prevented the arm from reaching its equilibrium position. What that equilibrium position was could be inferred from where the hand came to rest after the block was removed.

You can do this experiment yourself. Press your hand against the hand of a friend and ask your friend to withdraw their hand at some unpredictable time. Letting your hand go after their release, you'll see your hand go beyond where you pressed before; the harder you press, the farther your hand will go. Another way of doing the experiment is to hold a cup, glass, or filled soda can between your thumb and forefinger and have a friend pull the object out from your grasp. If you go with the flow—if you don't intervene—your fingers will close in on a new position somewhere "inside" the object. The harder you press, the deeper "inside the object" your fingers will go. (A little lubricant on your fingers helps the experiment go more smoothly.)

Why do this experiment with a friend? Why not do it by yourself? You can do it alone, but the result will be less dramatic because if you pull out the object yourself (with your other hand), you'll know about the withdrawal before it happens, so you'll be prepared for that event and will find it hard to go along with the do-not-intervene instruction. This is an example of feedforward control, the subject of the last chapter. You can't disregard what you yourself anticipate—an example of how your own mind can get in the way.[12]

Is the cup-removal demonstration an odd, unusual setup that says nothing about normal performance, or does it have ecological validity, a topic discussed earlier in this book? In my opinion, it has lots of ecological validity. According to the equilibrium point hypothesis, when you play the piano or type on a keyboard, your hands actually "want to go" to positions beneath or beyond the keyboard. Practitioners of karate learn this as part of their training. They're told to aim for points *beyond* the boards or blocks they plan on pummeling. Figure 8.16 illustrates the principle, although I can't be sure the gentleman in the picture was thinking explicitly about where he was directing his hand beneath the blocks he was bashing. No matter what a karate artist has in mind, it is almost certainly true that the farther beyond a board or block the hand aims for, the greater the force will be (Feld, McNair, & Wilk, 1979).[13]

Disequilibrium over the Equilibrium Point Hypothesis

What is the state of the equilibrium point hypothesis today? Is it universally embraced or do researchers have doubts about it? The hypothesis has been debated. A major question has been how equilibrium point (EP) control would actually be implemented by the nervous system. One view is that EP control relies on *preflexes*. The other view is that EP control relies rely on *reflexes*.

Emilio Bizzi and his colleagues, whose work has been used to introduce the EP hypothesis here, favored the preflex view, although Bizzi and his coauthors did not, to my knowledge, use that specific term, presumably because it was introduced in 1995 and Bizzi et al. did their work before that. Bizzi and coworkers suggested that muscle stiffnesses can be changed by varying the intensities of the motor signals sent to the muscles, with feedback input being more or less optional for the EP strategy to work. In support of this view, Bizzi et al. tested the EP hypothesis with deafferented monkeys, arguing that motor outflow alone alters the tensions of the muscles to change the equilibrium position of the joint, as per Figure 8.13.

Feldman and colleagues favored the reflex view. They argued that muscle stiffnesses can't be controlled directly, that reaching an equilibrium position requires feedback. According to the view of Feldman and colleagues, the mechanism used for implementing new equilibrium positions is the muscle stretch reflex (Figure 8.17), where the key regulated factor is the threshold length beyond which the stretch reflex turns on; below that length, the reflex remains quiet. Feldman and colleagues suggested that the threshold length for the muscle stretch reflex may be varied to modulate the way muscle tensions

Figure 8.16
A forceful karate chop aimed, by hypothesis, at a point beyond the surface of the blocks being smashed.

develop. They called this length λ or lambda, the Greek letter for the (lowercase) Roman letter "L" (for length). The idea is that setting the λ values for the muscles can affect the position at which the forces balance out, as seen in Figure 8.17, where the intercepts of the length–tension curves are varied but the slopes stay the same. This method differs from the one favored by Bizzi and colleagues, in which the slopes of the length–tension curves vary but the intercepts stay the same.[14]

Posture-Based Motion Planning

Although there is disagreement about the EP hypothesis with respect to neural implementation and other issues, the theory is worth knowing about because of the important role it has played in motor neuroscience. The EP hypothesis also has several attractions from an algorithmic perspective and can serve as a foundation for a cognitive–psychological model of motor control.

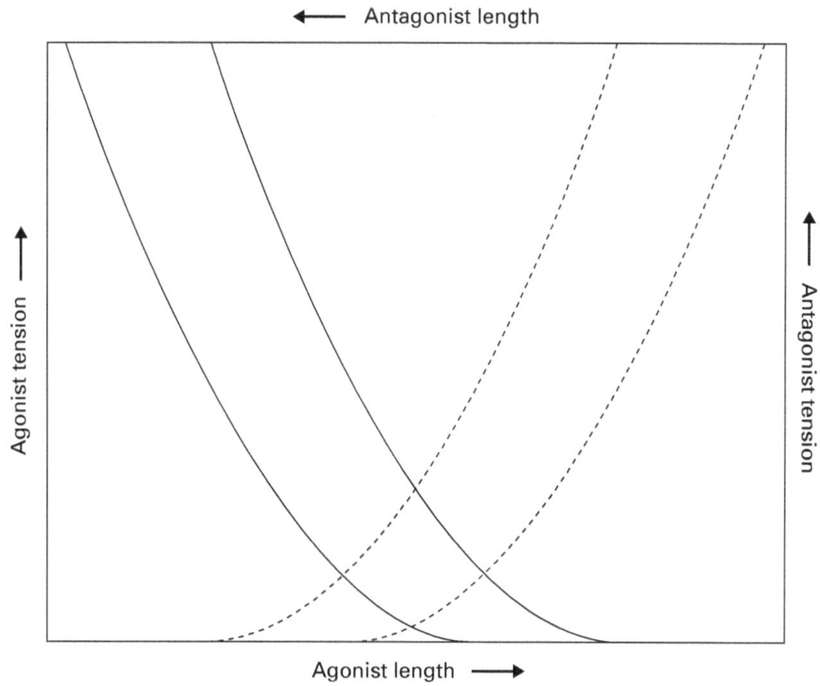

Figure 8.17
Method for modulating equilibrium positions by changing the muscle length at which the muscle stretch reflex is activated. Whereas agonist and antagonist length–tension curves had different slopes but identical intercepts in Figure 8.13, here the curves have identical slopes but different intercepts. As in Figure 8.13, the graph shows hypothesized length–tension curves both for muscle agonists (lengths increasing from left to right) and for muscle antagonists (lengths increasing from right to left). The muscles are in equilibrium where the agonist tensions (dashed lines) and antagonist tensions (solid lines) intersect.

What are those attractions? One is the finding that memory for limb positions is better than memory for limb movements. This result fits with the idea that goal positions are more important than the means of reaching them (Laabs, 1973; Rosenbaum, Meulenbroek, & Vaughan, 1999). In general, things that are important are better remembered than things that are not. Furthermore, solutions that are generated on the fly and used on a *pro tem* basis tend to be forgotten. Movements have this status vis-à-vis goal positions.

A second attraction of the EP hypothesis relates to the discovery that position variability increases and then decreases as limbs move from start positions to final positions. The decline occurs even without visual feedback, suggesting that there is convergence to a predefined target. Without such a predefined target, variability would keep growing as movement continues (Cooke, 1980).

A third attraction of the EP hypothesis is that movement kinematics scale with movement distance (Gordon & Ghez, 1994). When the hand goes from a starting position to another position some distance away, it tends to gradually speed up and then gradually slow down, as seen in Figure 8.18. The overall pattern relating speed to position is bell shaped, and that shape is preserved when the hand covers different

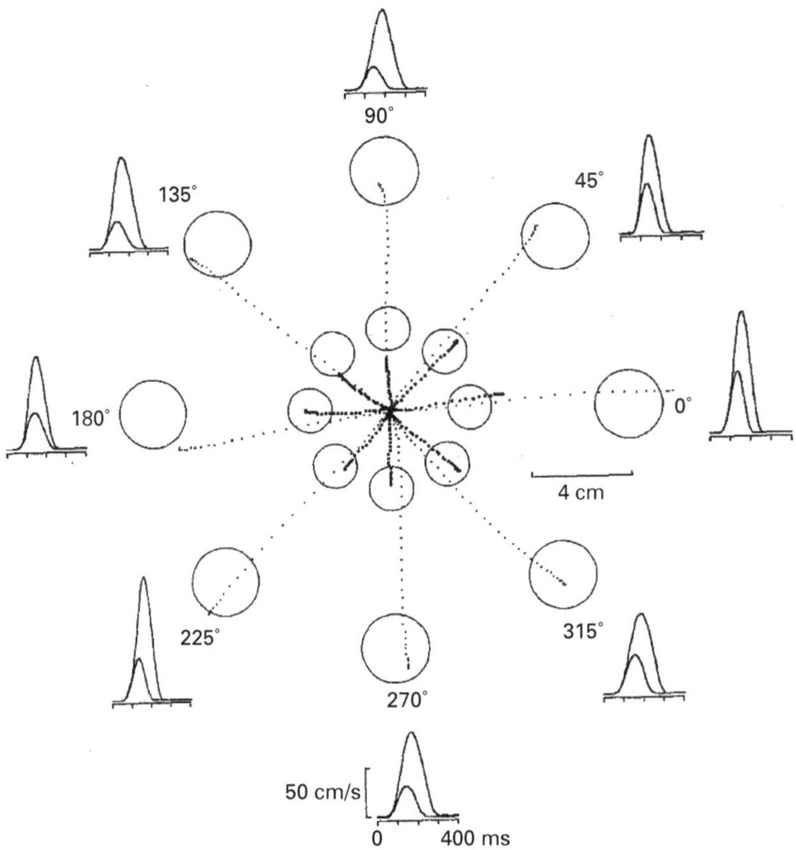

Figure 8.18
Speed profile invariance for center-out hand movements in different directions and over different distances.

distances in different directions, implying that the distances and directions are known in advance.

Fourth, consideration of the degrees of freedom suggests that there is an advantage to planning for final positions before planning paths to those final positions. Recall from the opening of this chapter that there are infinitely many final positions, even for a task as mundane as touching your nose; and for each final position, there are infinitely many paths; and for each of those infinitely many paths, there are infinitely many time patterns. With this layering of infinities, it makes sense to specify final positions first and then to specify paths and timing patterns. With thousands of trees in a forest, each of which has thousands of branches, it makes more sense to pick a tree and then find a branch than to first find a branch and then try to find the tree it belongs to.

An analogous idea is used in animation. Moving-picture cartoons starting in the days of Disney and others always used critical positions, or keyframes, which are established before transitions between the keyframes are filled in (Figure 8.19). High-level artists draw keyframes and lower-level artists connect the keyframes. This method is not only sensible from a hierarchical planning perspective; it is also economical from a computer-storage perspective. It is more efficient to specify goal states and compute

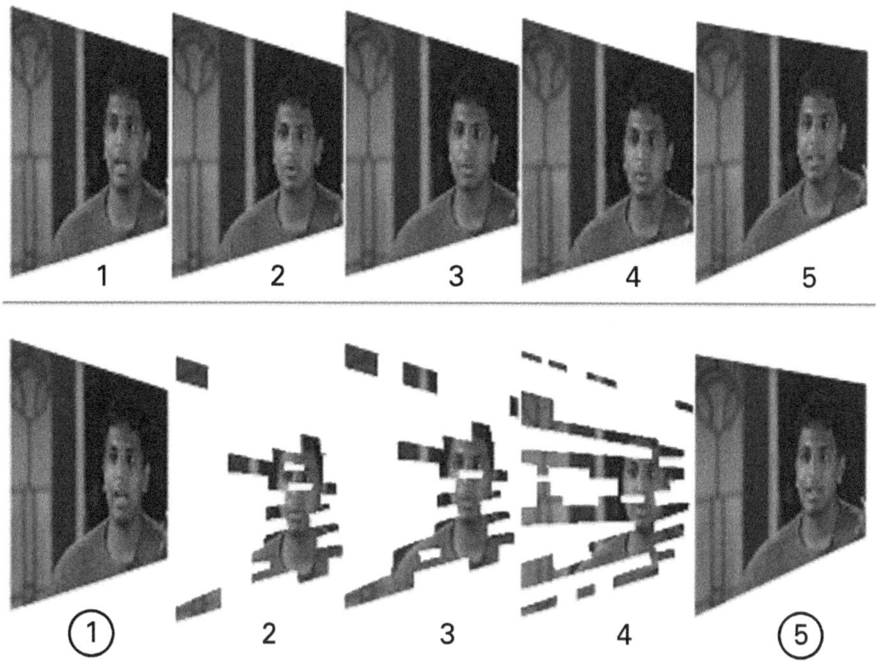

Figure 8.19
Two methods of computer animation. In the top panel, interframes are used, with every image fully represented (.avi format). In the bottom panel, keyframes are used, with a few sparsely represented interframes (.mpg format).

moves to them on the fly than to store every state along the way. Your computer is more likely to have .mpg files than .avi files.

How do these ideas come together in a psychological model of motor control? A model that my colleagues and I have developed relies on posture-based motion planning, as we call it (Jax, Rosenbaum, Vaughan, & Meulenbroek, 2003; Rosenbaum, 2017b; Rosenbaum et al., 1995, 2001). The idea is that movements are planned by first specifying goal positions and then by specifying movements to those goal positions. The movements have default properties that can be amended to permit avoidance of obstacles or follow paths with required curvature for purposes like drawing or dancing.[15] For any task, an infinite number of goal postures are possible, but acceptable candidates can be found by determining the costs of previously adopted goal postures. Those costs can be of different kinds, including how close a point along the limb-segment chain can be expected to come to a target and how expensive the relevant joint rotations will be. If no previously adopted goal posture is exactly right for the job, random search around the most promising goal posture can yield a better goal posture. Newly adopted goal postures can be stored while ones not used for a long time can be forgotten. The model has been used to simulate the way performers settle into a "groove" as they perform the same task repeatedly (Fischer, Rosenbaum, & Vaughan, 1997). It can also simulate compensation for changes in joint mobility, writing with different parts of the body, tool use, the benefits of performing the same tasks repeatedly, awkwardness in growth spurts, and the benefits of planning series of movements rather than just one movement at a time. The posture-based motion-planning model is just one of many computational models that have been developed to simulate voluntary movement patterns.

Soft Constraints

The previous sections covered two possible ways of addressing the degrees-of-freedom problem. One appealed to coupling. The other appealed to physical mechanics. Both approaches are attractive, but neither approach can be the full solution. What's also needed are *soft constraints*. If you can touch your nose in all sorts of creative ways, that means you can go beyond coupling and mechanics. You can rely on ad hoc preferences; that is what I mean by soft constraints.

Hard constraints, by contrast, are inescapable limitations imposed by physics or anatomy. Soft constraints are psychological limitations based on goals and emotions. If you strut down the street having just won the lottery, your gait signals your happiness, but if you trudge down the street having just been fired, your gait signals a very different emotional state. Physics alone doesn't determine your gait pattern, and neither does coupling. Soft constraints enter too, allowing for the expression of emotions and other states of mind such as respect for styles more appropriate in some circumstances than others (e.g., lifting your pinky while drinking tea with the queen but holding your mug with a power grip while slugging beer with your buddies). Soft constraints are especially interesting for cognitive psychologists and others concerned with knowledge and its expression. Some examples follow, along with personal anecdotes about their discovery.

The Grasp-Height Effect

I once walked into my home bathroom after a day at work, only to discover a toilet plunger standing on the bathroom sink counter. Upon seeing it, I remembered that I had left it there that morning before going to work. I had set it on the counter to dry on a paper towel after I had used it and washed it thoroughly. It needed to be dry before I could return it to the bathroom's carpeted floor. When I got home, having forgotten all about the plunger, I grabbed the plunger absentmindedly, intending to put it down on the floor. However, I stopped myself in midflight, noticing where I grabbed the shaft. I grasped it very high, near the top, not at mid-height, directly in front of me, which would have been a more direct reach and a more comfortable grasp. "Why did I grab it there?" I wondered.

It occurred to me that I might have grabbed the plunger high because I knew I would set it low. Had I grabbed it where my hand would have been most comfortable at grasp time, I would have had to bend down very low to put the plunger on the floor. Perhaps I had learned to tolerate some initial discomfort to avoid more massive discomfort at the end.

Later I brought the object-transport task to the laboratory (Cohen & Rosenbaum, 2004). Because plungers were convenient objects, being hardy things that could be picked up and set down nonchalantly, I decided to stick with them, so I got a fresh one for the research.

The university students in the resulting experiment (Figure 8.20) were asked to reach out and grab the plunger to bring it to each of a number of target shelves ranging in height from very low to very high. The heights were tested in a random order per subject. In the main experiment, the initial position of the plunger was the same at the beginning: a shelf at belly height.

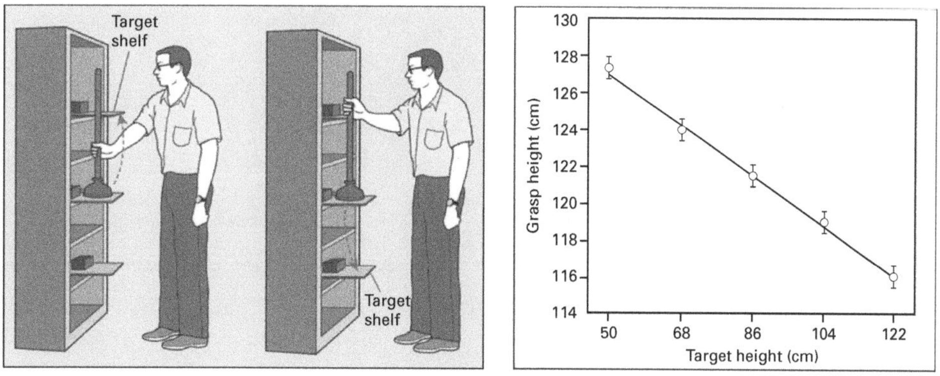

Figure 8.20
The grasp-height effect. Left panel: Two examples of the task. The subject reaches out and grasps the plunger low to bring it to the high target shelf or reaches out and grasps the plunger high to bring it to the low target shelf. Right panel: Grasp height as a function of target height.

Although none of the subjects claimed to have any awareness of what they did, all of them modulated their grasp heights depending on the heights to which they would bring the plunger, as seen in the right panel of Figure 8.20. The participants grasped the plunger high to place it low, and they grasped the plunger low to place it high. Over the range of target positions tested, all the subjects showed this inverse relation. This *grasp-height* effect, as we called it, indicated that a soft constraint applied to reaching for objects. That constraint, expressed verbally (although it was doubtless expressed nonverbally and unconsciously), was, "Don't end object manipulations in extreme positions."[16]

The End-State Comfort Effect

The same soft constraint was manifested in another everyday context (Figure 8.21). While eating at a restaurant, I saw a waiter filling glasses with water. Each glass stood with its open end down on a table cloth. The waiter grabbed each glass with his thumb near the table cloth and his little finger above, which was unusual because one normally picks up a glass with the thumb up higher than the pinky. After grabbing each glass in the unusual thumb-down position, the waiter flipped the glass to pour water into it. He filled each glass with water and then set it down on a tray for later transport to the dining area.

Why did the waiter flip the glasses by taking hold of them with his thumb down? By doing this, he could hold the glass steady in a thumb-up posture when he poured water into the glass and then set the glass down on the table. Had he grasped the glasses thumb *up* (the usual way), he would have found himself in an awkward posture when he needed the most control—when he had to hold the glass steady while pouring water into it and then when setting the filled glass down on the table.

To check that this observation held up more generally, my colleagues and I created a task modeled on the waiter's water-filling task. Rather than have subjects pick up inverted glasses to be filled with water, we invited subjects to reach out and turn a handle to rotate a pointer from one position to another (Figure 8.22). All the rotations

Figure 8.21
Grasping a glass with a thumb-up posture before flipping it (upper panels) and grasping the same glass with a thumb-down posture before flipping it (lower panels). The student who kindly posed for these pictures hams up the degree of discomfort/comfort experienced at the end of the flip.

covered 180°, and each trial began with the subjects standing with their hands by their sides. Once the subject was ready, he or she reached out with the hand that was supposed to be used in that trial as per prior instruction, grasping the handle and turning it to its target position. The subject was told to grasp the handle firmly, as in holding a tennis racquet (a so-called power grip). The question was, How would the subject take hold of the handle depending on its start and end positions?

As seen in Figure 8.22, subjects as a whole (and individually too) grasped the handle such that they avoided ending with the left thumb near position 6 or ending with the right thumb near position 4. The fact that different final positions were avoided with the left hand and right showed that the soft constraint was anatomical rather than spatial. Bringing the left thumb pointing to position 6 or bringing the right thumb to position 4 would have brought either arm close to the end of the range of motion, so those positions were avoided. Interestingly, the subjects adopted those positions when they first took hold of the handle, although they spent very little time in those positions.

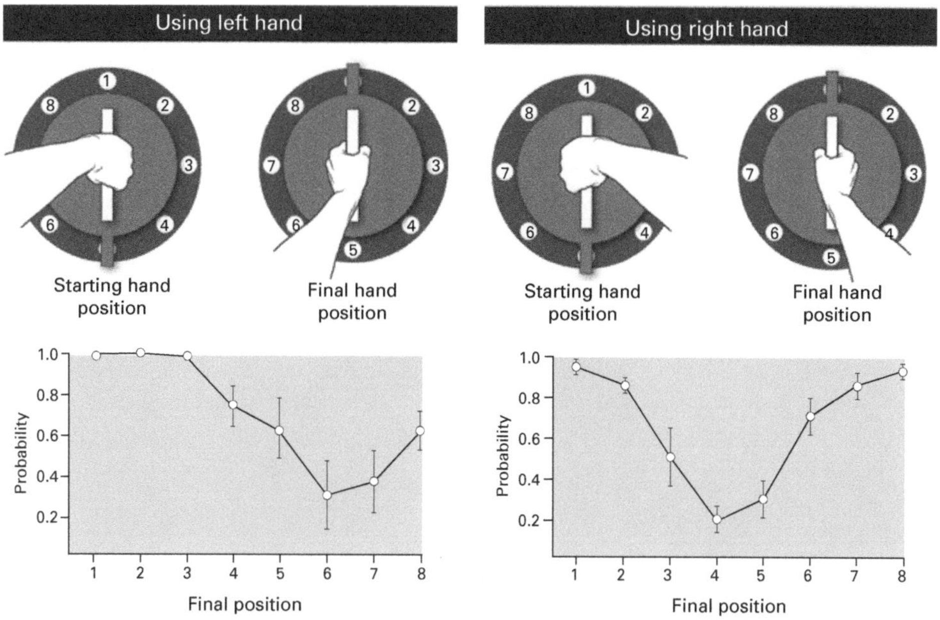

Figure 8.22
End-state comfort effect as shown by determining the probability of grabbing a handle to be turned 180° with the left hand (left side) or right hand (right side).

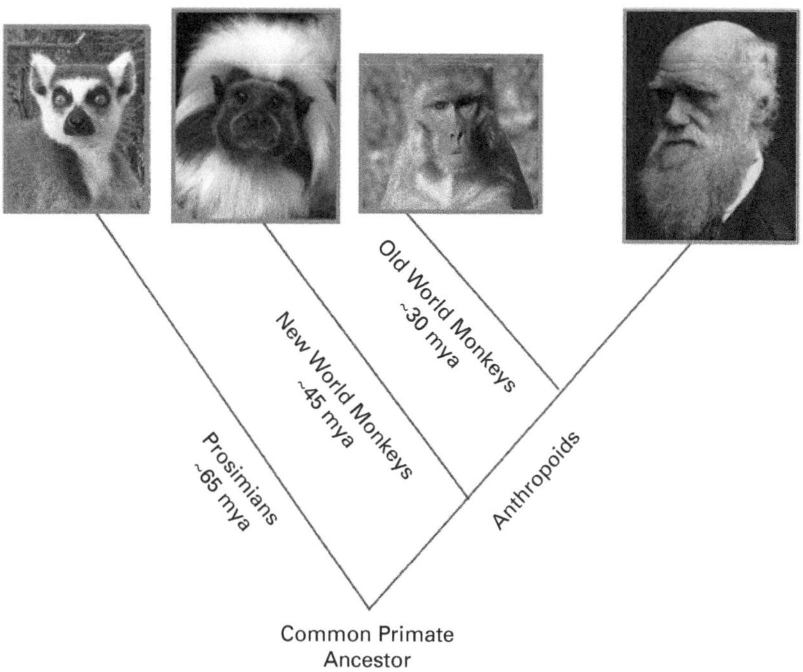

Figure 8.23
Species shown to exhibit second-order grasp planning and the times when their evolutionary lineage split off from the line leading to *Homo sapiens*.

Figure 8.24
A child enjoying spaghetti.

For the positions where more time was required, at the target sites where precision aiming would be needed, the extreme arm positions were avoided.

This behavioral pattern reflected the same soft constraint for object manipulation as seen in the grasp-height effect: "Don't end object manipulations in extreme body positions." Ending an object manipulation with joints at extreme positions is a bad idea. By contrast, ending at midrange positions gives more leeway, more power, and more comfort. My colleagues and I called this the *end-state comfort effect.*[17]

Second-Order Grasp Planning in Animals and Children

The tendency to adopt relatively uncomfortable grasp postures for the sake of greater comfort at the end of the task—the end-state comfort effect—reflects a more general principle: *second-order* grasp planning. The term refers to the idea that we go beyond planning for the immediate next state (*first-order* grasp planning), as in reaching around a barrier to get something on the other side.[18]

Because second-order grasp planning is fundamental for adaptive performance, it might be seen in infrahuman species. Indeed, that is the case (Figure 8.23). Cotton-top tamarin monkeys and lemurs show second-order grasp planning, as has been demonstrated by testing these animals on tasks where they could get food by overturning objects to retrieve food inside. These animals have dexterity similar to that of human adults. The fact that cotton-top tamarin monkeys and lemurs show second-order grasping suggests that such planning may have been in place as long ago as 65 million years ago, when the ancestors of lemurs and humans split off evolutionarily.[19] On the other hand, animals with *more* dexterity than human adults (capuchins, rhesus monkeys, and chimpanzees) *don't* show second-order grasping. These animals happily grab objects using canonical postures, ending up in postures that, to human adults, seem strange. These animals don't have to avoid extreme hand postures at the ends of their maneuvers because those hand positions aren't awkward for those animals. Evidently, if being at an extreme position has little cost, there is no reason to avoid it.

One kind of animal in this class is, interestingly enough, human children. One of the most unexpected findings from research on second-order grasp planning is that human children don't show the end-state comfort effect, at least when tested with tasks like picking up toy hammers or lifting toy swords. Kids pick up these objects in typical midrange hand postures, leaving their hands in extreme hand postures after bringing the objects to final positions for tasks such as hammering or engaging in mock sword fights (Crajé, Aarts, Nijhuis-van der Sanden, & Steenbergen, 2010; Keen, Lee, & Adolph, 2014; van Swieten et al., 2010; Wunsch, Henning, Aschersleben, & Weigelt, 2013).

Why do children adopt these strange final positions? Can't they plan ahead? That's doubtful, for kids can forecast beyond the immediate present. Otherwise, they wouldn't be able to reach around barriers, produce syntactically complex sentences, or fib as well as they do. Fibbing, though morally repugnant, takes computational sophistication.

Perhaps the answer to the question of why children adopt strange final positions is that those positions look uncomfortable for adults but are not uncomfortable for the children themselves. If children happily feed themselves while holding forks as they please (Figure 8.24), they may not find that way of forking painful or uncomfortable. It may not bother them in the least to be in those strange-looking poses, for their manual control may be no worse than it would be otherwise.[20]

Understanding how children make the transition from juvenile ways of behaving to more mature ways is an important challenge in action research. So too is the pattern of change that occurs on the other end of the developmental continuum, in old age. All through development, soft constraints as well as hard constraints affect the actions that are made (Wunsch et al., 2013).

9 Onward

You've come a long way by going through the eight chapters up to this point. You've learned about the intellectual background to this field, its applications and methods, its physical substrates, moving in space and time, learning, feedback, feedforward, and the degrees-of-freedom problem. In this last chapter, you'll read a few more things I hope you'll find interesting. First, you'll learn about some advances and challenges in medicine, robotics, and other hi-tech fields. Then you'll encounter some topics that deserve more coverage than they've gotten so far. Finally, you'll learn about some other topics and a surprising discovery that presages exciting times to come.

Medical Advances

If you keep your eye on the news, you've probably noticed stories about medical advances related to action, mind, and brain. I'll mention some that have caught my eye.

One concerned recording brainwaves from people who cannot produce language due to stroke. The patients can understand language but can't speak, write, type, or sign. The brain regions responsible for language production, but not language perception, have been impaired in these individuals. The advance that was announced allowed for translation of brain activity into audible speech.[1] The method is similar to that used to drive a robot with brain activity, as in the case of Jan Sherman, the woman with quadriplegia discussed in Chapter 3. Recorded brain activity is turned into output that the individual can modulate by thinking different thoughts. The patient and engineers work together to refine the transfer from brainwaves to soundwaves.

Another advance has been the development of artificial muscles. In Chapter 3, you learned about muscle fibers and how they generate forces. Muscles have properties that conventional mechanical devices don't. According to a review article about artificial muscles, "The desirable properties of . . . biological muscles are summarized as follows: they are energy efficient, they have a high contraction ratio, they are intrinsically compliant, their stiffness can be varied smoothly and dynamically and they . . . incorporate a sensory part" (Ariano et al., 2015). The article also said that "muscles are able to operate for billions of cycles over a period of a hundred years or more due to the ability to regenerate proteins in situ." The article, published in the *Journal of Applied Biomaterials & Functional Materials*, explained how progress is being made in the development of artificial muscles. Of course, it would be wonderful if people whose muscle fibers have been irreparably damaged from illness or injury could get artificial muscles. We are a long way off from that, but strides are being made to develop them.[2]

The aim of building artificial muscles is inspired by knowing that biological muscles exist and have desirable features. Bringing biology to engineering has become increasingly popular. The goal has been to develop *biomimetic* devices—that is, devices that imitate Nature. Engineers who take this approach reason that if Nature evolved working solutions, those solutions have a good chance of working in human-crafted devices. For example, spiders can walk on water and can walk upside down on ceilings. If people want devices that can do the same, spiders may provide useful lessons (Dickinson et al., 2000).

Another area where there has been cutting-edge work in biotechnology has been limb regeneration. Salamanders—or *axolotl* salamanders in particular (Figure 9.1)—can regrow lost limbs. Studies of these and other creatures have shown that limb regrowth depends on specific genes (Shastri, 2016).[3] Scientists with expertise in this area have a way to go to enable people to regrow limbs, but as knowledge in this field grows, limb regeneration will become more than just a pipe dream (more than a fantasy).

If limbs can't be regrown, an alternative treatment is limb transplantation. People have gotten hand transplants and, at least initially, had encouraging outcomes. There was even a case of a woman who got two new hands. Despite hope that she'd gain full use of both of these transplants, that hope faded with time. She suffered as her body's immune response took over. Ultimately, her newly attached hands had to be removed (Dobbs, 2019).[4] The fact that the procedure could be tried at all and that the results were promising at first is encouraging. More research is needed for such stories to end well.

Hand transplantation requires neurosurgery because nerves must be attached in order for newly installed limbs—hands or otherwise—to get efferent signals from the brain and spinal cord and to send afferent signals back to those structures. A story about neurosurgery that got some press recently concerned a young man who underwent brain surgery to alleviate *focal dystonia*—lack of muscle tone in a small set of muscles. During the surgery, the young man played the guitar (Figure 9.2).[5] He did this not just to have something do while surgeons poked around in his brain but rather because his

Figure 9.1
An axolotl salamander.

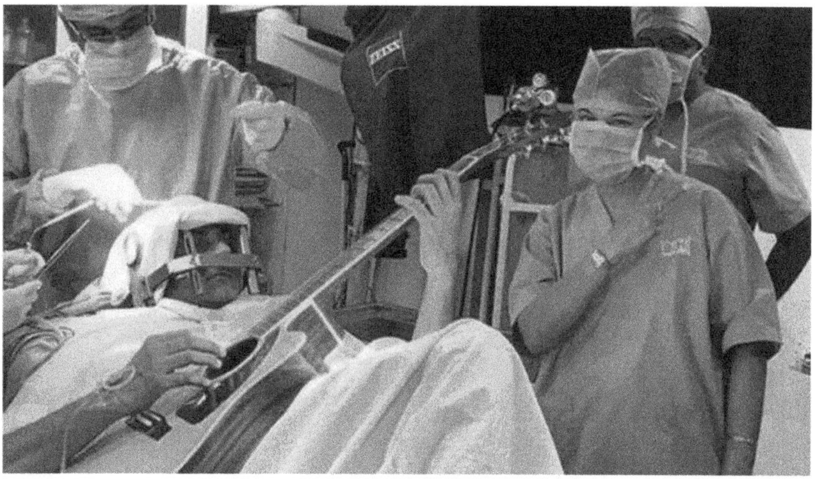

Figure 9.2
Patient suffering from focal dystonia in his left hand, playing the guitar during brain surgery aimed at relieving the focal dystonia.

focal dystonia took the form of his muscles freezing up when he attempted to played the guitar. Focal dystonia is seen in individuals who engage in the same physical task over long periods. One of the most famous focal dystonia patients was the classical pianist Leon Fleisher, who lost the use of his right hand as a result of the disorder. Later he performed pieces for left hand only. Fleisher made many recordings and gave many concerts both as a one- and two-handed pianist. Happily, he eventually was able to play with both hands again thanks to a variety of therapies, none of which involved brain surgery. Ironically, he died on the very day I was planning to write this text about him. His obituary appeared in the *New York Times* on August 2, 2020.

The cause of focal dystonia is unknown. The disorder has variously been ascribed to muscular factors, neural factors, and mental factors. In the case of the guitar player, his focal dystonia affected his fingering hand. When nothing else worked—massage, drugs aimed at relaxing the muscles, counseling, and so on—he resorted to brain surgery. The challenge for the surgeons was to quiet the messages in the young man's brain that caused his muscles to tense up. They needed to do this without deactivating the muscles themselves or hurting other functions. The best way to check the efficacy of the planned treatment was to have the patient play the guitar while candidate lesion sites were temporarily deactivated.

Another step forward in medical research has been aimed at helping patients stand or walk. Individuals who have lost or never had the ability to walk may be in that state for a variety of reasons. One of the most common is spinal cord injuries. People who cannot move their legs but can move their arms are *paraplegics*. People who cannot move their arms *or* their legs are *quadriplegics*. Jan Sherman, the woman discussed in Chapter 3, is a quadriplegic.

In the case of spinal cord injuries, it has long been hoped that a method could be developed to get nerves to grow where the injury occurred. Some progress has been made in the domain of nerve regrowth, but advances in this area, as in the others described above, have been limited (Arber & Costa, 2018).

Figure 9.3
A boy wearing an exoskeleton.

Another approach is to circumvent the injury by sending electrical signals directly to the muscles, in effect doing what Lucia and Luigi Galvani did with frog muscles (Figure 3.10) but in a more sophisticated, nuanced way. Here, too, there has been some progress, but the steps have been halting.

Yet another approach has been to send signals not to muscles (or not only to muscles) but also or instead to exoskeletons that can "walk for" patients or aid them in whatever walking they can do. Figure 9.3 shows a happy example.

Robotics

The previous section touched on a small set of medical advances. Another area where advances have been made is robotics. Predecessors to robots have been around for centuries. Royals in the late 1700s enjoyed mechanical dolls that played musical instruments and wrote letters (Figure 9.4).[6] Such devices performed in an open-loop fashion, repeating the same acts over and over again no matter what. Strictly speaking, they were not robots, for a true robot must be capable of closed-loop, adaptive control. Devices that always do the same thing with no possibility of feedback-based corrections are called *automatons* or *automata*.

With the growth of technology, actual robots became feasible. In fact, the word "robot" comes from a period—the 1920s—when industry and technology were taking off and true robots could be contemplated. The word *robot* was introduced in a play called *R.U.R.*, short for *Rossum's Universal Robots*, written in 1920 by a Czech author named Karel Čapek. In Čapek's story, people created robots to work for their human masters, but the robots rebelled and, in the end, extinguished the human race. "Be careful," Čapek warned us, foreseeing the possibility of ever more intelligent machines.

Robots with human forms, as Čapek's robots had, can be ominous, especially if, along with their humanoid bodies, they have unchanging facial expressions. A case in point was the alien robot from the science-fiction movie *The Day the Earth Stood Still* (Figure 9.5). The robot, called Gort, had a huge, rigid frame and no face. Its only face-like

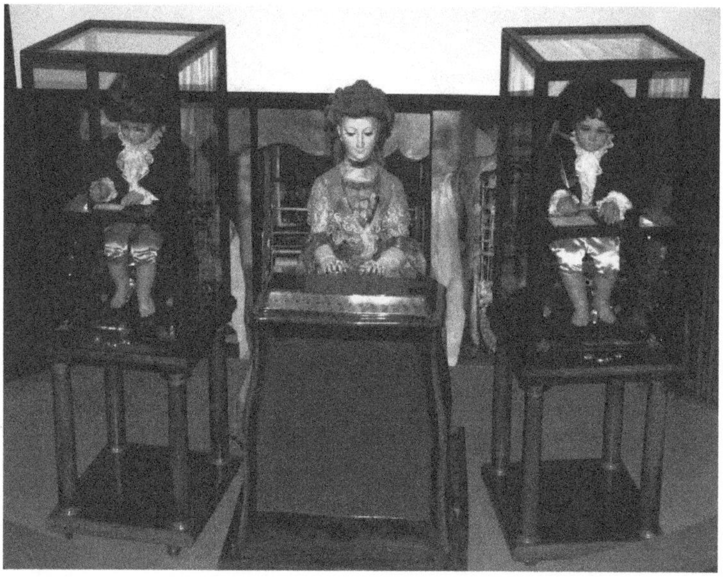

Figure 9.4
Automata from the 1700s at the Musée d'Art et d'Histoire of Neuchâtel, Switzerland.

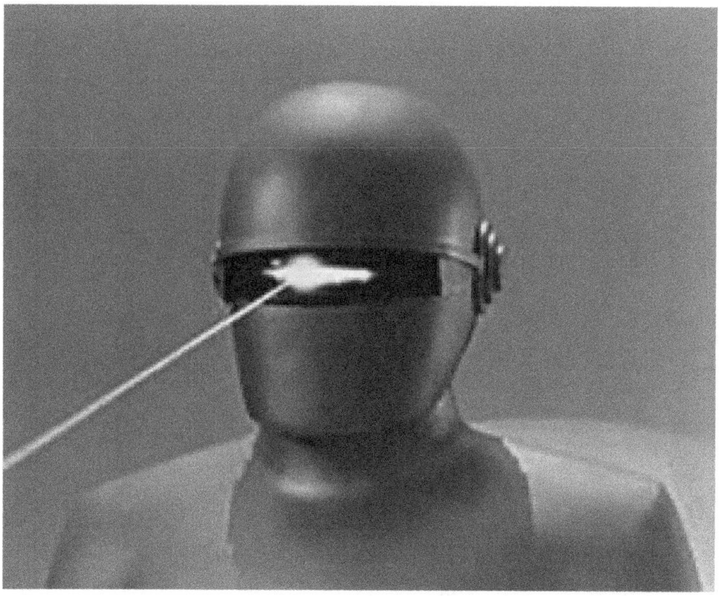

Figure 9.5
The robot Gort from the science fiction movie *The Day the Earth Stood Still*.

object was a visor that lifted when Gort would fire off death rays.[7] You wouldn't want to be in front of Gort when its visor rose!

Humanoid robots needn't have faces like Gort's. They can be socially engaging with faces capable of expressions. Even cartoon-like faces will suffice, as in the case of Pepper, a social robot (Figure 9.6) (Karniel, Nisky, Avraham, Peles, & Levy-Tzedek, 2010; Langer, Feingold-Polak, Mueller, Kellmeyer, & Levy-Tzedek, 2019).

Another reason to have humanoid robots is to have them provide a medium for testing one's hypotheses about action control. If a robot has arms, hands, and fingers and needs to carry out everyday physical tasks like folding laundry, it can be sobering if the robot takes much longer than people do to complete this household chore. Folding laundry comes to mind because a few years ago, videos of robots doing that task were embarrassingly slow, requiring excruciatingly many minutes to fold a towel. The fact that they could fold at all was a step forward at the time. More recently, progress has been made on this front. A team at the University of California, Berkeley has gotten robots to fold towels fairly quickly (Figure 9.7). Again, this may seem like a rather modest achievement, but considering what's needed to visually recognize a towel as such, developing a plan for folding the towel and carrying out the plan is admirable.

Folding towels is easier than folding shirts, especially when the shirts are part of a chaotic mess. In general, the greater the uncertainty about what's presented to a robot, the harder the problem. This is why robots can perform more ably when the same thing is done over and over again with only slight changes, like spray-painting automobiles on assembly lines. The positions of the cars or car parts coming down the assembly line vary a bit, but the robots can adjust to those variations. Over time, progress in robotics will be measured not just by how quickly robots can do the same tasks over and over again but by how well they can tackle new problems.

Figure 9.6
The social robot Pepper along with the robotics engineer, Shelly Levy-Tzedek (Ben Gurion University), who studies the use of robots as aids for physical and occupational therapy.

Figure 9.7
A towel-folding robot at the University of California, Berkeley.

A final remark about robotics pertains to the theme of exploiting physical mechanics. This theme has come up a number of times in this book. Relying on physical mechanics has impacted robotics. In early days, robots were often huge Gort-like devices. As said before, one robot was so huge that a human operator sat astride it and steered it, or tried to steer it, with levers. This behemoth was so hard to control that it had to be warehoused. The designers of the machine didn't exploit physical mechanics to make the control problem easier.

The idea of exploiting physics for robotic control came later. When it did, it revolutionized the field (Collins, Ruina, Tedrake, & Wisse, 2005). Engineers realized that robots can move in response to gravity and other mechanical forces, reducing the need for active control. Figure 9.8 is illustrative. It shows so-called passive walkers. These are not robots in the strict sense because they move ballistically. They have no electronics, yet they walk down slopes by virtue of how they're built. They amble down ramps with no more need for computation than apples falling from trees. By supplementing the devices with suitable electronics that permit feedback-based correction, they become full-fledged robots.[8]

Styles of Moving

In the early days of robotics, robots moved "robotically." They moved jerkily, coming to a stop at successive goal positions. The moves they made, at least stereotypically, were completed with quick starts and abrupt stops. That's not how people or animals move. We generally transition smoothly from one position to another, gliding through goal positions until coming to full stops where we want to end.

Movements generally have bell-shaped speed profiles, as seen in Figure 8.18. Furthermore, observed bell-shaped speed profiles have a special property: The ratio of maximum speed to mean speed tends to be close to 1.875 (Flash & Hogan, 1985; Hogan, 1984). As a result, the bells are neither sharply peaked nor overly flat. Bell-shaped speed profiles with the max-to-mean ratio around 1.875 tend to minimize mean squared jerk. Jerk is the

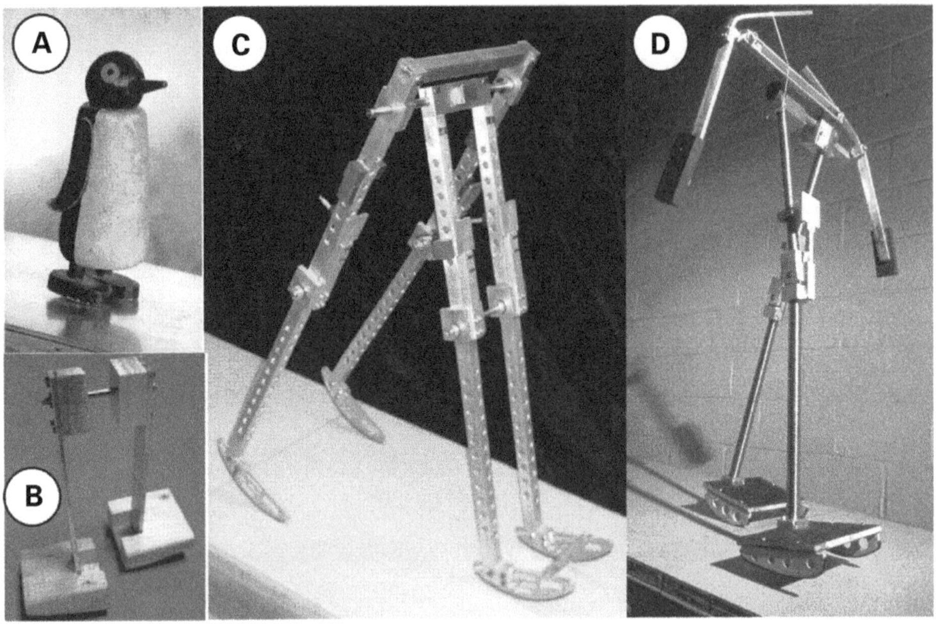

Figure 9.8
Passive walkers.

rate of change of acceleration, acceleration is the rate of change of velocity, and velocity is the rate of change of position. Velocity is a vector because it has a magnitude and a direction, so it is positive in one direction and negative in the opposite direction. Speed is a magnitude alone (so just one component of a velocity), and so is a scalar. The reason highways have speed limits rather than velocity limits is that the direction of motion is generally of no consequence. Because velocity can be positive or negative, acceleration can be positive or negative, and so can jerk. Squaring jerk rectifies the values (takes their absolute values, making all of them nonnegative), and taking the mean squared jerk gives the average of all the squared jerk values from the start of the movement to the end.

Moving smoothly isn't hardwired, however. You can move jerkily or smoothly at will. This reflects a remarkable ability to switch styles of movement. If an insect or a mouse couldn't switch instantly from relaxed strolls to frantic dashes, those animals wouldn't be as well adapted as they are. Similarly, if prowling cats couldn't switch from stalking their prey while barely moving to suddenly pouncing, cats might not have survived as long as they have.

Animals, including human animals, can switch modes of moving at the flick of a neural switch. How this happens is less mysterious than you might expect, for distinct neural circuits underlie different movement patterns. Central pattern generators (CPGs), which enable movement patterning without sensory feedback, exist for different gaits and can switch other neural circuits on and off.[9]

Although you can switch patterns of moving, you also have characteristic patterns that are yours alone. This was recently confirmed in a study of EMG patterns gathered while people rode a stationary bicycle and walked on a treadmill. Different people had different EMG patterns, although the gaits of all participants were identifiable as gaits for cycling and as gaits for walking (Reynolds, 2019). This result complements

Figure 9.9
Thermal image of a herd of cows in Amarillo, Texas.

one covered earlier in this text, where I shared with you that observers can distinguish acquaintances from moving dot patterns.

In another context involving visual motion perception, artificial intelligence (AI) vision systems have gotten so sophisticated that they can recognize styles of movement suggesting sickness in cattle herds. Based on a report I heard on the radio—I confess I don't read the *Western Livestock Journal*, where this work appeared—I learned that drones are being used to identify sick cows (Ledbetter, 2019). The drones take thermal images of cow herds (Figure 9.9), which are used to identify cows that appear, from their ongoing behavior, to be in bad "moo"-ds.

Emotions

Your style of moving at any given time may depend on your emotional state. You may swagger with a swelled head or drag yourself due to depression. Any comprehensive theory of action must allow for a role of emotions and other "soft factors" that affect movement styles.

What parts of the body express emotions most clearly? If you leap into the air, you may jump for joy, for you may have just won the lottery or heard that you got into medical school. But you could also leap for other reasons, like jumping to grab a rejection letter high on a shelf where you put it months ago and couldn't bring yourself to read until now. Your jumping might not reveal the nature of your emotion. But your face might. An observer checking your face might see you grinning in one case but grimacing in the other.

The face can reveal a lot, as this example suggests, but it doesn't reveal all. In fact, the same facial expression may mean different things in different contexts. A sour face may reflect shock at hearing an unexpected accusation, or it could be a response to a bite on a lemon. Looking at the body and surrounding context can help make sense of the emotion (Barrett, 2006). In fact, the face may not always be the most informative part of the body when it comes to identifying emotions. A study of poker players showed that the hands and torso were more informative than the face. Observers judging emotions of poker players did worse if the players' hands and torso were obscured than if the players' faces were hidden (Slepian, Young, Rutchick, & Ambady, 2013). Observations of lit dots on moving hands let observers identify anger from the way the dots dart about (Pollick, Paterson, Bruderlin, & Sanford, 2001).

Charles Darwin wrote an entire book on *The Expression of the Emotions in Man and Animals* (Darwin, 1872/1965). The book included a picture shown in Figure 9.10 to connect emotional expression to natural selection. The figure shows a cat in an agitated state. The cat is clearly scared and is doing what it can to protect itself. Its fangs are bared and its back is up. By engaging in these behaviors, the feline looks more ferocious than it would otherwise. A likely attacker would be unlikely to attack this particular cat if it could avoid doing so.

This example conveys Darwin's idea that creatures engage in behaviors that help them survive. Darwin suggested that the behaviors that animals engage in are simply ones that have tended to preserve the fecundity of the animals' forbears. If an ancestral cat bared his fangs and arched his back when approached by a snake, and if that act saved his life and enabled him to mate with a she-cat who behaved similarly, then their offspring may have the same behavioral tendencies. This is trial-and-error learning, *par excellence*, played out over generations.[10]

I'm not going to say much more about emotions and action here because the field of emotion is vast. I'll just mention some core questions, hoping to arouse your curiosity: What are emotions? How should emotions be categorized? What neural systems underlie various emotions? Are emotions universal or culturally determined? If they're

Figure 9.10
Cat in an agitated state.

not all universal, which ones are culturally bound and why? How can emotions be controlled when they should be put under wraps, and how can they be expressed more openly when they should be? What's the relation between emotions and moods? What are the triggers for both? Why are some people very emotional and/or moody and others aren't? How do emotions change over the lifetime? Are emotions experienced *after* actions are carried out or before as well?

The last question is one you may recognize. I will dwell on it for a moment. You may have heard or read the phrase, "You don't run from the tiger because you're scared; you're scared because you run from the tiger." A lot of work has been done on this idea. The jury is still out about it. Some work has supported the claim that emotions follow actions, but others have not. There has been considerable uproar over failures to replicate influential studies claiming that adoption of various body poses leads actors to feel associated emotions, a claim consistent with the action-first view. For example, it was claimed that adopting a power pose may make one feel more powerful, but that result wasn't replicated. Similarly, it was argued that pursing one's lips may make one feel sad whereas holding a pencil between one's teeth with one's cheeks far apart in a smile-like pose may make one feel happy, but that result hasn't been replicated either. These outcomes have cast doubt on the hypothesis that actions give rise to emotions rather than the other way around. Other results have supported the hypothesis.[11] I'll now offer some comments about this state of affairs.

One comment is that students often get upset when they hear that results don't replicate. Professional scientists may as well. But failures to replicate can arise for a number of reasons. The worst one is that researchers who reported a result lied. Detecting fraud has become a preoccupation of many scientists, and the open sharing of data has become the new normal. This is a healthy development in my opinion. Making data available to others can help skeptics go through the data themselves to judge their credibility. On the other hand, if a result doesn't hold up, that doesn't mean the result was fraudulent. Sometimes small changes of procedure can cause a result to appear in one lab and not in another. If a predicted result doesn't hold up, that doesn't necessarily kill the entire hypothesis. Sometimes, a variant of the hypothesis that preserves its main idea may still be correct, or the basis for the prediction may be shakier than was first realized. The feedback hypothesis of emotion—the proposal that emotions are experienced in reaction to completed actions—may be in this category.

A second comment pertains to everyday life experience. In everyday life, the feedback hypothesis of emotion may be quite helpful. The reason is that all of us may benefit from realizing that emotions can sometimes arise from interpretations of physiological states. If you drink a lot of coffee and get very jittery, it may be helpful to attribute the jitteriness to the coffee and not to self-blame about getting too anxious. If your boss doesn't smile at you when she passes you in the hall and you also had a lot of coffee a half hour earlier, your nervousness might just be a reaction to the coffee. Allowing for that possibility may make it easier for you to imagine that your boss didn't smile at you because of something going on in her life and not because she's thinking about you and dredging up, in an instant, a litany of complaints about you. It's helpful, then, to appreciate that feelings may have physical sources. (Morgan's canon comes to mind here once again.)

Third, I believe that the discussion of the feedback hypothesis will benefit from greater attention to the perception–action cycle (Figure 1.9). Studies like the ones referred to above concerning the adoption of poses and the subsequent evaluation of emotions are, in a sense, one-shot stimulus–response affairs. The stimulus is the feedback felt by the actor from his or her pose, and the response is the emotion that he or she reportedly feels. In everyday life, you don't typically do one thing, react to it, and then stop. You act, get feedback, act again, get more feedback, and so on. Saying that an emotion can be shaped by a single reaction to a single isolated action like adopting a power pose or holding a pencil in one's mouth misses the point that there is continual cycling between acting and perceiving. Trying to determine whether emotions precede or follow actions is, in a way, nonsensical, because it misses the cyclic nature of perception and action.

Fourth and finally, research on the feedback hypothesis, despite its flaws, calls to mind a concept that hasn't been mentioned here yet. William James is often credited with the action-first hypothesis, or what has also been called the James–Lange theory of emotion, named in honor of William James (discussed earlier in connection with ideomotor theory) and Carl Lange, another nineteenth-century scholar.[12] The concept to be brought to your attention also comes from William James. The idea is that we don't perceive *efferent* signals; we only perceive *afferent* signals or their effects. Said another way, you don't feel neural signals going out to your muscles. All you feel or perceive in other ways are the sensory consequences of what you do with your muscles and the effects they entail. This idea, which comes from James, accords with his theory of emotion.

Is there evidence for the hypothesis that afferent signals are perceived but efferent signals are not? A lovely experiment (figure 9.11) provided supportive evidence (Fessard & Paillard, 1948; see also Clarac, Massion, & Stuart, 2009). People tapped a finger and a toe in such a way that they perceived the two events as simultaneous. Opposing hypotheses made opposing predictions about the timing of the two events. If the internal guide for simultaneous performance were concurrent *efferent* signals, the finger tap would lead the toe tap because neural signals take less time to get to the finger than the toe. Conversely, if the internal guide for simultaneous performance were concurrent

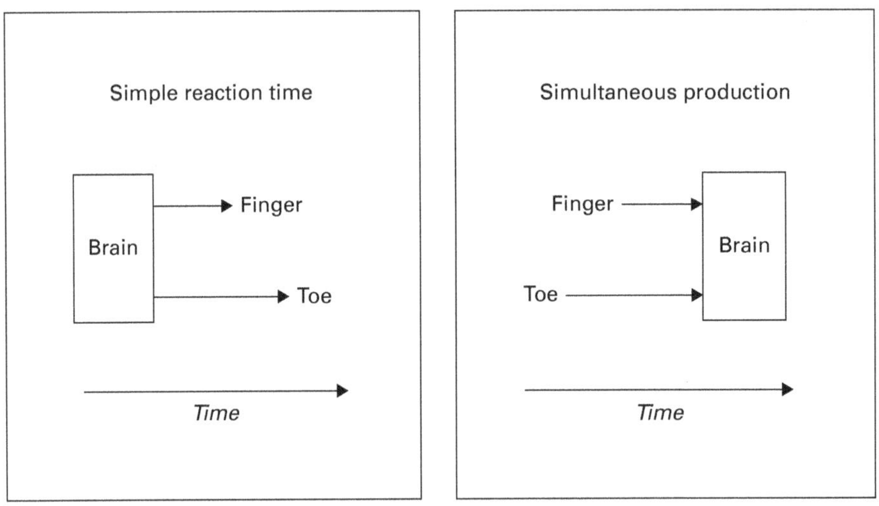

Figure 9.11
Sketch of observed time differences between the finger and toe when each effector is either tested on its own in simple reaction time tasks (left panel) or when the finger and toe are pressed on keys in simultaneous production tasks (right panel). The finger leads the toe in the simple reaction time task but follows the toe in the simultaneous production task.

afferent signals, the finger tap would occur *after* the toe tap because neural signals take longer to get from the toe to the brain than from the finger to the brain.

The obtained result favored the second prediction. When participants initiated finger-and-toe actions that they perceived as simultaneous, they pressed the toe before the finger. In fact, the lead time was very close to the greater time to respond with just the toe to an external signal than the time to respond with just the finger to an external signal. In other words, the lead time of the toe relative to the finger was what would be expected if the toe and finger were pressed so that sensory signals from the toe and finger arrived at the brain simultaneously. The delay was about 30 ms, a result consistent with the speed of nerve conduction estimated by Helmholtz, as discussed in Chapter 3. Thus, subjects cared about the timing of sensory feedback from the toe and finger but were oblivious to the timing of the motor commands sent to the two effectors. This is the outcome James would have predicted.[13]

Social Factors

Just as emotions affect actions, so do social factors. You saw some discussion of social aspects of action in this book. In the section on embodiment in Chapter 7, you saw that cells in the brain fire during observation of actions that observers can do themselves. Those cells occupy motor areas of the brain, not just sensory or perceptual areas, a fact that helps explain why imitative forces are so forceful (Figure 9.12). This may be why you feel the urge to yawn when others do or the urge to cross the street when your fellow pedestrians surge forward.

To these observations, two more can be added: *entrainment* and *mental modeling*. Entrainment means synchronization. When grandfather-clock pendulums or mechanical-metronome pendulums swing together, they entrain, as discussed in the last chapter. Such entrainment happens because of mechanical links, but other forms of entrainment arise without mechanical links. Fireflies blink synchronously, and people clap synchronously at the ends of concerts or plays in Eastern Europe and elsewhere. No leader tells the applauders when to come together or at what frequency. Their collective action provides an example of a self-organizing dynamical system, as does firefly entrainment (Néda, Ravasz, Brechet, Vicsek, & Barabási, 2000). Some species of fireflies don't entrain, and some audiences don't clap coincidentally. These observations indicate that dynamical systems are themselves subject to soft constraints.

The two cases of entrainment that were just mentioned—fireflies flashing and audiences applauding—come from naturalistic observations. Studies of entrainment in human behavior have also been conducted under experimental conditions and have shown that physical models of coupling, à la Christian Huygens, do a good job predicting the frequencies that individuals converge on when carrying out tasks simultaneously. These demonstrations have been obtained in studies of cyclic behaviors involving two people watching each other while rocking in rocking chairs. Demonstrations of coupling between people based on watching each other have also been obtained in studies of people swinging handheld batons (Richardson et al., 2006, 2007). Musicians, too, sway when playing together. An analysis of chamber-music players showed that members of a piano trio (piano, violin, and cello) swayed together more when they rated their joint emotional experience higher (Chang, Kragness, Livingstone, Bosnyak, & Trainor, 2019).

These results indicate that people's actions can become entrained as a result of physical and psychological factors. There are, of course, myriad other examples, including dancing with others, transporting objects with others, and so on. Acting together

Figure 9.12
Imitation is a powerful force.

with others doesn't require coupling based on deep reading of others' minds. When fireflies flash together and when members of an audience clap in unison, the members of those communities don't put themselves squarely in the minds of the others, scoping out, with intense psychoanalytic fervor, the innermost thoughts of their colleagues. The coordination is more automatic.

That said, there are action contexts in which others' mental states are taken into account. A demonstration was provided by Natalie Sebanz, Gunther Knoblich, and Wolfgang Prinz, working together at the time at a Max Planck Institute in Germany (Sebanz, Knoblich, & Prinz, 2003). They had subjects participate in a reaction time (RT) experiment in which the subjects performed on their own (bottom panel of Figure 9.13) and in pairs (top panel of Figure 9.13). The task in both cases was to press a button if the color of a ring worn on an observed finger was a designated color. Unpredictably from trial to trial, the finger either pointed to the left or right. Which way the finger pointed was uninformative; the direction of pointing was independent of the color, and color was the official cue for the required response.

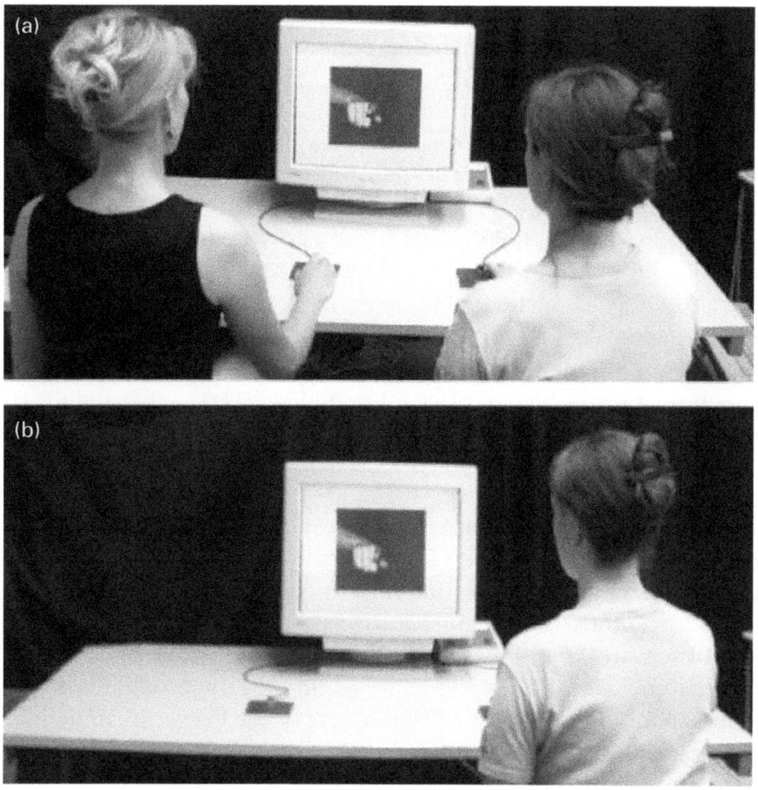

Figure 9.13
Two-person and one-person reaction time task.

In the solo condition, the RTs were no different when the finger pointed toward or away from the subject, but in the joint condition, the RTs were longer when the finger pointed away. The interpretation given by Sebanz et al. was that participants in the joint condition took the pointing finger into account, apparently treating it as an instruction to the other person to go ahead and respond. That interpretation needn't imply that the performers had deep insights into each other's minds, but it does suggest that the agency of the other person—his or her capacity to act—entered the minds of the participants.

Recognizing others' agency is seen in other activities (e.g., Schütz, Güldenpenning, Koester, & Schack, 2020). It becomes clear when people carry out actions in ways that show their appreciation of other people's needs. For example, if you hold a door open for others, that's an expression of courtesy. A naturalistic study of door-holding showed that people held doors longer when two people followed than when one person followed (Santamaria & Rosenbaum, 2011). This wasn't because the door holders were simply more likely to notice two followers than one. Instead, the willingness of people to wait longer holding a door open for two followers than for one was taken to suggest that the tendency to hold a door for others is designed to reduce the effort of the group as a whole. A given amount of extra time holding a door benefits more people when three people are involved than when two people are involved.[14]

Holding doors for others is an example of prosocial behavior (behavior that benefits others). Another example is handing someone an axe by the handle rather than the blade (Figure 9.14). Handing someone an axe by the handle indicates trust in the other person and a desire to make things easier for them.

Axe-passing hasn't been studied in the lab, as far as I know, but hammer-handing has. A study led by David Gonzalez at the University of Waterloo showed that people on the autism spectrum were less consistent than neurotypical people in handing a hammer to another person, holding the hammer by the handle rather than the head (Gonzalez, Glazebrook, Studenka, & Lyons, 2013). A related study concerned handing a dowel to another person who had to place the dowel in a high or low vertical position (Meyer, van der Wel, & Hunnius, 2013). Participants needed a few trials to learn to grasp the dowel where they otherwise would. Rather than grasping the dowel as individuals do on their own—high to place it low and vice versa—they did the opposite to accommodate the needs of the recipients.

A final comment for this section on social factors follows up on the remark that in the hammer-passing task of Gonzalez et al. (2013), it was reported that people on the autism spectrum were less consistent than neurotypical people in passing the hammer by the handle. It's easy to conclude from a result like this that people on the autism spectrum lack social interest. Given that such individuals may fail to make eye contact, may rock back and forth, or may repeat the same phrase over and over again in the presence of others, they may indeed seem socially uninterested.

But two psychologists (Figure 9.15) have questioned this assumption. Vikram Jaswal of the University of Virginia and Nameera Akhtar of the University of California, Santa Cruz have proposed that autistic individuals are in fact interested in social interactions. They based their suggestion on interviews with people on the autism spectrum as well as other evidence (Jaswal & Akhtar, 2019).[15] According to their research, people diagnosed as autistic may look away when spoken to not to *avoid* social contact but to *promote* it. The basis for this surprising claim is that looking *at* a speaker may make it harder for some autistic people to pay attention to the speaker's words. Looking *away* may help them listen more attentively. Not looking at a speaker may be misconstrued as lack of

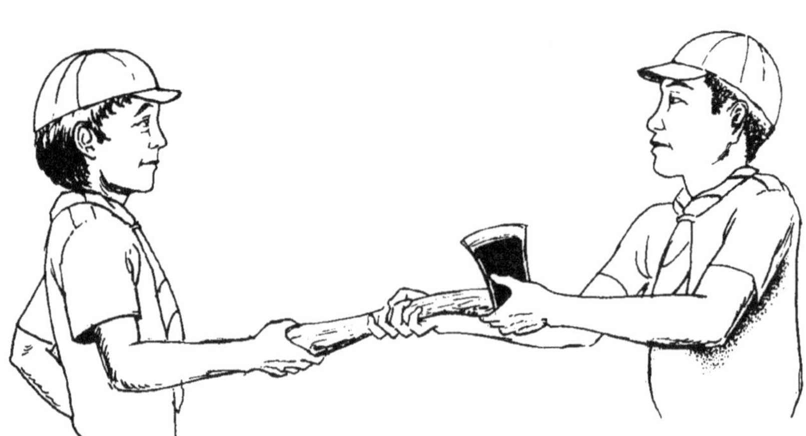

Figure 9.14
Passing an axe safely to another, as recommended by Scouts of South Africa.

Figure 9.15
Vikram Jaswal (left) and Nameera Akhtar (right).

interest or even dislike, but according to Jaswal and Akhtar, people on the autism spectrum may be far more interested in social action than their overt behavior may suggest.[16]

Inhibiting Actions

The end of this book is approaching, and when I get there, I will need to inhibit my further writing. In anticipation of that event, I should observe that there has been much discussion in this book about choosing and initiating action, but there has been much less about *inhibiting* actions. Stopping actions is, of course, very important. If you expressed every thought that crosses your mind, you'd get into a lot of trouble!

Inhibition was talked about earlier in this book in at least three contexts—in connection with neural inhibition, in connection with autoinhibition of just-completed parts of action and stuttering (Figure 4.7), and in connection with damping down interneurons in speech production (Figure 8.8). Two more aspects of inhibition will now be mentioned. One concerns the point of no return. The other concerns doing two things at once.

Regarding the point of no return, experiments have been done in which people have been asked to prepare and perform an action after a go signal but then to refrain from doing the action if a stop signal appears. A key independent variable in these studies has been the delay between the go signal and the stop signal. The studies have shown that, beyond a certain time, the stop signal has no effect; motor commands heading to the musculature will have an impact no matter what. Determining the point-of-no-return time and what factors influence it has been a topic of interest (de Jong, Coles, Logan, & Gratton, 1990).

A surprising discovery was that impulsive people aren't quicker to trigger actions in response to "go" signals but are instead slower to inhibit actions in response to "stop" signals. The delay between the go signal and the stop signal that was needed for impulsive individuals to stop their responses was longer than the delay that was needed for less impulsive individuals to stop theirs (Logan, Schachar, & Tannock, 1997).

Knowing more about the inhibition of actions can inform personal choices, training, and policymaking. For example, knowing more about the ability to stop what one is on the verge of doing may prove important in domains like policing, where police officers must make decisions about when, if ever, to use deadly force (Di Nota & Huhta, 2019).

Another aspect of inhibiting actions is, in my opinion, one of the most striking discoveries that has ever been made in the field of action, mind, and brain. It is the discovery that inhibiting actions doesn't necessarily mean canceling actions but instead entails adding actions to one another. If an added action fully opposes an action in progress, the two actions cancel. A horse pulling a rope will be brought to a stop by a horse pulling the rope just as hard in the opposite direction. Thus, actions can be stopped by adding more actions, not just halting actions that are under way.

Evidence for adding actions to other actions was obtained in a study (Figure 9.16) in which people aimed the hand toward a target light and then were supposed to change course and head to a new light that could come on at some other unpredictable location (Flash & Henis, 1991). A critical independent variable was the delay between the first stimulus and second, as in the studies of the point of no return. In contrast to the no-return studies, where the second stimulus signaled the need to stop, in the study I'm telling you about now, the second stimulus signaled the need to change course.

The main results are shown in Figure 9.16. Analysis of the movement paths indicated that the first move wasn't terminated. Instead, the first move was completed and a new movement was added. The new move went from the first target to the second. Adding this move to the move from the start position to the first target yielded a composite trajectory that veered toward the second target. Forming a composite trajectory epitomizes vector summation, which has been mentioned often in this book, beginning with the classic work of Georgopoulos and colleagues, discussed in Chapter 3, and continuing with the work of Bizzi and colleagues, discussed in Chapter 8.[17]

Many other studies support the idea of superposition of movements (doing two or more moves at once) with the net output being the vector sum of an initial move toward a first target plus a second move from the first target to the second. For example, ongoing movements were used in an influential study of speech perception (Figure 9.17). Here people prepared to move a mouse to bring a cursor to one picture or another based on

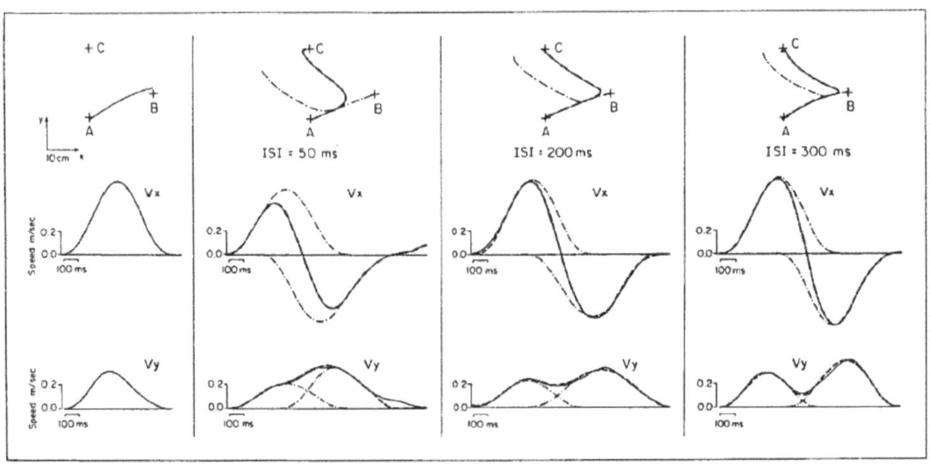

Figure 9.16
Results of moving the hand from target A to target B alone (left panel) or redirecting the hand to target C (the three right panels) if the target C light comes on 50 ms, 100 ms, or 200 ms after the B light. The top row shows the observed handpath as a solid curve and the hypothesized added handpath to the new target (C) as a dashed curve. Beneath the recorded handpath is the velocity in the left–right dimension, *Vx*. Below that is the velocity in the in–out dimension, *Vy*. Both velocity profiles are bell shaped. Solid curves and dashed curves in the bottom two rows are, respectively, observed and hypothesized velocity profiles.

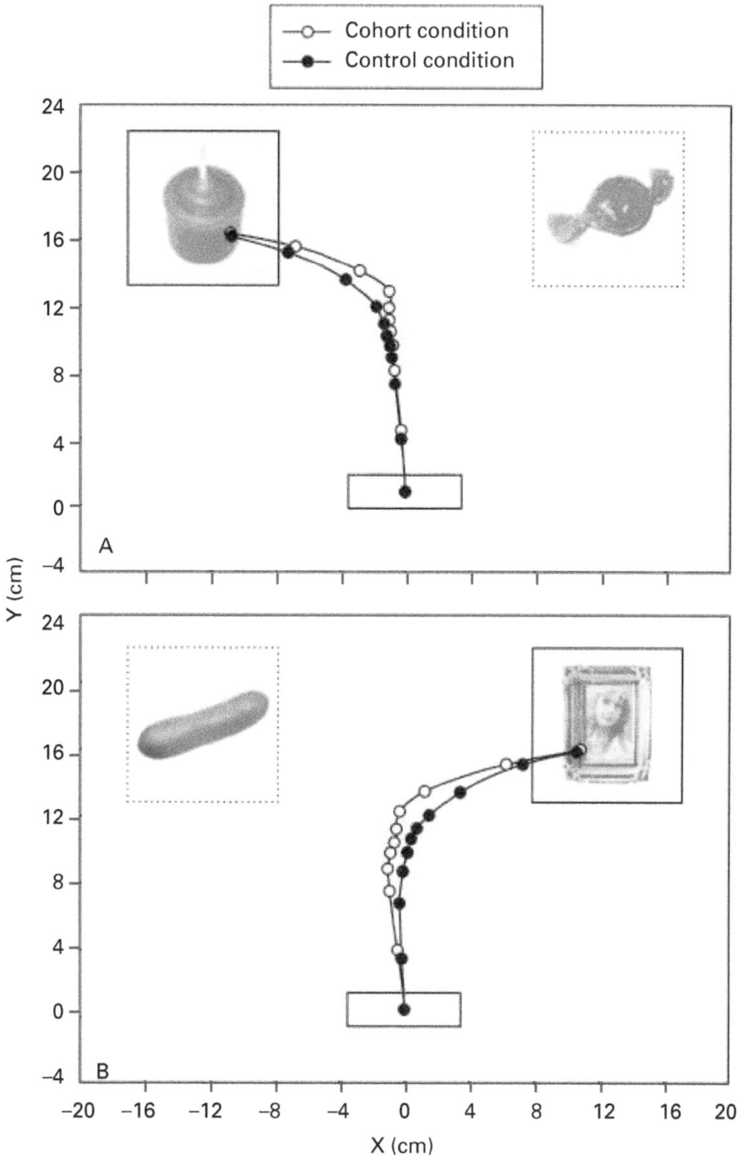

Figure 9.17
Handpaths toward a picture of a candle when the word that could be heard to signal the cursor movement to that target was "candle," and the other possible word was "candy," which starts with the same sound (cohort condition). Alternatively, when the candle on the left was the target, the other possible word (not shown in the top panel) was "picture," which starts with a different sound (control condition). An analogous condition is shown in the bottom panel, where the target is on the right rather than left. In this case, the named target is "picture" and the alternative is "pickle" (cohort condition). Not shown for the bottom panel is the case where the picture on the right is paired with a target that starts with a different sound, like "candle" (control condition).

which picture was named aloud (Spivey, Grosjean, & Knoblich, 2005; Tanenhaus et al., 1995). It was found that people curved the path to the named picture sooner if its name differed immediately from the name of the other possible picture. For example, there was a more curved handpath to a picture of a candle when the other possible target was a portrait than when the other possible word was a candy. The words "candle" and "candy" start off sounding the same, but the words "candle" and "portrait" start off sounding differently. The fact that the move to the target is more direct when the two possible names differed right from the start suggests that word identification can occur before words (or their sound patterns) end. Continuously measuring computer mouse positions made it possible to discover this. Other studies that have used continuous monitoring of ongoing behavior have also yielded new discoveries about action, mind, and brain (Song & Nakayama, 2009; Spivey, 2007; Tipper, Lortie, & Baylis, 1992; Welsh & Elliott, 2004).

Precrastination

You have now arrived at the last section of this book. This section will concern more macroscopic forms of behavior than you have read about recently in this text. Much of the work that has been described so far has involved small-scale acts like moving a computer mouse from one place to another, as in the last study described. Now I'd like you to consider larger-scale actions like walking down the aisle of a supermarket and picking up items for purchase. Walking and reaching are well-coordinated acts, although in the research literature, they have typically been treated separately.

Consider the problem of approaching a table with an object on it. Suppose you want to pick up the object and carry it somewhere. If the object is near the edge of the table on the side from which you approach, you will, unsurprisingly, walk up to the table, pick up the object, and carry it away. But if the object is far from the edge of the table close to your start position and near the edge on the opposite side, you will probably walk all the way around the table to get the object, or you might stretch over the table to get it. Which would you do? In an experiment on this question (Figure 9.18), it was found that people were generally willing to walk long distances to enable short reaches. In fact, they

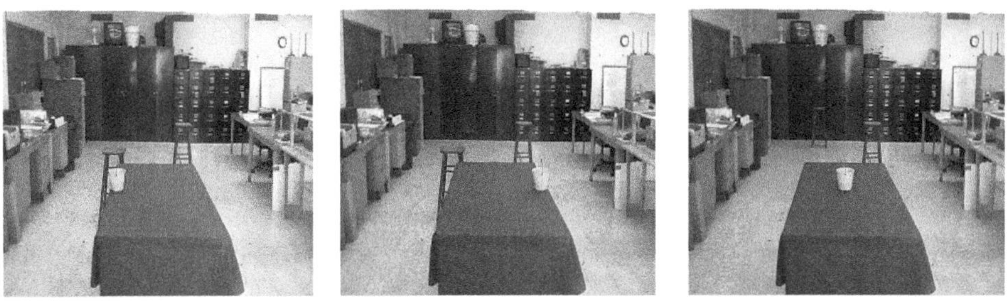

Figure 9.18
Examples of bucket and target-stool arrangements used to determine whether people would walk around the left or right side of the table to pick up and carry the bucket to the left or right stool beyond the table. As seen here, the factors that were varied were the bucket position (left, right, or middle) and the distances of the left and right stools from the ends of the table. The pictures are taken from the spot where the subjects stood before heading one way or the other. Walks along the left side of the table required use of the right hand, whereas walks along the right side of the table required use of the left hand.

were willing to walk about 11 meters to avoid a 1-meter reach (Rosenbaum, 2012; Rosenbaum, Brach, & Semenov, 2011).

The study just covered concerned evaluations of different kinds of action costs—those related to walking and those related to reaching. In other work, the question concerned the trade-offs people make between cognitive and perceptual-motor demands. As in the walk-and-reach studies, it was found that there are regular trade-offs. Subjects were willing to walk through narrow gaps to avoid having to memorize more digits. Narrowing the gap by a certain amount was equivalent to memorizing some number of additional digits (Feghhi & Rosenbaum, 2019, 2021).

Another study of action decisions led to a surprise discovery. In fact, it was so surprising that it got worldwide attention, including coverage in the *New York Times* and other outlets (DeMelo, 2019; Richtel, 2014). In 2014, two students and I asked lab participants to walk down an alley (Figure 9.19) and pick up a bucket on the left or right side of the alley (Rosenbaum et al., 2014). The participants (undergraduates at Penn State University) were told simply to pick up whichever bucket seemed easier. We expected participants to pick up the bucket closer to the finish line; that way, the bucket that was selected would be the one that had to be carried a shorter distance. Our long-term plan was to determine how far people were willing to walk with a light load compared to a heavy load.

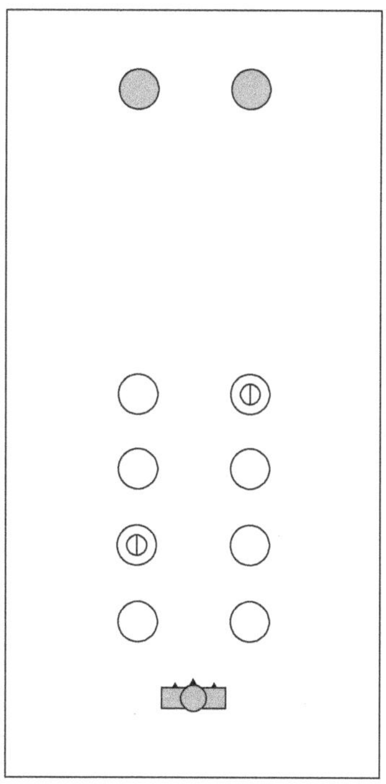

Figure 9.19
Setup for experiments where precrastination was discovered. The subject (gray figure) is asked to walk down the alley and do whatever is easier: either pick up the bucket on the left, shown here on the platform second closest to the subject, and carry it to the platform on the left (gray circle) or pick up the bucket on the right, shown here on the platform farthest from the subject, and carry it to the platform on the right (the other gray circle). The left bucket was picked up with the left hand. The right bucket was picked up with the right hand. The left and right buckets appeared on all possible left and right platforms.

To our astonishment, the result we got was exactly the opposite of what we expected. Most of our subjects picked up the bucket near the start line, not the one near the finish line. This meant that they chose the bucket that had to be carried farther, not the one that had to be carried less far.

At first, we thought our instructions were off, that our participants misunderstood what we were asking or that they were responding to some inadvertent signal we may have been communicating where the "right answer," they thought, was to pick the bucket requiring longer carries. Through nine experiments, we ruled out these artifactual explanations and came to realize that by picking up the near bucket, participants acted in a way that let them have one less thing to think about. Picking up a bucket sooner rather than later lowered the mental load. Although picking up a bucket may not seem like a very mentally loading requirement, the truth of the matter, as seen in this book, is that carrying out a physical act like that is not cognitively trivial.

In later experiments, Lisa Fournier and others tested this hypothesis. Their participants walked down an alley (Figure 9.20) and brought back two buckets arrayed in a line before them (Fournier et al., 2018). One bucket was nearby; the other was farther way. Surprisingly, most people picked up the near bucket, carried it out to the farther bucket, and then returned with both buckets. Picking up the near bucket on the way out meant carrying it further than needed, as in the initial experiment.

Why did people do this? Why didn't they pass the near bucket and then pick it up on the way back, retrieving the far bucket first? Fournier et al. hypothesized that the

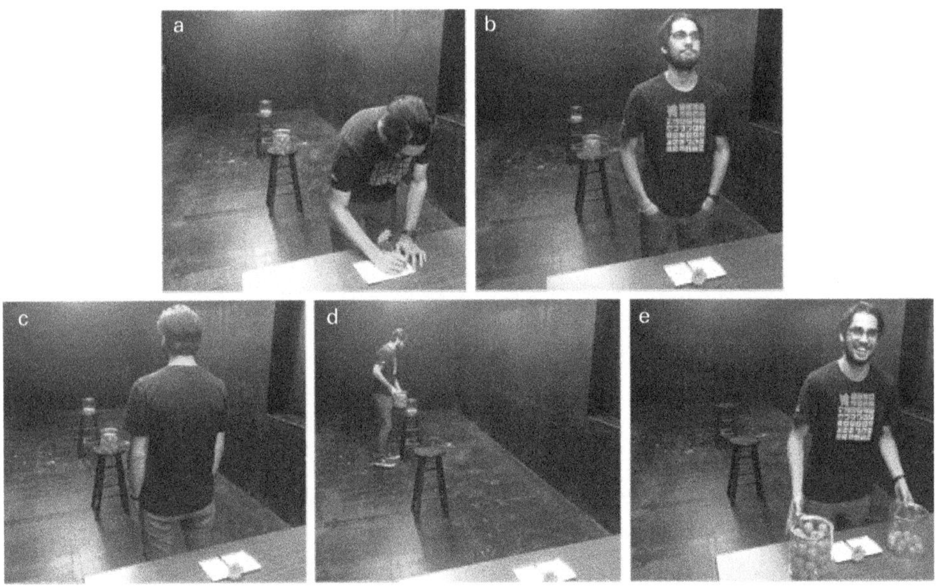

Figure 9.20
Another setup where precrastination has been demonstrated. (a) The subject engages in a filler task involving the alphabet and arithmetic. (b) The subject memorizes a list of digits or does nothing. (c) The subject turns around and looks at the buckets, which contain varying numbers of balls (not discussed here). (d) The subject heads out and chooses to pick up the near bucket before picking up the far bucket. (e) Finally, the subject returns to the home position with both buckets and tries to recall the list of digits if there was one.

reason was to reduce the load on working memory. Remembering to do things in the future—*prospective* memory—adds to mental load (McDaniel & Einstein, 2007). That mental load can be reduced by getting things done sooner rather than later.

Consistent with this hypothesis, Fournier et al. showed that people were even more likely to pick up the near bucket first if they memorized digits beforehand, knowing they would have to recall the digits later, after returning with both buckets. Picking up the near bucket on the way out helped lower the memory load, which was all the more important when memory had an extra burden.[18]

Results like these have made it possible to identify a general tendency that wasn't apparent before. My students and I called this tendency *precrastination*. We chose that word because it is, roughly, the opposite of procrastination. We defined precrastination as the tendency to complete tasks as soon as possible, even at the expense of extra effort (Rosenbaum et al., 2014).

Precrastination is a common tendency (Blinch & DeWinne, 2019; Rosenbaum et al., 2019; Vangsness & Young, 2020; VonderHaar, McBride, & Rosenbaum, 2019; Wasserman, 2018). It is manifested in a wide range of circumstances: answering emails immediately, paying bills as soon as they come in rather than waiting to collect interest by leaving money in the bank, submitting papers and proposals well before they are due, getting surgeries before second opinions are obtained, and convicting people prematurely to get the cases over and done with.

Recognizing that these behaviors are of a piece came to light in studies of walking and reaching. It was surprising that research on mundane tasks like walking and reaching could reveal this wider tendency, but in a way that is the main message of this book: no matter how small scale the actions, they can reveal a lot about much wider scales. What you do depends on what you know, and what you know depends on what you do at all virtually scales.

This is why the study of action holds so much promise for revealing more about the mind and brain. Continued work in this field will surely contribute to society in ways that can scarcely be imagined. It has already shaped the world in ways that are more widespread than is generally appreciated, as so much of this book has been meant to show.

I hope this book will help you follow a course of action that is rewarding for you and for everyone you affect, directly or indirectly.

Notes

Preface

1. There are a number of fine books on the topics to be covered here, although with different emphases from this one's: Graziano (2008); Hommel, Brown, and Nattkemper, (2016); Jeannerod (2006); Magill and Anderson (2010); Schmidt, Lee, Winstein, Wulf, and Zelaznik (2018); Tresilian (2012); and Tversky (2019). Two other books with different emphases than that of the present work, whose quality or lack thereof can be judged by others, are by me (Rosenbaum, 2010, 2017b).

Chapter 1

1. Contrary to Descartes's belief that animals have no free will and cannot think, modern research has shown that animals can reason to an impressive degree (e.g., Dona & Chittka, 2020; Wasserman & Zentall, 2006).

2. Here's the joke: Descartes goes to a café. He sits down and orders a coffee. "Would you like cream in your coffee, Monsieur?" the waiter asks him. "I think not," Descartes replies, and then he disappears.

3. I am very grateful to Iman Feghhi for telling me about Omar Khayyam's earlier development of analytic geometry.

4. Quote from Locke's *Essay Concerning Human Understanding* excerpted in https://wwnorton.com/college/history/ralph/workbook/ralprs24.htm. Locke's blank-slate hypothesis and the term itself are so famous that another book used that phrase as part of its title: *The Blank Slate—The Modern Denial of Human Nature* (Pinker, 2002).

5. The expression "until you are blue in the face" is a figurative, not a literal, expression. It means continuing with something for so long that, if you were holding your breath while doing it, you would turn blue.

6. The phrase, "To know is not enough," is the motto of Hampshire College in Amherst, Massachusetts, where I taught from 1981 to 1986.

7. This call for action has been issued before: Heuer (2003), Rosenbaum (2005, 2017b), Rosenbaum and Feghhi (2019), Schack (2010), and Schack and Frank (2020).

8. Important works by these investigators include Beilock (2015) and Witt (2011).

Chapter 2

1. It took some time for psychologists to see that this form of learning is as important or, in some cases, more important than learning by trial and error. Socially based learning is so efficient that it provides a way of circumventing the need for long training based on reinforcement of individual acts (Bandura, 1977).

2. Further testing of the hypothesis would be desirable. Two predictions can be checked. First, tall people should make their last glance down earlier (at a higher step) than short people because, for short people, the last step vanishes closer to the bottom. Second, the last look down should only depend on how far a walker is from the bottom, not on how many steps they have descended. As a result, there should be no effect of the number of steps on the staircase on the position of the last downward look.

3. The scientist who introduced the concept of ecological validity was Egon Brunswik (1947).

4. According to Wikipedia (https://en.wikipedia.org/wiki/Doubting_Thomas), "A doubting Thomas is a skeptic who refuses to believe without direct personal experience—a reference to the Apostle Thomas, who refused to believe that the resurrected Jesus had appeared to the ten other apostles, until he could see and feel the wounds received by Jesus on the cross."

5. An even more precise method would be to use an eye-movement recording apparatus (Ioannidou, Hermens, & Hodgson, 2017), although then there is the potential worry that the apparatus affects the behavior.

6. The Moses reference is to the movie *The Ten Commandments*, and the reference to singing kids is to the movie *The Sound of Music*.

7. Students who graduate from Stanford, as I did when I got my PhD in 1977, smile at the notion that their diplomas say they got their degrees from a junior university. The quote in the text above comes from https://en.wikipedia.org/wiki/Leland_Stanford_Jr.

8. In the town where I live, Culver City, California, there is a kinescope in a public space in the restaurant district. People wonder at the motion even today, which is especially ironic considering that Culver City, the home of Sony Studios and the place where the *Wizard of Oz* and other movie classics were filmed, calls itself The Heart of Screenland. The town kinescope is across the street from the hotel where the little people who played the Munchkins stayed. Evidently, they were a rowdy bunch. Another old-fashioned device for generating movies is the flipbook. You may have one of these. They are available in toy stores, novelty shops, and check-out areas of bookstores. By holding the flipbook with your thumb and index finger, you can riffle the pages past your thumb and see a little movie, a person hop on a bike and ride away, for example, or a ballet dancer doing pliés. From Amazon.com, I learned while working on this footnote that you can now create flipbooks yourself with suitable software.

9. The movies that Edison and other early movie producers created flickered a lot, so these movies came to be called "flicks." From this term, combined with "net" from the internet, came the name Netflix, the internet entertainment giant that had revenues of $20.15 billion in 2019.

10. There is some controversy about this hypothesis (Tresilian, 1999).

11. On the Wikipedia site devoted to high-speed photography (https://en.wikipedia.org/wiki/High-speed_photography), you can see such pictures. One picture that caught my eye did so for ironic reasons. It was the image of an electric fan that looked like it was turned off. According to the Wikipedia site, the fan was turning at full speed when the shot was taken. A fan viewed in a strobe light flashing at critical frequencies can give the illusion of standing still. Be careful if you find yourself in that situation!

12. https://www.sciencemag.org/news/2018/07/video-reveals-secret-dragonfly-s-backward-flight

13. Even young babies can recognize moving dot patterns as originating from living creatures, as shown by their looking preferences (Simion, Regolin, & Bulf, 2008).

14. http://www.motioncapturesociety.com/resources/industry-history

15. Hz stand for cycles per second and is a tribute to Heinrich Rudolf Hertz (1857–1894), the first physicist to prove the existence of electromagnetic waves (https://en.wikipedia.org/wiki/Heinrich_Hertz).

16. There are video-based systems that can do a reasonably good job of automatically picking up marker positions, such as the Simi Reality Motion Systems GmbH (Germany), used, for example, in a study by Yang and Wang (2020).

17. Surface electrodes can also be placed on muscles (i.e., on the flesh over muscles), where they can pick up electrical activity in the muscles. This technique is called electromyography. The signals it yields are called EMGs. As you might expect, electrodes that have been inserted into the brain and into muscles are more sensitive than those that lie on the surface.

18. I have invented the conversation directed to the patient.

19. A possible way to use fMRI to help locked-in patients was suggested by a student who heard me lecture about this work. His idea was to have the patient imagine one kind of activity, like playing tennis, to signal "yes" in response to questions she was asked and to imagine another kind of activity, like walking through her house, to signal "no." It turned out that this method had already been tried, but I was so proud of the student that I proclaimed his ingenuity and his name (with his permission) in a blog for Psychology Today (Rosenbaum, 2017a). At the time of this writing, the student is in medical school.

20. Hebb's principle appears in his book, The Organization of Behavior: A Neuropsychological Theory (1949), but the phrase about neurons wiring together if they fire together wasn't contained in the book. Rather, it was coined later by someone else.

21. This chapter did not cover methods of externally activating or inhibiting areas of the brain to study the effects of those interventions. Relevant studies will be discussed in the next chapter, relying on electrodes, but because this chapter discussed cutting-edge technologies, it is worth mentioning a technological advance over traditional electrode stimulation methodology. The technique involves optogenetics. Here neurons are genetically modified so they can be controlled with pulses of light. Important advances have been made with this technique (https://en.wikipedia.org/wiki/Optogenetics).

Chapter 3

1. The anatomy of the eye, the ear, and other perceptual organs is commonly discussed in works that privilege perception, so will I refrain from delving into those subjects here.

2. The lizard muscle article that I read online was at https://www.bbc.com/news/health-49876827. The article (October 1, 2019) is entitled "Babies in the Womb Have Lizard-Like Hand Muscles."

3. https://en.wikipedia.org/wiki/Human_skeleton

4. For pursuing the argument, it doesn't really matter if 2D:4D ratios are larger in females than in males or vice versa. It turns out that the hypothesis that has been most widely considered is that 2D:4D ratios are larger in females than in males—that is, by hypothesis, that males have longer ring fingers relative to index fingers than females do, but again, all that really matters here is that there is a difference in either direction.

5. Alexander wrote many books. His volume, *Exploring Biomechanics* (1992), is one of the most beautiful books you could ever hope to see.

6. Some of these musings appeared elsewhere (Rosenbaum, 2017b).

7. As far as I know, there has been little research on the neural control of tail or trunk movements. This seems like an extremely interesting topic to investigate because both appendages are in the midline of the body, so both sides of the brain may be more involved in controlling them than is the case for arms and legs.

8. For discussion of whether Donders's law applies to the arm, see Gielen, Vrijenhoek, and Flash (1997).

9. https://www.arthritis.org/about-arthritis/types/gout/what-is-gout.php

10. https://www.mayoclinic.org/diseases-conditions/lupus/symptoms-causes/syc-20365789

11. https://en.wikipedia.org/wiki/Charge_conservation

12. There is uncertainty about whether Franklin actually did the kite experiment. He described it in a publication without saying who the experimenter was, but he often failed to credit himself in his writings.

13. Franklin also discovered the Gulf Stream, which flows from the Gulf of Mexico along the East Coast of the United States and then east toward Europe, keeping the west side of Europe warmer than it would be otherwise. Franklin discovered the Gulf Stream, or the possibility of a stream within the ocean, by noticing what looked like a river in the ocean as he sailed between America and Europe (2003). To my way of thinking—and I don't know whether Franklin scholars have suggested this before—here again, we see Franklin's receptivity to the idea of flowing currents. Could it be that his invention of the public library similarly reflected an openness to the free flow of books and that his belief in democracy reflected an openness to the free flow of ideas? Less grandly, his invention of swimming fins and the urinary catheter also fits with this flow-centered hypothesis. Other inventions might as well.

14. https://en.wikipedia.org/wiki/Lucia_Galeazzi_Galvani

15. https://en.wikipedia.org/wiki/Rosalind_Franklin

16. This book doesn't include many pictures of the neuromuscular junction, interneuronal junctions, or neurons because those images are quite familiar and readily available. Likewise for pictures of the brain.

17. You can see Jacqueline du Pré playing the cello, in her signature piece, Elgar's *Cello Concerto*, by searching with relevant keywords on YouTube. A movie about her, along with her relationship to her sister Hilary, is called *Hilary and Jackie*. I recommend it highly.

18. Motor neurons can die as a result of diseases whose names may be familiar to you. Polio myelitis is a viral disease that can leave a person paralyzed in one or more parts of the body. The great classical violinist Itzhak Perlman is unable to walk because of this disease, and President Franklin Roosevelt was also unable to walk because of it. Amyotrophic lateral sclerosis (ALS) or Lou Gehrig's disease is a disorder that leads to progressive degeneration of motor neurons, causing weakness, paralysis, and finally death. Lou Gehrig was a famous baseball player who succumbed to the disease in 1941. The cause of ALS is unknown and is so far untreatable. There are other motor neuron diseases as well.

19. The term "final common pathway" was coined by Sir Charles Sherrington, the same person who coined the word "synapse." Sherrington's work will be discussed in more detail in the next chapter.

20. The word "cortex" means "shell." The cerebral cortex forms a shell around the midbrain.

21. Jackson's deduction was so insightful that for this insight and other contributions, he received many prestigious awards. A building in a medical school in England is named after him, and Oliver Sacks, the popular neurologist-writer, declared that Jackson profoundly affected and inspired him.

22. It is tempting to think that in the many years that have passed since Penfield's surgery to help epileptic patients, drugs and other treatments have come along that obviate (make unnecessary) surgery as a treatment for this disease. Sadly, that is not the case. Strides have been made in the treatment of epilepsy, but the malady is still poorly understood and not as easily treated as one would hope (Couzin-Frankel, 2019).

23. Watch "Jan Sherman moves robotic arm with her mind": https://www.youtube.com/watch?v=Vg6b00VDjvM. For more about brain–computer interfaces, see Velasquez-Manoff (2020).

24. Taking a weighted sum is the same as doing vector summation, which you may have learned about in a math class. As you may remember, or perhaps are learning here for the first time, it is possible to add vectors by placing them head to tail from the starting point to the end of the vector chain. In the present case, where the axes for the vector summation can be thought of as left–right and in–out, if each vector goes in a particular direction and over some length within the space, you can have a first vector pointing out from the origin, and then you can put the start of the next vector at the end of the first vector, the start of the third vector at the end of the second vector, and so on. If you do this for all the vectors in the population, you can go from the start point to an endpoint. If each of the vectors in this conceptual scheme corresponds to each neuron's firing rate, the location that is finally reached will depend on the weighted sum of all the neurons' activations.

25. Population coding is also appealing from the standpoint of avoiding the need for a neural chief executive. Cognitive psychologists study executive processes (high-level choice processes), but it's essential not to postulate a chief executive (a homunculus) in the brain. Population coding achieves this by instantiating democracy (Rosenbaum, 2014).

26. More recently, a great deal of research has been done on motor imagery using fMRI and other techniques (e.g., Guillot, Di Rienzo, MacIntyre, Moran, & Collet, 2012; Vogt et al., 2007).

Chapter 4

1. It was recently shown through similar logic that bats have cognitive maps (Fenton, 2020).

2. Another result integrates the two spatial scales (Marteniuk & Bertram, 2001). Participants lifted a cup from one platform and transported it to another. In one condition, the participants stood still. In the other, they moved the cup while walking. In the standing-still condition, the handpath looked the same when plotted in extrinsic (external spatial) coordinates and in intrinsic (joint-based) coordinates. But in the walking condition, the graph was chaotic when plotted in intrinsic coordinates but orderly when plotted in extrinsic coordinates.

3. O'Keefe and Nadel (1978) reviewed the discovery of place cells. In an article in a publication of the Association for Psychological Science, the *APS Observer*, the second author of the book, Lynn Nadel (https://www.psychologicalscience.org/observer/nadel), reported that younger children are less apt to form cognitive maps than older children and that the developmental period when cognitive maps can be formed coincides with the period when children overcome childhood amnesia, the long-term forgetting of personal events (before age 2.5 years or so). Nadel hypothesized that both changes can be traced to maturation of the hippocampus, the part of the brain where place cells are located. The hippocampus is also the part of the brain that is important for the formation of new declarative memories.

Consistent with Nadel's hypothesis, Buzsáki and Llinás (2017) linked the hippocampus's capacity for representing series of places in space to the capacity for recalling series of personal events.

4. http://www.scholarpedia.org/article/Head_direction_cells

5. A particularly dramatic demonstration of the centrality of graded activity of location-related neurons involved recording from hippocampal place cells from rats running in a wheel. The rats had just gone through the right side of a maze to get water and then had to run in a wheel before going through the left side of the maze. Left-side place cells became activated during the wheel running but did not become activated in the wheel if the order of events was reversed and the rats were biding their time in the wheel before heading to the right side of the maze after going through the left side (Buzsáki & lLinás, 2017).

6. "Worst stutter in the world" on YouTube: https://www.youtube.com/watch?v=uULRVq1qj5U

7. https://en.wikipedia.org/wiki/P%C4%81%E1%B9%87ini

8. Recall or learn for the first time that correlations reflect the correspondence between sets of numbers. If there is no relation, the correlation is zero or close to zero. If there is a positive correlation, low numbers are associated with low numbers and high numbers are associated with high numbers. If there is a negative correlation, low numbers are associated with high numbers and high numbers are associated with low numbers. If long intertap times are followed by short intertap times and short intertap times are followed by long intertap times, the correlation between successive intertap times (also known as the lag 1 autocorrelation) is negative. An autocorrelation is a correlation of one variable with itself. The word "auto" means self. An automobile is a self-driving, rather than horse-driven, vehicle. The term "lag-1" means that the autocorrelation is taken for variables just 1 apart, lag 2 is 2 apart, and so on.

9. For reviews, see Wing (1980, 1993).

10. Further elegant work pursuing the mathematical analysis of rhythmic finger tapping was reported by Ivry and Hazeltine (1995) and Spencer, Zelaznik, Diedrichsen, and Ivry (2003).

11. Skinner (1969) focused on behaviors that could influence the likelihood of their provoking stimuli, so-called operant or instrumental conditioning. John B. Watson (e.g., Watson, 1930), the father of behaviorism in America, had earlier focused on reactions that didn't necessarily affect the likelihood of external stimuli but instead influenced the body's response to external stimuli, such as salivation in response to food or a stimulus such as a bell that comes to be associated with the food. The latter form of learning is called classical conditioning. Classical conditioning was discovered by Ivan Pavlov (1928).

12. A note about terminology: the term "negative reinforcement," which was not included in the main text, is not the same as punishment. Negative reinforcement is the cessation of feedback that has value (either positive or negative). Being told over and over again that your baseball pitches are good is positive reinforcement. Not being told anything about your baseball pitches after that is negative reinforcement (the negation or removal of feedback given before).

13. A possible neural basis for producing movements without sensory feedback was established around the time that Lashley published these observations. Brown (1911, 1917) reported that in deafferented cats, it was possible to observe stepping. Later work identified central pattern generators that emit rhythmic outputs entirely on their own. These were first found in insects and later were found in other animals, including mammals. For a review, see https://en.wikipedia.org/wiki/Central_pattern_generator.

14. Sherrington was actually aware that spinal mechanisms could permit sequences of movements without sensory input. In fact, to his credit, he advocated publication of work by a junior colleague (Brown, 1911, 1914) showing that series of efferent signals for walking could be obtained from the spinal cord without afferent input. Later research confirmed and clarified the nature of these central pattern generators, as they came to be called (Grillner, 1981, 1996).

15. Gentner, Grudin, and Conway (1980) showed that the hand and finger moves toward keys on a keyboard often begin before the previous keystroked has been completed.

16. In mentioning the "reciprocal innervation among the sensory-motor centers," he is referring directly to Sherrington's work.

17. According to Wikipedia (https://en.wikipedia.org/wiki/Pig_Latin), "Pig Latin is a language game or argot in which words in English are altered, usually by adding a fabricated suffix or by moving the onset or initial consonant or consonant cluster of a word to the end of the word and adding a vocalic syllable to create such a suffix."

18. Translation achieved via http://www.snowcrest.net/donnelly/piglatin.html.

19. Chomsky and Lashley were apparently ignorant of each other, although they were at the same institution (Harvard University) at the same time (Bruce, 1994).

20. The last point about unconscious mechanisms deserves special comment because Sigmund Freud, it turns out, was not the originator of the idea in any formal, scientific sense. The idea actually came from Hermann von Helmholtz, the German scientist who measured the speed of nerve conduction, as discussed in Chapter 2, and who argued for unconscious inferences in perception, as discussed in Chapter 8.

21. The researchers who most actively kept track of, and catalogued, action slips have been Donald Norman (1981) and James Reason (1990).

22. The word "parameter" is more general than feature because "feature" connotes a discrete (present or absent) value, whereas a parameter can be a continuous variable.

23. Thankfully, I don't have Parkinson's disease, or not yet at the time of this writing. Noting new phenomena related to one's field illustrates what cognitive psychologists call the *prepared mind.*

24. Further evidence for putting a premium on reducing cognitive or computational costs comes from work by Kool, McGuire, Rosen, and Botvinick (2010).

25. This is called a *tree-traversal* process. See Abrahamse, Jiménez, Verwey, and Clegg (2010); Povel and Collard (1982); and Rosenbaum, Kenny, and Derr (1983).

26. Taking longer to start action sequences with more elements was shown before my team and I showed this with choice reaction times. Henry and Rogers (1960.) showed that simple RTs are longer for complex (multielement) action sequences than for simple action sequences. That finding was extended by Sternberg, Monsell, Knoll, and Wright (1978). Abrahamse et al. (2010), and Klapp (1977). Klapp and Maslovat (2020) have reviewed this literature. The idea that action sequences rely on hierarchically organized plans was introduced by Bryan and Harter (1897, 1899) and has been supported in studies using a wide range of methods including, but not limited to, recording reaction times. See, for example, Logan and Crump (2011) and Schack and Frank (2020).

27. Estes's (1972) model served as a basis for a model of typing (Rumelhart & Norman, 1982). For a detailed review and critique of that model, see Rosenbaum (2010).

28. These monkeys were *deafferented*; that is, their afferent fibers we cut. Remember that afferent fibers carry sensory signals. Afferent fibers that enter the spinal cord enter the dorsal side of the spinal cord. A mnemonic to use to remember the term "afferent" is that these fibers go *at* the spinal cord. Nerve fibers that carry motor commands from the spinal cord to muscles are called *efferent* fibers. A mnemonic to use to remember the term "efferent" is that these fibers *exit* the spinal cord.

29. This conclusion does not bear on the claims made by Sherrington about the anatomy and physiology of the spinal reflex. It only bears on the necessity of reflexes for basic sequencing and timing. Movements made *without* feedback are generally less graceful than movements made *with* feedback, suggesting that somesthetic feedback—feedback from the body—plays some role in motor control, even if it is not always necessary for basic sequencing and timing. See, for example, Keele (1968).

Chapter 5

1. The subject in the memory experiment was a runner. He related numbers he heard to track times and other data he knew from his personal life.

2. Of course, motivation and persistence are required to devote 10,000 hours of deliberate practice to any skill. Guidelines for cultivating those attitudes have been advanced by Wulf and Lewthwaite (2016). Being wise to the fact that long practice may never pay off is useful as well.

3. The learning curve is attributable to Robert Rescorla and Alan Wagner. The underlying principle, that degree of improvement is proportional to the remaining distance to ultimate learning, is called the Rescorla–Wagner law of learning (Rescorla & Wagner, 1972). The Rescorla–Wagner law was formulated in connection with classical (Pavlovian) conditioning.

4. I have accepted my mediocrity as a violinist and enjoy it as a hobby. As a friend of my wife's once said when she and Judy were finishing their coffees in our kitchen and I joined them after practicing a Bach partita in another room, "That sounded hard!" It was one of the funniest lines I've ever heard. If it's not clear why I found the line funny, it's because truly skilled performance should make things seem easy, not hard. Others have come to terms with the fact that music may better serve them as an avocation (hobby) than as a vocation (job). On this score, see the amusing piece by Jennifer Weiner entitled "I'll Never be Rachmaninoff" (*New York Times Sunday Review*, December 30, 2018, p. 9).

5. I was aided in writing this section by a nicely done *Psychology Today* blog called "The YouTube Overconfidence Effect," by Ira Hyman: https://www.psychologytoday.com/us/blog/mental-mishaps/201804/the-youtube-overconfidence-effect

6. I've made up the dialog to bring it to life.

7. For a more nuanced account of this work, see MacKay (2019).

8. This is not Ericsson of 10,000 hours fame. Similar results were reported by Basak, Voss, Erickson, Boot, and Kramer (2011).

9. Beilock has a wonderful TED talk on this topic, found easily on the internet.

10. Gareth Gaskell has a TED talk on this topic as well, easily found on the internet.

11. Transfer is also easier between tasks with many shared elements than between tasks with fewer shared elements. The more elements two activities have in common, the better the transfer between them. This is known as the principle of identical elements (Thorndike & Woodworth, 1901).

12. For reviews, see Anderson (2015), Newell and Rosenbloom (1981), and Rosenbaum (2010).

13. There has been debate about the power law formulation (Evans, Brown, Mewhort, & Heathcote, 2018; Heathcote, Brown, & Mewhort, 2000; Rosenbaum & Janczyk, 2019).

14. https://en.wikipedia.org/wiki/Moro_reflex

15. For a more detailed discussion of affordances, see Warren (1984) and Warren and Whang (1987).

16. For a recent book on this perspective, see Wagman and Blau (2019). For a discussion of why the perspective is less popular than one might expect, see Chong and Proctor (2020).

17. A fuller statement is that resonance "describes the phenomena of amplification that occurs when the frequency of a periodically applied force is in harmonic proportion to a natural frequency of the system on which it acts" (https://en.wikipedia.org/wiki/Resonance).

Chapter 6

1. As noted in the last chapter, there is some debate about whether a power law equation is the best description. Some authors, cited before, have argued that a negative exponential equation is better. In a power law equation, amount of practice is raised to some power. In an exponential equation, amount of practice is in the power term (in the exponent).

2. Note that I am using a quote of the quote by a later author.

3. For another estimate of the number of chunks that can be remembered, see Cowan (2001).

4. Rosenbaum and Janczyk (2019) described a computational model of trial-and-error learning. Variations of the theory can be imagined. Methods that have been tried (behaviors that are performed) can get graded marks rather than just "good" or "bad" marks.

5. One might think that feedback cycle times are positively related to action completion times, but that may not be the case. A huge amount of time can go into preparing for a trip to outer space, but the mission can fail in an instant, or a small amount of time can be spent preparing for and carrying out an action and it takes years to get feedback about it, as in committing a crime of passion and then spending years in prison for the offense.

6. As mentioned in Chapter 5, toddlers fall thousands of times (Adolph et al., 2012).

7. The distinction between step tracking and ramp tracking is more than just didactic (for teaching). It turns out that the nervous system has different neural systems for the two kinds of tracking. Multiple sclerosis may lead to problems with smooth pursuit eye movements but not saccadic eye movements, and some drugs affect one kind of eye movement but not the other (Alpern, 1972; Gouras, 1985).

8. I name Richard Pew in full here as a humble expression of thanks for his help to me early in my career when he invited me to join him in writing a review chapter for Stevens's *Handbook of Experimental Psychology*, second edition (Pew & Rosenbaum, 1988). Working on that chapter helped me organize in my own mind a useful way of organizing concepts and lines of work in the field of motor performance. Dick graciously went along with that organization for the chapter.

9. A great deal of follow-up work has been done with the method developed by Nissen and Bullemer, including theoretical insights into the course of skill learning (Willingham, 1998). For example, it has been shown that implicit learning can compensate for failed explicit learning (Miyamoto, Wang, & Smith, 2020).

10. Later, Woodworth became known as the dean of American experimental psychology, achieving that name because of his monumental treatise (Woodworth, 1938) called *Experimental Psychology*. This was the first English-language book based on laboratory research about mental life and behavior. A second edition appeared later (Woodworth & Schlosberg, 1954). William James's monumental treatise, *The Principles of Psychology*, published in 1890, described less experimental work simply because that work hadn't been done yet. James's musings did much to inspire experiments that were later done to test his intuitions and reasoning. Several such studies will be described in the next chapter.

11. Daniel Kahneman, the winner of the Nobel Prize for his establishment of the field of behavioral economics along with Amos Tversky, wrote about these two processes in a book called *Thinking, Fast and Slow* (Kahneman, 2011).

12. Drawing on the logic that Woodworth (1899) used, the visual feedback loop time is generally thought to be on the order of 200 ms (Elliott, Helsen, & Chua, 2001; Keele & Posner, 1968).

13. I had hoped to show the apparatus that Woodworth used but couldn't find a picture of it, so I consulted with an expert who would know if one existed. Romeo Chua of the University of British Columbia, coauthor of a major review article about Woodworth's research (Elliott et al., 2001), was kind enough to double-check and confirm my sense that none existed. On December 16, 2020, he generously emailed me as follows: "I don't have a picture of Woodworth's setup. The best I can come up with is this website that has an archive of images of old instruments. There is an image of a vertical and horizontal kymograph, but I'm not sure whether they would be the type that Woodworth used. http://homepage.ntu.edu.tw/~linth/index.101.htm#No.22."

14. Reviews of the research spawned by Woodworth's findings have been provided by Digby Elliott and colleagues (Elliott et al., 2001, 2010). A biographical memoir about Woodworth by a colleague of his who was a noted vision scientist (Graham, 1967) scarcely mentioned Woodworth's work on movement control.

15. Here the word "arithmetic" is pronounced with the emphasis on "met" rather than "rith."

16. For more on the relation between Fitts's law and the Hick–Hyman law, see Wright, Marino, Chubb, and Mann (2019).

17. When people move cursors to targets on computer screens, they don't usually move as quickly as possible. Rather, they move at comfortable speeds. The Fitts aiming task is done under the instruction to move to the target as quickly as possible, so an open question is whether Fitts's law applies to spontaneously adopted movement times. Young, Pratt, and Chau (2009) have tackled this question.

18. The reason for the slash in the citation date was that the paper only appeared as an informal report until it was brought to publication twenty years later by Professor Alan Wing of Birmingham University. Markus Janczyk and I (Rosenbaum & Janczyk, 2019) did some detective work to learn who Crossman was, because nothing was known about his (or her identity) before then. Crossman did not provide a first name in his publications, only initials. Thanks to Markus, we were able to learn that Crossman's first name was Ted and other things about him, including the fact that he was on the faculty at UC Berkeley.

19. There is also some variability along the axis perpendicular to the main axis of movement, but it tends to be smaller. Typically, an ellipse can be drawn around the endpoints of the moves. The ellipse is much longer than it is wide; that is, its long axis aligns with the line connecting the starting point to the target, while its shorter axis aligns with the perpendicular to that line.

20. David Meyer is best known for his discovery, with Roger Schvaneveldt, of semantic priming, the finding that words with similar meanings prime each other in memory (Meyer & Schvaneveldt, 1971).

21. A reflection of the sophistication that is brought to bear in aiming is that people aim for targets in a way that maximizes gains and minimizes losses when presented with landing areas marked with gain and loss zones (Trommershäuser, Maloney, & Landy, 2008). Performance in these tasks is not susceptible to strong loss aversion of the sort that can lead to suboptimal economic decisions (Tversky & Kahneman, 1974).

22. Neurotypical people do so as well, of course, and detailed analysis of hand position as a function of time following the presentation of an oriented target indicates that the hand tilts even before leaving the start gate in anticipation of the final orientation (Brown, Moore, & Rosenbaum, 2002).

23. Ebbinghaus (1885/1964) is best known for devising one of the first scientific methods for studying memory.

24. For more about these topics, see Milner and Goodale (2006) and Ganel and Goodale (2019).

25. The relation between these two amplitudes is call the *gain* of the system.

26. This is not a quote from the article. It is one I've made up to try to give a sense of how the experiment was conducted for participants (university students) in the explicit instruction condition.

27. Wooster (1923) was cited 73 times according to Google Scholar on February 27, 2020. Richard Held, who, with Alan Hein, essentially rediscovered Wooster's effects (Held & Hein, 1963), was cited 24,529 times according to Google Scholar on the same day. Held and Hein (1963) did not cite Wooster (1923), although it is possible they did not know about her work. For more about Margaret Wooster, see http://www.feministvoices.com/margaret-wooster-curti/.

28. https://youtu.be/DphlhmtGRqI

29. This article deeply inspired me as a young experimental psychologist because it showed how simple experiments could bear on deep questions of longstanding interest. I felt privileged later to be in the same department as Charlie Harris (the Human Information Processing Research Department at Bell Labs, headed by Saul Sternberg) and to interact with Irv Rock at Rutgers Newark when my wife, Judith Kroll, and he were on the faculty there.

30. This paper also described many other illusions of body shape and position induced by muscle vibration.

Chapter 7

1. These terms come from Gregory (1966).

2. Other terms for outflow are efferent copy (von Holst & Mittelstaedt, 1950) and corollary discharge (Sperry, 1950).

3. Two aspects of this result are noteworthy. One is that the illusion provides support for modularity, the notion that there are separate faculties in the brain that carry out their functions largely oblivious to what the others are doing (Fodor, 1983). Your hand is no less a part of you than your eye is, yet when you move your hand, your visual system interprets the incoming visual signal in a way that is ignorant of the hand movement.

The other noteworthy aspect of the result concerns Bayesian statistics and the mounting evidence that the brain makes intelligent use of prior probabilities to compute posterior probabilities—in other words, to determine how likely the source of some event is given the prior odds of its occurring. Even though it's very unlikely that the house you're looking at could bounce up and down, you may see it doing so if you press on your eyeball with your finger. It's interesting to consider this outcome vis-à-vis Bayesian views of vision.

4. Similarly, some people who get migraine headaches experience scintillating visual patterns that move with their eyes (Richards, 1971).

5. I often turn this into a class demonstration with two students taking turns looking at each other as each one looks into a handheld mirror trying to see their own saccades.

6. The suppression of sensory input due to one's own activity turns out to be widespread. When male crickets chirp to attract mates or scare off rivals, their chirps are incredibly loud (100 dB) and would deafen the crickets producing the sounds if some extra neural discharge didn't suppress the neural response to the intense auditory signals. Such auxiliary neural signals have been found (Poulet & Hedwig, 2002). In crickets, when efferent commands are sent to the wings to make the chirps, additional signals inhibit the auditory afferents.

7. Even if some saccades are made while the lids are closed, the time that the visual system is shut down is probably five hours a day or more.

8. It has been discovered that rats seem to enjoy being tickled (Ishiyama & Brecht, 2016).

9. Evidence consistent with this feedforward control hypothesis has been obtained in brain-imaging studies. People diagnosed with schizophrenic symptoms have been found to have atypical circuits associated with feedforward control (Frith, Blakemore, & Wolpert, 2000; Fu & McGuire, 2003).

10. https://en.wikipedia.org/wiki/Erich_von_Holst. There are many Max Planck Institutes in Germany and a few outside Germany as well. They are named for the developer of quantum mechanics, Max Planck.

11. von Holst and Mittelstaedt were less dramatic in their characterization of the neural control of the housefly than I have been here. They said that there is a copy of the efferent commands to the visual centers, or some sort of corollary discharge. I believe that it makes more sense to think in terms of a perceptual representation formed in advance of the efferent commands, both because that account offers a basis for motivating the efferent commands in the first place and because it provides a format for evaluating reafference (the perceptual input that results from the movement).

12. https://en.wikipedia.org/wiki/Hermann_Lotze

13. This outcome was expected from earlier work (Liberman, Cooper, Shankweiler, & Studdert-Kennedy, 1967).

14. Molto, Nalborczyk, Palluel-Germain, and Morgado (2020) provided a meta-analysis of studies on action effects on distance perception and confirmed these effects.

Chapter 8

1. Bernstein's focus on physical action makes sense given the context in which he worked, which was in the Soviet Union, where putting things into practice (*praxis*) was the order of the day. Karl Marx (Figure 1.7), the father of Marxism, had developed the philosophical, historical, and economic ideology that led to the formation of the Soviet Union. For Marx, praxis was central. Bernstein, working in that context and spending most of his career under Stalin, focused on praxis either by choice or by necessity. Videos of Bernstein can be seen on the internet. In seeing him in action for the first time, I was shocked by his missing teeth, something a high-level scientist in the West wouldn't have. Bernstein worked under difficult conditions (Talis, 2015).

2. While researching this famous picture for this book, I ran across the following stupefying quote in the Wikipedia article (https://en.wikipedia.org/wiki/The_Creation_of_Adam) on *The Creation of Adam*: "In 1990 in Anderson, Indiana, physician Frank Meshberger noted in the Journal of the American Medical Association that the background figures and shapes portrayed behind the figure of God appeared to be an anatomically accurate picture of the human brain. On close examination, borders in the painting correlate with major sulci of the cerebrum in the inner and outer surface of the brain, the brain stem, the frontal lobe, the basilar artery, the pituitary gland and the optic chiasm."

3. Another example of an extreme form of performance is playing the piano by lying down on a piano bench, extending one's hands over one's head and crossing them. A delightful video that shows this can be found on YouTube: Salut Salon "Wettstreit zu viert" | "Competitive Foursome" (https://youtu.be/BKezUd_xw20).

4. Search for Alvin Law—Drumming at 21 Years Old, or go to https://www.youtube.com/watch?v=Bjwuwi7lgLY.

5. The adjective that is used is indeed "dynamical," not "dynamic."

6. There is a wonderful YouTube on this equation called "This equation will change how you see the world (the logistic map)": https://www.youtube.com/watch?v=ovJcsL7vyrk.

7. Mechsner et al. (2001) also showed that visual symmetry is more important than muscle symmetry in a study involving turning cranks with two hands beneath a table, causing two flags to turn above the table. The authors used an ingenious gear system to modify the relations between crank turns and observed flag rotations. They found that the simplicity of the perceptual (visual representation) is what determined quality of performance. The two hands could produce complex patterns if the visual goal was simple. The same message came from other studies of bimanual control (Kunde & Weigelt, 2005), and a similar point was made in a study where people merely kept their hands in contact with felt moving surfaces (Rosenbaum, Dawson, & Challis, 2006).

8. https://en.wikipedia.org/wiki/Christiaan_Huygens

9. In YouTube, search for "synchronization of metronomes" or go to https://www.youtube.com/watch?v=aaxw4zbulms or https://www.youtube.com/watch?v=t58lgkreubo.

10. https://en.wikipedia.org/wiki/Morgan%27s_Canon#cite_note-2

11. An example of changing the stiffness of a joint while the joint remains in the same position is preparing to catch a baby that will be dropped from a window of a burning building. That is the example I use in my teaching to help make the concept as vivid as possible.

12. When I taught about this one year in my Action, Mind, and Brain class, a student said (in so many words), "Think of a waiter holding a tray. If the waiter places a glass on the tray himself, you won't see the tray drop to a lower position, or at most you'll see it drop a tiny little bit. But if someone else puts the glass on the tray, you'll see a larger drop." The student's point was that the waiter had a feedforward representation of what would happen. The feedforward representation helped the waiter resist the placement of the glass on the tray by his own hand. The student who came up with that observation is now a tenured professor at the University of Illinois: Allejandro Lleras. One of his papers (Thomas & Lleras, 2009) is cited near the end of Chapter 7.

13. No one, to my knowledge, has actually tried to record where hands go when piano or computer keyboards are suddenly yanked away. That would be an informative and entertaining experiment. Conceivably, typists and pianists can do "air keyboarding," that is, typing or playing the piano for extended periods in the air. Their hand and finger movements could be recorded with motion-tracking equipment to test this hypothesis. If the equilibrium point hypothesis is correct, and if the performers can get over the lack of normal feedback, they should move their hands and fingers to where their extremities would go if the keyboards were present. I hope someone will try this. It would be a fun thing to watch on YouTube.

14. According to Feldman and Latash (2005), deafferentation was not as complete or as effective as Polit and Bizzi (1978) assumed. Other questions have been raised about other aspects of the equilibrium point hypothesis, including the extent to which movements to equilibrium points run off on their own without the need or possibility of movement shaping (Gomi & Kawato, 1996). Defenders of EP control argue that nothing in the EP hypothesis precludes movement shaping.

15. Desired trajectories can be obtained by superimposing back-and-forth movements onto the main movement. The back-and-forth movements can go from the starting posture back to another suitably chosen "bounce posture" and then back to the starting posture. Vector summation can be used to shape the movement by choosing a suitable bounce posture.

16. Children as young as three years of age begin to show the grasp height effect (Jovanovic & Schwarzer, 2017).

17. For reviews, see Rosenbaum et al. (2012, 2014, 2017b).

18. A milestone of human development is learning to inhibit direct reaches when objects of interest are on the other side of a clear barrier (Diamond, 1990).

19. For a video of a cotton tamarin monkey engaging in second-order grasp planning, go to https://sites.google.com/site/rosenb13/davidrosenbaum.

20. I wondered whether "pinky" is a racist term. Had it been, I would not have used it. The research I have done has said it is not.

Chapter 9

1. Search on the internet for "Scientists Create Speech from Brain Signals" (*New York Times*, April 14, 2019). https://www.nytimes.com/2019/04/24/health/artificial-speech-brain-injury.html?smid=nytcore-ios-share

2. A related advance has been the development of new shape-morphing, self-healing, intelligent material for soft robotics (Ford et al., 2019). https://scitechdaily.com/new-shape-morphing-self-healing-intelligent-material-developed-for-soft-robotics/. Note the word "soft." In contrast to the rigid material of traditional robots or their fictional counterparts such as the one that will be shown in Figure 9.6, soft materials can make robots more resilient. For more on this, see https://science.sciencemag.org/content/365/6449/145.abstract?et_rid=49127623&et_cid=2929163.

3. https://www.sciencemag.org/news/2016/06/how-some-salamanders-regrow-their-limbs

4. https://www.wired.com/story/devastating-allure-of-medical-miracles/

5. https://www.bbc.com/news/world-asia-india-40678481#:~:text=An%20Indian%20musician%20played%20the,known%20as%20%22musician's%20dystonia%22

6. https://en.wikipedia.org/wiki/Jaquet-Droz_automata

7. https://en.wikipedia.org/wiki/Gort_(The_Day_the_Earth_Stood_Still)

8. Videos can be found on the internet by searching for "passive dynamics."

9. Caggiano et al. (2018) identified midbrain structures that modulate cat gait patterns. As a proof of concept that simple switching is possible for gait patterns, a fully worked-out electronic circuit was developed for this purpose in an insect robot (Ohara et al., 2019). This development may be edifying with respect to the seemingly impossible feat of switching from one language to another, as fluent bilinguals do all the time (Kroll & de Groot, 2005). In principle, the same sort of switching mechanism that is used to switch gaits could also be used to switch languages.

10. A critical feature of trial-and-error learning in this context is that the behaviors that emerge needn't be chosen deliberately at the behest of an inner chief executive (CEO). Positing such an inner CEO invites infinite-regress problems: who tells the CEO what to decide? I have written about this, attempting to use Darwin's theory as a basis for a theory of mental function (Rosenbaum, 2014).

11. Wikipedia has fine articles about the facial feedback hypothesis and the James–Lange theory.

12. For a review, see Barrett (2017).

13. The researchers who did the finger-and-toe study also estimated the time for nerve signals to get from the brain to the finger and from the brain to the toe by recording simple reaction times for the finger and the toe separately. The simple reaction time for the toe was longer than the simple reaction time for the finger by about 30 ms. For more about synchronization, see Aschersleben and Prinz (1995).

14. The door-holding study was based on naturalistic observations at the student union building on the Penn State University (University Park) campus. No attempt was made to analyze the data in terms of gender, race, or other factors. Of course, data of this kind could be analyzed that way, and cultural differences could be studied as well by making analogous observations in other places.

15. Hughes (1996) provided further evidence of motor-planning challenges in autism.

16. For succinct coverage of Jaswal and Akhtar's work, search for "Important new work on autism could change interventions" on the internet. This should bring you to a site run by the UC Santa Cruz News Center: https://news.ucsc.edu/2018/06/akhtar-autism.html.

17. Reliance on vector summation (neural population coding) can be viewed as democracy in the brain: everyone votes all the time, without a central executive who makes decisions (Rosenbaum, 2014).

18. This result was replicated by Patterson and Kahan (2020). See Raghunath, Fournier, and Kogan (2020) for some qualifications.

References

Abbs, J. H. (1986). Invariance and variability in speech production: A distinction between linguistic intent and its neuromotor implementation. In J. S. Perkell & D. H. Klatt (Eds.), *Invariance and variability in speech processes* (pp. 202–219). Hillsdale, NJ: Erlbaum.

Abrahamse, E. L., Jiménez, J., Verwey, W. B., & Clegg, B. A. (2010). Representing serial action and perception. *Psychonomic Bulletin and Review, 17,* 603–623.

Adolph, K. E., Cole, W. G., Komati, M., Garciaguirre, J. S., Badaly, D., Lingeman, J. M., Gladys, L. Y., & Sotsky, R. B. (2012). How do you learn to walk? Thousands of steps and dozens of falls per day. *Psychological Science, 23,* 1387–1394.

Aglioti, S., DeSouza, J. F., & Goodale, M. A. (1995). Size-contrast illusions deceive the eye but not the hand. *Current Biology, 5,* 679–685.

Alexander, R. M. (1984). Walking and running. *American Scientist, 72,* 348–354.

Alexander, R. M. (1992). *Exploring biomechanics*. New York, NY: W. H. Freeman.

Alpern, M. (1972). Eye movements. In D. Jameson & L. Hurvich (Eds.), *Handbook of sensory physiology* (Vol. VII/4, pp. 303–330). Berlin, Germany: Springer.

Anderson, J. R. (2015). *Cognitive psychology and its implications* (8th ed.). New York, NY: Worth.

Arber, S., & Costa, R. M. (2018). Connecting neuronal circuits for movement. *Science, 360,* 1403–1404.

Ariano, P., Accardo, D., Lombardi, M., Bocchini, S., Draghi, L., De Nardo, L., & Fino, P. (2015). Polymeric materials as artificial muscles: An overview. *Journal of Applied Biomaterials & Functional Materials, 13,* 1–9.

Asatryan, D. G., & Feldman, A. G. (1965). Functional tuning of the nervous system with control of movement or maintenance of a steady posture: 1. Mechanographic analysis of the work of the joint on execution of a postural task. *Biophysics, 10,* 925–935.

Aschersleben, G., & Prinz, W. (1995). Synchronizing actions with events: The role of sensory information. *Perception & Psychophysics, 57,* 305–317.

Baddeley, A. D., & Longman, D. J. A. (1978). The influence of length and frequency of training session on the rate of learning to type. *Ergonomics, 21,* 627–635.

Bandura, A. (1977). *Social learning theory*. Englewood Cliffs, NJ: Prentice-Hall.

Bandura, A., Ross, D., & Ross, S. A. (1961). Transmission of aggression through imitation of aggressive models. *Journal of Abnormal and Social Psychology, 63,* 575–582.

Baron-Cohen, S., Wheelwright, S., Hill, J., Raste, Y., & Plumb, I. (2001). The "reading the mind in the eyes" test revised version: A study with normal adults, and adults with Asperger syndrome or high-functioning autism. *Journal of Child Psychology and Psychiatry and Allied Disciplines, 42*, 241–251.

Barrett, L. F. (2006). Solving the emotion paradox: Categorization and the experience of emotion. *Personality and Social Psychology Review, 10*, 20–46.

Barrett, L. F. (2017). *How emotions are made: The secret life of the brain*. Boston, MA: Houghton Mifflin Harcourt.

Basak, C., Voss, M. W., Erickson, K. I., Boot, W. R., & Kramer, A. F. (2011). Regional differences in brain volume predict the acquisition of skill in a complex real-time strategy videogame. *Brain and Cognition, 76*, 407–414.

Bassett, D. S., Wymbs, N., Porter, M. A., Mucha, P., Carlson, J. M., & Grafton, S. T. (2011). Dynamic reconfiguration of human brain networks during learning. *Proceedings of the National Academy of Sciences, 108*, 7641–7646.

Beilock, S. (2015). *How the body knows its mind*. New York, NY: Simon & Schuster.

Beilock, S. L., Carr, T. H., MacMahon, C., & Starkes, J. L. (2002). When paying attention becomes counterproductive: Impact of divided versus skill-focused attention on novice and experienced performance of sensorimotor skills. *Journal of Experimental Psychology: Applied, 8*, 6–16.

Beilock, S. L., & Goldin-Meadow, S. (2010). Gesture changes thought by grounding it in action. *Psychological Science, 21*, 1605–1610.

Bernstein, N. (1967). *The coordination and regulation of movements*. London, UK: Pergamon.

Bick, P. A., & Kinsbourne, M. (1987). Auditory hallucinations and subvocal speech in schizophrenic patients. *American Journal of Psychiatry, 144*, 222–225.

Bizzi, E., Mussa-Ivaldi, F. A., & Giszter, S. (1991). Computations underlying the execution of movement: A biological perspective. *Science, 253*, 287–291.

Blakemore, S. J., Wolpert, D. M., & Frith, C. D. (1998). Central cancellation of self-produced tickle sensation. *Nature Neuroscience, 1*, 635–640.

Blinch, J., & DeWinne, C. R. (2019). Pre-crastination and procrastination effects occur in a reach-to-grasp task. *Experimental Brain Research, 237*, 1129–1139.

Blundon, J. A., Roy, N. C., Teubner, B. J., Yu, J., Eom, T. Y., Sample, K. J., & Rose, D. C. (2017). Restoring auditory cortex plasticity in adult mice by restricting thalamic adenosine signaling. *Science, 356*, 1352–1356.

Bobbert, M. F., Huijing, P. A., & van Ingen Schenau, G. J. (1986). An estimation of power output and work done by the human triceps surae muscle-tendon complex in jumping. *Journal of Biomechanics, 19*, 899–906.

Botvinick, M., & Cohen, J. (1998). Rubber hands "feel" touch that eyes see. *Nature, 391*, 756.

Brashers-Krug, T., Shadmehr, R., & Bizzi, E. (1996). Consolidation in human motor memory. *Nature, 382*, 252–255.

Brosch, S., & Pirsig, W. (2001). Stuttering in history and culture. *International Journal of Pediatric Otorhinolaryngology, 59*, 81–87.

Brown, L. E., Moore, C. M., & Rosenbaum, D. A. (2002). Feature-specific processing dissociates action from recognition. *Journal of Experimental Psychology: Human Perception and Performance, 28*, 1330–1344.

Brown, T. G. (1911). The intrinsic factors in the act of progression in the mammal. *Proceedings of the Royal Society, London, 84,* 308–319.

Brown, T. G. (1914). On the nature of the fundamental activity of the nervous centres; together with an analysis of the conditioning of rhythmic activity in progression, and a theory of the evolution of function in the nervous system. *Journal of Physiology, 48,* 18.

Bruce, D. (1994). Lashley and the problem of serial order. *American Psychologist, 49,* 93–103.

Brunswik, E. (1947). Systematic and representative design of psychological experiments. In *Proceedings of the Berkeley Symposium on Mathematical Statistics and Probability* (pp. 143–202). The Regents of the University of California.

Bryan, W. L. & Harter, N. (1897). Studies in the physiology and psychology of the telegraphic language. *Psychological Review, 4,* 27–53.

Bryan , W. L. Harter , N. (1899). Studies on the telegraphic language: The acquisition of a hierarchy of habits. *Psychological Review, 1899,* 6, 345–378.

Buzsáki, G., & Llinás, R. (2017). Space and time in the brain. *Science, 358,* 482–485.

Buzsáki, G., & Mizuseki, K. (2014). The log-dynamic brain: How skewed distributions affect network operations. *Nature Reviews Neuroscience, 15,* 264–278.

Caggiano, V., Leiras, R., Goñi-Erro, H., Masini, D., Bellardita, C., Bouvier, J., Caldeira, V., Fisone, G., & Kiehn, O. (2018). Midbrain circuits that set locomotor speed and gait selection. *Nature, 553,* 455–460.

Chang, A., Kragness, H. E., Livingstone, S. R., Bosnyak, D. J., & Trainor, L. J. (2019). Body sway reflects joint emotional expression in music ensemble performance. *Scientific Reports, 9,* 205.

Chase, W. G., & Ericsson, K. A. (1981). Skilled memory. In J. R. Anderson (Ed.), *Cognitive skills and their acquisition* (pp. 141–189). Hillsdale, NJ: Erlbaum.

Chomsky, N. (1968). *Language and mind.* New York, NY: Harcourt Brace Jovanovich.

Chong, I., & Proctor, R. W. (2020). On the evolution of a radical concept: Affordances according to Gibson and their subsequent use and development. *Perspectives on Psychological Science, 15,* 117–132.

Churchland, P. S. & Seijnowski, T. J. (2016). *The computational brain.* Cambridge, MA: MIT Press.

Clarac, F., Massion, J., & Stuart, D. G. (2009). Reflections on Jacques Paillard (1920–2006)—A pioneer in the field of motor cognition. *Brain Research Reviews, 61,* 256–280.

Clark, A. (1997). *Being there: Putting brain, body, and world together again.* Cambridge, MA: MIT Press.

Clausen, J., Fetz, E., Donoghue, J., Ushiba, J., Spörhase, U., Chandler, J., Bierbaumer, N., & Soekadar, S. R. (2017). Help, hope, and hype: Ethical dimensions of neuroprosthetics. *Science, 356,* 1338–1339.

Coates, M. (2005). Why do most species have five digits on their hands and feet? *Scientific American.* Retrieved from https://www.scientificamerican.com/article/why-do-most-species-have/.

Cohen, R. G., & Rosenbaum, D. A. (2004). Where objects are grasped reveals how grasps are planned: Generation and recall of motor plans. *Experimental Brain Research, 157,* 486–495.

Collins, S., Ruina, A., Tedrake, R. & Wisse, M. (2005). Efficient bipedal robots based on passive-dynamic walkers. *Science, 307,* 1082–1085.

Cooke, J. D. (1980). The organization of simple, skilled movements. In G. E. Stelmach & J. Requin (Eds.), *Tutorials in motor behavior* (pp. 199–212). Amsterdam, Netherlands: North-Holland.

Cooper, L. A., & Shepard, R. N. (1984). Turning something over in the mind. *Scientific American, 251*(6), 106–114.

Costill, D. L., Daniels, J., Evans, W., Fink, W., Krahenbuhl, G., & Saltin, B. (1976). Skeletal muscle enzymes and fiber composition in male and female track athletes. *Journal of Applied Physiology, 40,* 149–154.

Couzin-Frankel, J. (2019). Epilepsy's next frontier. *Science, 366,* 1300–1304.

Cowan, N. (2001). The magical number 4 in short-term memory: A reconsideration of mental storage capacity. *Behavioral and Brain Sciences, 24,* 87–114.

Crajé, C., Aarts, P., Nijhuis-van der Sanden, M., & Steenbergen, B. (2010). Action planning in typically and atypically developing children (unilateral cerebral palsy). *Research in Developmental Disabilities, 31,* 1039–1046.

Crossman, E. R. F. W. (1959). A theory of the acquisition of speed skill. *Ergonomics, 2,* 153–166.

Crossman, E. R. F. W., & Goodeve, P. J. (1983). Feedback control of hand-movement and Fitts' law. *Quarterly Journal of Experimental Psychology, 35A,* 251–278. (Original work published 1963)

Csikszentmihályi, M. (1990). *Flow: The psychology of optimal experience.* New York, NY: Harper & Row.

Czyż, S. H., & Moss, S. J. (2016). Specificity vs. generalizability: Emergence of especial skills in classical archery. *Frontiers in Psychology, 7.* https://doi.org/10.3389/fpsyg.2016.01178.

Daley, M. A., & Biewener, A. A. (2006). Running over rough terrain reveals limb control for intrinsic stability. *Proceedings of the National Academy of Sciences, 103,* 15681–15686.

Darwin, C. R. (1965). *The expression of the emotions in man and animals.* Chicago, IL: University of Chicago Press. (Original work published 1872)

Deecke, L., Scheid, P., & Kornhuber, H. H. (1969). Distribution of readiness potential, premotion positivity, and motor potential of the human cerebral cortex preceding voluntary finger movements. *Experimental Brain Research, 7,* 158–168.

de Jong, R., Coles, M. G., Logan, G. D., & Gratton, G. (1990). In search of the point of no return: The control of response processes. *Journal of Experimental Psychology: Human Perception and Performance, 16,* 164–182.

DeMelo, J. (2019, March 25). Precrastination: When the early bird gets the shaft. *New York Times,* Smart Living Section. Retrieved from https://www.nytimes.com/2019/03/25/smarter-living/precrastination-when-the-early-bird-gets-the-shaft.html.

Dewey, J. (1896). The reflex arc concept in psychology. *Psychological Review, 3,* 357–370.

Diamond, A. (1990). Developmental time course in human infants and infant monkeys, and the neural bases of inhibitory control of reaching. *Annals of the New York Academy of Sciences, 608,* 637–676.

Dickinson, M. H., Farley, C. T., Full, R. J., Keohl, M. A., Kram, R., & Lehman, S. (2000). How animals move: An integrative view. *Science, 288,* 100–106.

Di Nota, P. M., & Huhta, J. M. (2019). Complex motor learning and police training: Applied, cognitive, and clinical perspectives. *Frontiers in Psychology, 10,* 1797.

Dobbs, D. (2019, February 18). The devastating allure of medical miracles. *Wired (Science).* Retrieved from https://www.wired.com/story/devastating-allure-of-medical-miracles/.

Dodge, R. (1900). Visual perception during eye movement. *Psychological Review, 7,* 454–465.

Dona, H. S. G., & Chittka, L. (2020). Charles H. Turner, pioneer in animal cognition. *Science, 370,* 530–531.

Donders, F. C. (1847). Beitrag zur Lehre von den Bewegungen des menschlichen Auges [Contribution to the study of the movements of the human eye]. In *Hollhdischen Beitragen zu den Anatomischen und Physiologischen Wissenschaften* (Vol. 1, pp. 104–145). Utrecht, Netherlands: Bötticher'sche Buchhandlung.

Duhamel, J.-R., Colby, C. L., & Goldberg, M. E. (1992). The updating of the representation of visual space in parietal cortex by intended eye movements. *Science, 255,* 90–92.

Dumay, N., & Gaskell, M. G. (2007). Sleep-associated changes in the mental representation of spoken words. *Psychological Science, 18,* 35–39.

Durgin, F. H., Klein, B., Spiegel, A., Strawser, C. J., & Williams, M. (2012). The social psychology of perception experiments: Hills, backpacks, glucose, and the problem of generalizability. *Journal of Experimental Psychology: Human Perception and Performance, 38,* 1582–1595.

Easton, T. A. (1972). On the normal use of reflexes. *American Scientist, 60,* 591–599.

Ebbinghaus, H. E. (1964). *Memory: A contribution to experimental psychology.* New York, NY: Dover. (Original work published 1885)

Ehrsson, H. H., Spence, C., & Passingham, R. E. (2004). That's my hand! Activity in the premotor cortex reflects feeling of ownership of a limb. *Science, 305,* 875–877.

Ekman, P., & Friesen, W.V. (1975). *Unmasking the face: A guide to recognizing emotions from facial clues.* Oxford, England: Prentice-Hall.

Elliott, D., Hansen, S., Grierson, L. E., Lyons, J., Bennett, S. J. & Hayes, S. J. (2010). Goal-directed aiming: Two components but multiple processes. *Psychological Bulletin, 136,* 1023–1044.

Elliott, D., Helsen, W. F., & Chua, R. (2001). A century later: Woodworth's (1899) two component model of goal-directed aiming. *Psychological Bulletin, 127,* 342–357.

Elsner, B., & Hommel, B. (2001). Effect anticipation and action control. Journal of Experimental Psychology: *Human Perception and Performance, 27,* 229–240.

Engelkamp, J. (1998). *Memory for actions.* Hove, East Sussex, UK: Psychology Press/Taylor & Francis.

Erickson, K. I., Boot, W. R., Basak, C., Neider, M. B., Prakash, R. S., Voss, M. W., Graybiel, A. M., Simons, D. J., Fabiani, M., Gratton, G., & Kramer, A. F. (2010). Is bigger better? Striatal volume predicts level of video game skill acquisition. *Cerebral Cortex, 20,* 2522–2530.

Ericsson, K. A. Krampe, R. T., & Tesch-Romer, C. (1993). The role of deliberate practice in the acquisition of expert performance. *Psychological Review, 100,* 363–406.

Estes, W. K. (1972). An associative basis for coding and organization in memory. In A. W. Melton & E. Martin (Eds.), *Coding processes in human memory* (pp. 161–190). Washington, DC: Winston.

Evans, N. J., Brown, S. D., Mewhort, D. J., & Heathcote, A. (2018). Refining the law of practice. *Psychological Review, 125,* 592–605.

Feghhi, I., & Rosenbaum, D. A. (2019). Judging the subjective difficulty of different kinds of tasks. *Journal of Experimental Psychology: Human Perception and Performance, 45*, 983–994.

Feghhi, I., & Rosenbaum, D. A. (2021). Effort-avoidance is not simply error-avoidance. *Psychological Research, 85*, 1462–1472.

Feld, M. S., McNair, R. E., & Wilk, S. R. (1979). The physics of karate. *Scientific American, 240*(4), 150–158.

Feldman, A., & Latash, M. L. (2005). Testing hypotheses and the advancement of science: Recent attempts to falsify the equilibrium point hypothesis. *Experimental Brain Research, 161*, 91–103.

Fenton, M. B. (2020). Bats navigate with cognitive maps. *Science, 369*, 142.

Fessard, A., & Paillard, J. (1948). Recherche sur la simultanéité dans l'exécution des mouvements volontaires [Research on simultaneity in the execution of voluntary movements]. *Comptes Rendus des Seances de la Societe de Biologie et des ses Filiales (Paris), 132*, 933–935.

Fischer, M. H., Rosenbaum, D. A., & Vaughan, J. (1997). Speed and sequential effects in reaching. *Journal of Experimental Psychology: Human Perception and Performance, 23*, 404–428.

Fischer, M. H., & Zwaan, R. A. (2008). Embodied language: A review of the role of the motor system in language comprehension. *Quarterly Journal of Experimental Psychology, 61*, 825–850.

Fitts, P. M. (1954). The information capacity of the human motor system in controlling the amplitude of movement. *Journal of Experimental Psychology, 47*, 381.

Fitts, P. M. (1964). Perceptual-motor skill learning. In A. W. Melton (Ed.), *Categories of human learning* (pp. 243–285). New York, NY: Academic Press.

Flash, T., & Henis, E. (1991). Arm trajectory modification during reaching towards visual targets. *Journal of Cognitive Neuroscience, 3*, 220–230.

Flash, T., & Hogan, N. (1985). The coordination of arm movements: An experimentally confirmed mathematical model. *Journal of Neuroscience, 5*, 1688–1703.

Fodor, J. A. (1983). *Modularity of mind: An essay on faculty psychology*. Cambridge, MA: MIT Press.

Ford, M. J., Ambulo, C. P., Kent, T. A., Markvicka, E. J., Pan, C., Malen, J., & Majidi, C. (2019). A multifunctional shape-morphing elastomer with liquid metal inclusions. *Proceedings of the National Academy of Sciences, 116*, 21438–21444.

Forssberg, H., Grillner, S., & Rossignol, S. (1975). Phase dependent reflex reversal during walking in chronic spinal cats. *Brain Research, 55*, 247–304.

Fournier, L. R., Coder, E., Kogan, C., Raghunath, N., Taddese, E., & Rosenbaum, D. A. (2018). Which task will we choose first? Precrastination and cognitive load in task ordering. *Attention, Perception & Psychophysics, 81*, 489–503.

Franck, N., Farrer, C., Georgieff, N., Marie-Cardine, M., Daléry, J., d'Amato, T., & Jeannerod, M. (2001). Defective recognition of one's own actions in patients with schizophrenia. *American Journal of Psychiatry, 158*, 454–459.

Frith, C. D., Blakemore, S. J., & Wolpert, D. M. (2000). Abnormalities in the awareness and control of action. *Philosophical Transactions of the Royal Society of London, 355*, 1771–1788.

Fu, C., & McGuire, P. (2003). Hearing voices or hearing the self in disguise? Revealing the neural correlates of auditory hallucinations in schizophrenia. In T. Kircher (Ed.), *The self in neuroscience and psychiatry* (pp. 425–435). Cambridge, UK: Cambridge University Press.

Ganel, T., & Goodale, M. A. (2019). Still holding after all these years: An action-perception dissociation in patient DF. *Neuropsychologia, 128*, 249–254.

Gentner, D. R., Grudin, J., & Conway, E. (1980). *Finger movements in transcription typing* (Technical Report 8001). La Jolla: University of California, San Diego, Center for Human Information Processing.

Georgopoulos, A. P., Lurito, J. T., Petrides, M., Schwartz, A. B., & Massey, J. T. (1989). Mental rotation of the neuronal population vector. *Science, 243*, 234–236.

Georgopoulos, A. P., Schwartz, A. B., & Kettner, R. E. (1986). Neuronal population coding of movement direction. *Science, 233*, 1416–1419.

Gibson, J. J. (1950). *Perception of the visual world.* Boston, MA: Houghton-Mifflin.

Gielen, C. C. A. M., Vrijenhoek, E. J., & Flash, T. (1997). Principles for the control of kinematically redundant limbs. In M. Fetter, H. Misslisch, & D. Tweed (Eds.), *Three-dimensional kinematics of eye-, head-, and limb-movements* (pp. 285–297). Chur, Switzerland: Harwood Academic.

Gladwell, M. (2008). *Outliers: The story of success.* New York, NY: Little, Brown.

Glenberg, A. M., & Kaschak, M. P. (2002). Grounding language in action. *Psychonomic Bulletin & Review, 9*, 558–565.

Glover, S. R. (2002). Visual illusions affect planning but not control. *Trends in Cognitive Sciences, 6*, 288–292.

Goldfield, E. C., Kay, B. A., & Warren, W. H. (1993). Infant bouncing: The assembly and tuning of action systems. *Child Development, 64*, 1128–1142.

Goldin-Meadow, S., & Wagner, S. M. (2005). How our hands help us learn. *Trends in Cognitive Sciences, 9*, 234–241.

Gomi, H., & Kawato, M. (1996). Equilibrium-point control hypothesis examined by measured arm stiffness during multijoint movement. *Science, 272*, 117–120.

Gonzalez, D. A., Glazebrook, C. M., Studenka, B. E., & Lyons, J. (2013). Motor interactions with another person: Do individuals with autism spectrum disorder plan ahead? *Frontiers in Integrative Neuroscience, 7*, 1–9.

Goodale, M. A., & Milner, A. D. (1992). Separate visual pathways for perception and action. *Trends in Neuroscience, 15*, 20–25.

Goodall, J. (1964). Tool-using and aimed throwing in a community of free-living chimpanzees. *Nature, 201*, 1264–1266.

Goodwin, G. M., McCloskey, D. J., & Matthews, P. B. C. (1972). The contribution of muscle afferents to kinaesthesia shown by vibration-induced illusions of movement and by the effect of paralysing joint afferents. *Brain, 95*, 705–748.

Gordon, J., & Ghez, C. (1994). Accuracy of planar reaching movements: I. Independence of direction and extent variability. *Experimental Brain Research, 99*, 97–111.

Gould, S. J. (1980). *The panda's thumb: More reflections in natural history.* New York, NY: W. W. Norton.

Gouras, P. (1985). Oculomotor system. In E. R. Kandel & J. H. Schwartz (Eds.), *Principles of neural science* (2nd ed., pp. 571–583). New York, NY: Elsevier.

Grady, D. (2019, June 20). About the idea that you're growing horns from looking down at your phone. *New York Times*. Retrieved from https://www.nytimes.com/2019/06/20/health/horns-cellphones-bones.html.

Graham, C. H. (1967). *Robert Sessions Woodworth 1869–1962: A biographical memoir by Clarence H. Graham*. Washington, DC: National Academy of Sciences.

Gray, P. (2011). Psychology (Sixth Ed.). New York, NY: Worth.

Graziano, M. S. (2008). *The intelligent movement machine: An ethological perspective on the primate motor system*. Oxford, UK: Oxford University Press.

Graziano, M. S., & Gross, C. G. (1994). Mapping space with neurons. *Current Directions in Psychological Science, 3*, 164–167.

Graziano, M. S., Taylor, C. S. R., & Moore, T. (2002). Complex movements evoked by microstimulation of precentral cortex. *Neuron, 34*, 841–851.

Greenwald, A. G. (1970). A choice reaction time test of ideo-motor theory. *Journal of Experimental Psychology, 86*, 20–25.

Gregory, R. L. (1966). *Eye and brain*. New York, NY: McGraw-Hill.

Grillner, S. (1981). Control of locomotion in bipeds, tetrapods, and fish. In V. B. Brooks (Ed.), *Handbook of physiology, Section 1: The nervous system, Vol. II, Motor control* (pp. 1179–1236). Bethesda, MD: American Physiological Society.

Grillner, S. (1996). Neural networks for vertebrate locomotion. *Scientific American, 274*(1), 64–69.

Guillot, A., Di Rienzo, F., MacIntyre, T., Moran, A., & Collet, C. (2012). Imagining is not doing but involves specific motor commands: A review of experimental data related to motor inhibition. *Frontiers in Human Neuroscience, 6*, 247.

Hamilton, J. (2020, June 18). A scientist's pink cast leads to discovery about how the brain responds to disability. Retrieved from https://www.npr.org/sections/health-shots/2020/06/18/877621475/a-scientists-pink-cast-leads-to-discovery-about-how-the-brain-responds-to-disabi.

Hard, B. M., Recchia, G., & Tversky, B. (2011). The shape of action. *Journal of Experimental Psychology: General, 140*, 586–604.

Heathcote, A., Brown, S., & Mewhort, D. J. K. (2000). The power law repealed: The case for an exponential law of practice. *Psychonomic Bulletin and Review, 7*, 185–207.

Hebb, D. O. (1949). *The organization of behavior: A neuropsychological theory*. New York: Wiley.

Held, L. I. (2014). *How the snake lost its legs: Curious tales from the frontier of evo-devo*. Cambridge, UK: Cambridge University Press.

Held, R. (1965). Plasticity in sensory-motor systems. *Scientific American, 213*(5), 84–94.

Held, R., & Hein, A. (1963). Movement-produced stimulation in the development of visually guided behavior. *Journal of Comparative and Physiological Psychology, 56*, 872–876.

Henry, F. M., & Rogers, D. E. (1960). Increased response latency for complicated movements and a "memory drum" theory of neuromotor reaction. *Research Quarterly*, 31, 448–458.

Herbort, O., Büschelberger, J., & Janczyk, M. (2018). Preschool children adapt grasping movements to upcoming object manipulations: Evidence from a dial rotation task. *Journal of Experimental Child Psychology, 167*, 62–77.

Hermer, L., & Spelke, E. (1996). Modularity and development: The case of spatial reorientation. *Cognition, 61*, 195–232.

Heuer, H. (2003). Motor control. In A. Healy & R. Proctor (Eds.), *Handbook of psychology: Vol. 4. Experimental psychology* (pp. 317–354). Hoboken, NJ: Wiley.

Hogan, N. (1984). An organizing principle for a class of voluntary movements. *Journal of Neuroscience, 4*, 2745–2754.

Hollerbach, J. M., & Atkeson, C. G. (1986). Characterization of joint-interpolated arm movements. In H. Heuer & C. Fromm (Ed.), *Generation and modulation of action patterns* (pp. 41–54). Berlin, Germany: Springer-Verlag.

Holst, E. von (1939). Die relative Koordination als Phänomenon und als Methode zentralnervöse Funktionsanalyze. *Ergebnisse der Physiologie, biologischen Chemie und experimentellen Pharmakologie., 42*(1), 228–306. [English translation in Holst, E. von. (1973)]. Relative coordination as a phenomenon and as a method of analysis of central nervous functions. In *The behavioural physiology of animal and man: The collected papers of Erich von Holst (Vol. 1)* [R. Martin, Translator]. London, England: Methuen.

Holst, E. von, & Mittelstaedt, H. (1950). Das Reafferenzprinzip. Wechselwirkungen zwischen Zentralnervensystem und Peripherie. *Naturwissenschaften, 37*, 464–476.

Hommel, B., Brown, S. B., & Nattkemper, D. (2016). *Human action control.* Basel, Switzerland: Springer.

Hommel, B., Müsseler, J., Aschersleben, G., & Prinz, W. (2001). The theory of event coding (TEC): A framework for perception and action planning. *Behavioral and Brain Sciences, 24*, 849–937.

Hughes, C. (1996). Brief report: Planning problems in autism at the level of motor control. *Journal of Autism & Developmental Disorders, 26*, 99–107.

Ioannidou, F., Hermens, F., & Hodgson, T. L. (2017). Mind your step: The effects of mobile phone use on gaze behavior in stair climbing. *Journal of Technology in Behavioral Science, 2*, 109–120.

Isaacson, W. (2003). *Benjamin Franklin: An American life*. New York, NY: Simon & Schuster.

Ishikawa, T. & Montello, D. R. (2006). Spatial knowledge acquisition from direct experience in the environment: Individual differences in the development of metric knowledge and the integration of separately learned places. *Cognitive Psychology, 52*, 93–129.

Ishiyama, S., & Brecht, M. (2016). Neural correlates of ticklishness in the rat somatosensory cortex. *Science, 354*, 757–760.

Ivry, R. B., & Hazeltine, R. E. (1995). Perception and production of temporal intervals across a range of durations: Evidence for a common timing mechanism. *Journal of Experimental Psychology: Human Perception and Performance, 21*, 3–18.

James, W. (1890). *Principles of psychology*. New York, NY: Holt.

Jaswal, V., & Akhtar, N. (2019). Being versus appearing socially uninterested: Challenging assumptions about social motivation in autism. *Behavioral and Brain Sciences, 42*, E82.

Jax, S. A., & Rosenbaum, D. A. (2007). Hand path priming in manual obstacle avoidance: Evidence that the dorsal stream does not only control visually guided actions in real time. *Journal of Experimental Psychology: Human Perception and Performance, 33*, 425–441.

Jax, S. A., Rosenbaum, D. A., Vaughan, J., & Meulenbroek, R. G. J. (2003). Computational motor control and human factors: Modeling movements in real and possible environments. *Human Factors, 45*, 5–27.

Jeannerod, M. (1981). Intersegmental coordination during reaching at natural objects. In J. Long & A. Baddeley (Eds.), *Attention and performance IX* (pp. 153–169). Hillsdale, NJ: Erlbaum.

Jeannerod, M. (2006). *Motor cognition: What action tells the self.* New York, NY: Oxford University Press.

Jeka, J. J., & Lackner, J. R. (1994). Fingertip contact influences human postural control. *Experimental Brain Research, 100*, 495–502.

Jenkins, J. G., & Dallenbach, K. M. (1924). Obliviscence during sleep and waking. *American Journal of Psychology, 35*, 605–612.

Johansson, G. (1973). Visual perception of biological motion and a model for its analysis. *Perception & Psychophysics, 14*, 201–211.

Johnson, G. (2008). *The ten most beautiful experiments*. New York, NY: Knopf.

Jokl, E. (1981). The human hand. *International Journal of Sport Psychology, 12*, 140–148.

Jovanovic, B., & Schwarzer, G. (2017). The development of the grasp height effect as a measure of efficient action planning in children. *Journal of Experimental Child Psychology, 153*, 74–82.

Kahneman, D. (2011). *Thinking, fast and slow*. New York, NY: Farrar, Straus and Giroux.

Kardas, M., & O'Brien, E. (2018). Easier seen than done: Merely watching others perform can foster an illusion of skill acquisition. *Psychological Science, 29*, 521–536.

Karniel, A., Nisky, I., Avraham, G., Peles, B. C., & Levy-Tzedek, S. (2010). A Turing-like handshake test for motor intelligence. In *Haptics: Generating and perceiving tangible sensations* (pp. 197–204). Berlin, Germany: Springer.

Keele, S. W. (1968). Movement control in skilled motor performance. *Psychological Bulletin, 70*, 387–403.

Keele, S. W., & Posner, M. I. (1968). Processing of visual feedback in rapid movements. *Journal of Experimental Psychology, 77*, 155–158.

Keen, R., Lee, M. H., & Adolph, K. (2014). Planning an action: A developmental progression in tool use. *Ecological Psychology, 26*, 98–108.

Keetch, K. M., Schmidt, R. A., Lee, T. D., & Young, D. E. (2005). Especial skills: Their emergence with massive amounts of practice. *Journal of Experimental Psychology: Human Perception and Performance, 31*, 970–978.

Kelso, J. A. S. (1984). Phase transitions and critical behavior in human bimanual coordination. *American Journal of Physiology: Regulatory, Integrative and Comparative, 246*, 1000–1004.

Klapp, S. T. (1977). Reaction time analysis of programmed control. *Exercise and Sport Sciences Reviews, 5*, 231–253.

Klapp, S. T., & Maslovat, D. (2020). Programming of action timing cannot be completed until immediately prior to initiation of the response to be controlled. *Psychonomic Bulletin & Review, 27*, 821–832.

Kool, W., McGuire, J. T., Rosen, Z. B., & Botvinick, M. M. (2010). Decision making and the avoidance of cognitive demand. *Journal of Experimental Psychology: General, 139*, 665–682.

Kornmuller, A. E. (1930). Eine experimentelle Anesthesie der auberen Augenmuskein am Menschen und ihre Auswirkungen [Experimental anesthesia of the external eye muscles in humans and its effects]. *Journal für Psycyhologie und Neurologie, 41*, 354–366.

Kosslyn, S. M., Pick, H. L., Jr., & Fariello, G. R. (1974). Cognitive maps in children and men. *Child Development, 45*, 707–716.

Kots, Y. M., & Syrovegnin, A. V. (1966). Fixed set of variants of interactions of the muscles of two joints in the execution of simply voluntary movements. *Biophysics, 11*, 1212–1219.

Kroll, J. F., & de Groot, A. M. B. (2005). *Handbook of bilingualism: Psycholinguistic approaches.* New York, NY: Oxford University Press.

Krupic, J., Bauza, M., Burton, S., & O'Keefe, J. (2018). Local transformations of the hippocampal cognitive map. *Science, 359*, 1143–1146.

Kunde, W., & Weigelt, M. (2005). Goal congruency in bimanual object manipulation. *Journal of Experimental Psychology: Human Perception and Performance, 31*, 145–156.

Kwakkel, G., Veerbeek, J. M., van Wegen, E. E., & Wolf, S. L. (2015). Constraint-induced movement therapy after stroke. *The Lancet Neurology, 14*, 224–234.

Laabs, G. J. (1973). Retention characteristics of different reproduction cues in motor short-term memory. *Journal of Experimental Psychology, 100*, 168–177.

Lackner, J. R. (1988). Some proprioceptive influences on the perceptual representation of body shape and orientation. *Brain, 111*, 281–297.

Lakoff, G., & Johnson, M. (1999). *Philosophy in the flesh—The embodied mind and its challenge to Western thought.* New York, NY: Basic Books.

Lametti, D. (2010, December 21). "Can't tickle yourself? That's a good thing." *Scientific American.*

Langer, A., Feingold-Polak, R., Mueller, O., Kellmeyer, P., & Levy-Tzedek, S. (2019). Trust in socially assistive robots: Considerations for use in rehabilitation. *Neuroscience & Biobehavioral Reviews, 104*, 231–239.

Langman, V. A., Roberts, T. J., Black, J., Maloiy, G. M., Heglund, N. C., Weber, J. M., Kram, R., & Taylor, C. R. (1995). Moving cheaply: Energetics of walking in the African elephant. *Journal of Experimental Biology, 198*, 629–632.

Lashley, K. (1917). The accuracy of movement in the absence of excitation from the moving organ. *American Journal of Physiology, 43*, 169–194.

Lashley, K. S. (1951). The problem of serial order in behavior. In L. A. Jeffress (Ed.), *Cerebral mechanisms in behavior* (pp. 112–131). New York, NY: Wiley.

Ledbetter, K. (2019). Using drones to spot sick cows from the sky. *Western Livestock Journal*, August 6. https://www.wlj.net/top_headlines/using-drones-to-spot-sick-cows-from-the-sky/article_eb62c34e-405b-11e9-aa08-03244e32bc84.html.

Lee, D. N., & Aronson, E. (1974). Visual proprioceptive control of standing in human infants. *Perception & Psychophysics, 15*, 529–532.

Lee, D. N., & Reddish, P. E. (1981). Plummeting gannets: A paradigm of ecological optics. *Nature, 293*, 293–294.

Leed, J. E., Chinn, L. K., & Lockman, J. J. (2019). Reaching to the self: The development of infants' ability to localize targets on the body. *Psychological Science, 30*, 1063–1073.

Leslie, M. (2019). The mismeasure of hands? *Science, 364,* 923–925.

Liberman, A. M., Cooper, F. S., Shankweiler, D. P., & Studdert-Kennedy, M. G. (1967). Perception of the speech code. *Psychological Review, 74,* 431–461.

Libet, B., Gleason, C. A., Wright, E. W., & Pearl, D. K. (1983). Time of conscious intention to act in relation to onset of cerebral activity (readiness-potential). *Brain, 106,* 623–642.

Linkenauger, S. A., Witt, J. K., Stefanucci, J. K., Bakdash, J. Z., & Proffitt, D. R. (2009). The effects of handedness and reachability on perceived distance. *Journal of Experimental Psychology: Human Perception and Performance, 35,* 1649–1660.

Loeb, G. E. (1995). Control implications of musculoskeletal mechanics. *Proceedings of the 17th International Conference of the Engineering in Medicine and Biology Society, 2,* 1393–1394.

Logan, G. D., & Crump, M. J. C. (2011). Hierarchical control of cognitive processes: The case for skilled typewriting. In B. Ross (Ed.), *The Psychology of Learning and Motivation,* Vol. 54 (pp.1–27). Burlington, MA: Academic Press.

Logan, G. D., Schachar, R. J., & Tannock, R. (1997). Impulsivity and inhibitory control. *Psychological Science, 8,* 60–64.

MacDonald, M. C., Koranda, M. J., Bulgarelli, F., & Weiss, D. J. (2020). Is language production planning emergent from action planning? A preliminary investigation. *Frontiers in Psychology, 11,* 1193.

MacKay, D. G. (2019). *Remembering: What 50 years of research with famous amnesia patient HM can teach us about memory and how it works.* Amherst, New York: Prometheus Books.

Magill, R., & Anderson, D. (2010). *Motor learning and control.* New York, NY: McGraw-Hill.

Maguire, E. A., Gadian, D. G., Johnsrude, I. S., Good, C. D., Ashburner, J., Frackowiak, R. S., & Frith, C. D. (2000). Navigation-related structural change in the hippocampi of taxi drivers. *Proceedings of the National Academy of Sciences, 97,* 4398–4403.

Maguire, E. A., Woollett, K., & Spiers, H. J. (2006). London taxi drivers and bus drivers: A structural MRI and neuropsychological analysis. *Hippocampus, 16,* 1091–1101.

Marr, D. (1982). *Vision.* San Francisco, CA: W. H. Freeman.

Marteniuk, R. G., & Bertram, C. P. (2001). Contributions of gait and trunk movements to prehension: Perspectives from world- and body-centered coordinates. *Motor Control, 5,* 151–165.

Matin, E. (1974). Saccadic suppression. *Psychological Bulletin, 81,* 899–918.

Mazzoni, P., & Krakauer, J. W. (2006). An implicit plan overrides an explicit strategy during visuomotor adaptation. *Journal of Neuroscience, 26,* 3642–3645.

McDaniel, M. A., & Einstein, G. O. (2007). *Prospective memory: An overview and synthesis of an emerging field.* Thousand Oaks, CA: SAGE.

McMahon, T. A. (1984). *Muscles, reflexes, and locomotion.* Princeton, NJ: Princeton University Press.

Mechsner, F., Kerzel, D., Knoblich, G. & Prinz, W. (2001). Perceptual basis of bimanual coordination. *Nature, 414,* 69–73.

Merleau-Ponty, M. (1945). *Phénoménologie de la perception* [*Phenomenology of perception*]. Paris, France: Gallimard.

Meyer, D. E., Abrams, R. A., Kornblum, S., Wright, C. E., & Smith, J. E. K. (1988). Optimality in human motor performance: Ideal control of rapid aimed movements. *Psychological Review, 95*, 340–370.

Meyer, D. E., & Schvaneveldt, R. W. (1971). Facilitation in recognizing pairs of words: Evidence of a dependence between retrieval operations. *Journal of Experimental Psychology, 90*, 227–234.

Meyer, D. E., Smith, J. E. K., Kornblum, S., Abrams, R. A. & Wright, C. E. (1990). Speed-accuracy tradeoffs in aimed movements: Toward a theory of rapid voluntary action. In M. Jeannerod (Ed.), *Attention and performance XIII: Motor representation and control* (pp. 173–226). Hillsdale, NJ: Erlbaum.

Meyer, D. E., Smith, J. E. K., & Wright, C. E. (1982). Models for the speed and accuracy of aimed movements. *Psychological Review, 89*, 449–482.

Meyer, M., van der Wel, R. P. R. D., & Hunnius, S. (2013). Higher-order action planning for individual and joint object manipulations. *Experimental Brain Research, 225*, 579–588.

Michel, A. (2020, April). Mind over body. *APS Observer*. Retrieved from https://www.psychologicalscience.org/observer/mind-over-body?utm_source=APS+Emails&utm_campaign=f9073b151d-OBSERVER_APRIL_2020&utm_medium=email&utm_term=0_d2c7283f04-f9073b151d-62631487.

Miller, G. A. (1956). The magical number seven plus or minus two: Some limits on our capacity for processing information. *Psychological Review, 63*, 81–97.

Miller, G. A., Galanter, E., & Pribram, K. H. (1960). *Plans and the structure of behavior*. New York, NY: Holt, Rinehart, & Winston.

Milner, A. D., & Goodale, M. A. (2006). *The visual brain in action*. Oxford, UK: Oxford University Press.

Miyamoto, Y. R., Wang, S., & Smith, M. A. (2020). Implicit adaptation compensates for erratic explicit strategy in human motor learning. *Nature Neuroscience 23*, 443–455.

Molto, L., Nalborczyk, L., Palluel-Germain, R., & Morgado, N. (2020). Action effects on visual perception of distances: A multilevel Bayesian meta-analysis. *Psychological Science, 31*, 488–504.

Morasso, P. (1981). Spatial control of arm movements. *Experimental Brain Research, 42*, 223–227.

Moser, E. I., Kropff, E., & Moser, M.-B. (2008). Place cells, grid cells, and the brain's spatial representation system. *Annual Review of Neuroscience, 31*, 69–89.

Nabavinik, M., Abaszadeh, A., Mehranmanesh, M., & Rosenbaum, D. A. (2017). Especial skills in experienced archers. *Journal of Motor Behavior, 50*, 249–253.

Néda, Z., Ravasz, E., Brechet, Y., Vicsek, T., & Barabási, A.-L. (2000). Self-organizing processes: The sound of many hands clapping. *Nature, 403*, 849–850.

Newell, A. M., & Rosenbloom, P. S. (1981). Mechanisms of skill acquisition and the law of practice. In J. R. Anderson (Ed.), *Cognitive skills and their acquisition* (pp. 1–55). Hillsdale, NJ: Erlbaum.

Nissen, M. J., & Bullemer, P. (1987). Attentional requirements of learning: Evidence from performance measures. *Cognitive Psychology, 19*, 1–32.

Norman, D. A. (1981). Categorization of action slips. *Psychological Review, 88*, 1–15.

Ohara, M., Kurosawa, M., Sasaki, T., Tanaka, T., Kawamura, S., Hayakawa, Y., & Saito, K. (2019, January). Development of hardware neural networks IC with switchable gait pattern for insect-type microrobot. In *2019 IEEE/SICE International Symposium on System Integration (SII)* (pp. 663–668).

O'Keefe, J., & Nadel, L. (1978). *The hippocampus as a cognitive map*. Oxford, UK: Oxford University Press.

Owen, A. M., Coleman, M. R., Boly, M., Davis, M. H., Laureys, S., & Pickard, J. D. (2006). Detecting awareness in the vegetative state. *Science, 313*, 1402.

Pan, S. C., & Rickard, T. C. (2015). Sleep and motor memory: Is there room for consolidation? *Psychological Bulletin, 141*, 812–834.

Patterson, E. E., & Kahan, T. A. (2020). Precrastination and the cognitive-load-reduction (CLEAR) hypothesis. *Memory, 28*, 107–111.

Pavlov, I. P. (1928). *Lectures on conditioned reflexes*. New York, NY: International Publishers.

Penfield, W., & Jasper, H. H. (1954). *Epilepsy and the functional anatomy of the human brain*. New York, NY: Little, Brown.

Pew, R. W. (1966). Acquisition of hierarchical control over the temporal organization of a skill. *Journal of Experimental Psychology, 71*, 764–771.

Pew, R. W., & Rosenbaum, D. A. (1988). Human motor performance: Computation, representation, and implementation. In R. C. Atkinson, R. J. Herrnstein, G. Lindzey, & R. D. Luce (Eds.), *Stevens' handbook of experimental psychology* (2nd ed., pp. 473–509). New York, NY: Wiley.

Pinker, S. (2002). *The blank slate*. New York, NY: Viking.

Pirsig, R. M. (1974). *Zen and the art of motorcycle repair*. New York, NY: William Morrow.

Polanyi, M. (1958). *Personal knowledge: Towards a post-critical philosophy*. Chicago, IL: University of Chicago Press. [Reprinted as Polanyi, M. (1964). *Personal knowledge*. New York: Harper & Row.]

Polit, A., & Bizzi, E. (1978). Processes controlling arm movements in monkeys. *Science, 201*, 1235–1237.

Pollick, F. E., Paterson, H. M., Bruderlin, A., & Sanford, A. J. (2001). Perceiving affect from arm movement. *Cognition, 82*, B51–B61.

Poulet, J. F. A., & Hedwig, B. A (2002). Corollary discharge maintains auditory sensitivity during sound production. *Nature, 418*, 872–876.

Povel, D.-J., & Collard, R. (1982). Structural factors in patterned finger tapping. *Acta Psychologica, 52*, 107–124.

Prinz, W. (2012). *Open minds: The social making of agency and intentionality*. Cambridge, MA: MIT Press.

Prinz, W., & Hommel, B. (Eds.). (2002). *Common mechanisms in perception and action: Attention and performance* (Vol. XIX). Oxford, UK: Oxford University Press.

Proctor, R. W., & Vu, K.-P. L. (2006). *Stimulus-response compatibility principles: Data, theory, and application*. Boca Raton, FL: CRC Press.

Proffitt, D. R., Bhalla, M., Gossweiler, R., & Midgett, K. (1995). Perceiving geographical slant. *Psychonomic Bulletin & Review, 2*, 409–428.

Radman, Z. (Ed.). (2013). *The hand, an organ of the mind.* Cambridge, MA: MIT Press.

Raghunath, N., Fournier, L. R., & Kogan, C. (2020). Precrastination and individual differences in working memory capacity. *Psychological Research.* Advance online publication.

Ramirez, N. F., & Kuhl, P. (2017). The brain science of bilingualism. *Young Children, 72,* 38–44.

Reason, J. (1990). *Human error.* Cambridge, UK: Cambridge University Press.

Rescorla, R. A., & Wagner, A. R. (1972). A theory of Pavlovian conditioning: Variations in the effectiveness of reinforcement and nonreinforcement. In A. H. Black & W. F. Prokasy (Eds.). *Classical conditioning II: Current research and theory* (pp. 64–99). New York, NY: Appleton-Century-Crofts.

Reynolds, G. (2019, October 23). Something in the way we move. *New York Times.* Retrieved from https://www.nytimes.com/2019/10/23/well/move/something-in-the-way-we-move.html.

Richards, W. (1971). The fortification illusions of migraines. *Scientific American, 224(5),* 88–96.

Richardson, M. J., Marsh, K. L., Isenhower, R. W., Goodman, J. R., & Schmidt, R. C. (2007). Rocking together: Dynamics of intentional and unintentional interpersonal coordination. *Human Movement Science, 26,* 867–891.

Richardson, M. J., Marsh, K. L., & Schmidt, R. C. (2006). Effects of visual and verbal interaction on unintentional interpersonal coordination. *Journal of Experimental Psychology: Human Perception and Performance, 31,* 62–79.

Richtel, M. (2014, July 19). Sometimes, early birds are too early. *New York Times,* p. BU3. Retrieved from http://www.nytimes.com/2014/07/20/business/sometimes-early-birds-are-too-early.html?src=twr&_r=1.

Rock, I. (1966). *The nature of perceptual adaptation.* New York, NY: Basic Books.

Rock, I., & Harris, C. S. (1967). Vision and touch. *Scientific American, 216,* 5, 96–104.

Roland, P. E., Larsen, B., Lassen, N. A., & Skinhoj, E. (1980). Supplementary motor area and other cortical areas in organization of voluntary movements in man. *Journal of Neurophysiology, 43,* 118–136.

Rosenbaum, D. A. (1980). Human movement initiation: Specification of arm, direction, and extent. *Journal of Experimental Psychology: General, 109,* 444–474.

Rosenbaum, D. A. (2005). The Cinderella of psychology: The neglect of motor control in the science of mental life and behavior. *American Psychologist, 60,* 308–317.

Rosenbaum, D. A. (2009). Walking down memory lane: Where walkers look as they descend stairs provides hints about how they control their walking behavior. *American Journal of Psychology, 122,* 425–430.

Rosenbaum, D. A. (2010). *Human motor control* (2nd ed.). San Diego, CA: Academic Press/Elsevier.

Rosenbaum, D. A. (2012). The tiger on your tail: Choosing between temporally extended behaviors. *Psychological Science, 23,* 855–860.

Rosenbaum, D. A. (2014). *It's a jungle in there: How competition and cooperation in the brain shape the mind.* New York, NY: Oxford University Press.

Rosenbaum, D. A. (2017a, August 10). Inner tennis. *Psychology Today.* Retrieved from https://www.psychologytoday.com/blog/perception-and-performance/201708/inner-tennis.

Rosenbaum, D. A. (2017b). *Knowing hands: The cognitive psychology of manual control.* New York, NY: Cambridge University Press.

Rosenbaum, D. A., Brach, M., & Semenov, A. (2011). Behavioral ecology meets motor behavior: Choosing between walking and reaching paths. *Journal of Motor Behavior, 43*, 131–136.

Rosenbaum, D. A., Chapman, K. M., Weigelt, M., Weiss, D. J., & van der Wel, R. (2012). Cognition, action, and object manipulation. *Psychological Bulletin, 138*, 924–946.

Rosenbaum, D. A., Dawson, A. M., & Challis, J. H. (2006). Haptic tracking permits bimanual independence. *Journal of Experimental Psychology: Human Perception and Performance, 32*, 1266–1275.

Rosenbaum, D. A., & Feghhi, I. (2019). The time for action is at hand. *Attention, Perception & Psychophysics, 81*, 2123–2138.

Rosenbaum, D. A., Fournier, L. R., Levy-Tzedek, S., McBride, D. M., Rosenthal, R., Sauerberger, K., VonderHaar, R. L., Wasserman, E. A., & Zentall, T. R. (2019). Sooner rather than later: Precrastination rather than procrastination. *Current Directions in Psychological Science, 28*, 229–233.

Rosenbaum, D. A., Gong, L., & Potts, C. A. (2014). Pre-crastination: Hastening subgoal completion at the expense of extra physical effort. *Psychological Science, 25*, 1487–1496.

Rosenbaum, D. A., Inhoff, A. W., & Gordon, A. M. (1984). Choosing between movement sequences: A hierarchical editor model. *Journal of Experimental Psychology: General, 113*, 372–393.

Rosenbaum, D. A., & Janczyk, M. (2019). Who is or was E. R. F. W. Crossman, the champion of the Power Law of Learning and the developer of an influential model of aiming? *Psychonomic Bulletin & Review, 26*, 1449–1463.

Rosenbaum, D. A., Kenny, S., & Derr, M. A. (1983). Hierarchical control of rapid movement sequences. *Journal of Experimental Psychology: Human Perception and Performance, 9*, 86–102.

Rosenbaum, D. A., Loukopoulos, L. D., Meulenbroek, R. G. M., Vaughan, J., & Engelbrecht, S. E. (1995). Planning reaches by evaluating stored postures. *Psychological Review, 102*, 28–67.

Rosenbaum, D. A., Marchak, F., Barnes, H. J., Vaughan, J., Slotta, J., & Jorgensen, M. (1990). Constraints for action selection: Overhand versus underhand grips. In M. Jeannerod (Ed.), *Attention and performance XIII: Motor representation and control* (pp. 321–342). Hillsdale, NJ: Erlbaum.

Rosenbaum, D. A., van Heugten, C., & Caldwell, G. C. (1996). From cognition to biomechanics and back: The end-state comfort effect and the middle-is-faster effect. *Acta Psychologica, 94*, 59–85.

Rosenbaum, D. A., Weber, R. J., Hazelett, W. M., & Hindorff, V. (1986). The parameter remapping effect in human performance: Evidence from tongue twisters and finger fumblers. *Journal of Memory and Language, 25*, 710–725.

Rothwell, J. C. (1987). *Control of human voluntary movement.* London, UK: Croom-Helm.

Ryle, G. (1949). *The concept of mind.* New York, NY: Barnes & Noble.

Rumelhart, D. E., & Norman, D. A. (1982). Simulating a skilled typist: A study of skilled cognitive-motor performance. *Cognitive Science, 6*, 1–36.

Santamaria, J. P., & Rosenbaum, D. A. (2011). Etiquette and effort: Holding doors for others. *Psychological Science, 22*, 584–588.

Schack, T. (2010). *Die kognitive Architektur menschlicher Bewegungen* [The cognitive architecture of human movements]. Osnabrück, Germany: Meyer & Meyer.

Schack, T., & Frank, C. (2020). Mental representation and the cognitive architecture of skilled action. *Review of Philosophy and Psychology*. Advance online publication.

Schmidt, R. A., Lee, T., Winstein, C., Wulf, G., & Zelaznik, H. (2018). *Motor control and learning* (6th ed.). Champaign, IL: Human Kinetics.

Schmidt, R. A., Zelaznik, H. N., Hawkins, B., Frank, J. S., & Quinn, J. T., Jr. (1979). Motor output variability: A theory for the accuracy of rapid motor acts. *Psychological Review, 86*, 415–451.

Schütz, C., Güldenpenning, I., Koester, D., & Schack, T. (2020). Social cues can impact complex behavior unconsciously. *Scientific Reports, 10*, 21017.

Scoville, W. B., & Milner, B. (1957). Loss of recent memory after bilateral hippocampal lesions. *Journal of Neurology, Neurosurgery, and Psychiatry, 20*, 11–21.

Sebanz, N., Knoblich, G., & Prinz, W. (2003). Representing others' actions: Just like one's own? *Cognition, 88*, 11–21.

Shastri, D. (2016, June 29). How some salamanders regrow their limbs. *Science*. Retrieved from https://upload.wikimedia.org/wikipedia/commons/f/f6/AxolotlBE.jpg.

Shea, J. B., & Morgan, R. L. (1979). Contextual interference effects on acquisition, retention, and transfer of a motor skill. *Journal of Experimental Psychology: Human Learning and Memory, 5*, 179–187.

Shepard, R. N., & Metzler, J. (1971). Mental rotation of three-dimensional objects. *Science, 171*, 701–703.

Shergill, S. S., Bays, P. M., Frith, C. D., & Wolpert, D. M. (2003). Two eyes for an eye: The neuroscience of force escalation. *Science, 301*, 187.

Sherrington, C. S. (1947). *The integrative action of the nervous system*. New Haven, CT: Yale University Press. (Original work published 1906)

Simion, F., Regolin, L., & Bulf, H. (2008). A predisposition for biological motion in the newborn baby. *Proceedings of the National Academy of Sciences, 105*, 809–813.

Skinner, B. F. (1948). 'Superstition' in the pigeon. *Journal of Experimental Psychology, 38*, 168–172.

Skinner, B. F. (1969). *Contingencies of reinforcement: A theoretical analysis*. New York, NY: Appleton-Century-Crofts.

Slepian, M. L., Young, S. G., Rutchick, A. M., & Ambady, N. (2013). Quality of professional players' poker hands is perceived accurately from arm motions. *Psychological Science, 24*, 2335–2338.

Song, J.-Y. & Nakayama, K. (2009). Hidden cognitive states revealed in choice reaching tasks. *Trends in Cognitive Science, 13*, 360–366.

Spelke, E. (1994). Initial knowledge: Six suggestions. *Cognition, 50*, 431–445.

Spelke, E. S., & Kinzler, K. D. (2007). Core knowledge. *Developmental Science, 10*, 89–96.

Spencer, J. P., Clearfield, M., Corbetta, D., Ulrich, B., Buchanan, P., & Schöner, G. (2006). Moving toward a grand theory of development: In memory of Esther Thelen. *Child Development, 77*, 1521–1538.

Spencer, R. M. C., Zelaznik, H. N., Diedrichsen, J., & Ivry, R. B. (2003). Disrupted timing of discontinuous but not continuous movements by cerebellar lesions. *Science, 300*, 1437–1439.

Sperry, R. (1950). Neural basis of the spontaneous optokinetic response produced by visual inversion. *Journal of Comparative and Physiological Psychology, 43*, 482–489.

Spivey, M. (2007). *The continuity of mind*. New York, NY: Oxford University Press.

Spivey, M. J., Grosjean, M., & Knoblich, G. (2005). Continuous attraction toward phonological competitors. *Proceedings of the National Academy of Sciences, 102*, 10393–10398.

Sternberg, S., Monsell, S., Knoll, R. L., & Wright, C. E. (1978). The latency and duration of rapid movement sequences: Comparisons of speech and typewriting. In G. E. Stelmach (Ed.), *Information processing in motor control and learning* (pp. 117–152). New York, NY: Academic Press.

Straka, H., Simmers, J., & Chagnaud, B. P. (2018). A new perspective on predictive motor signaling. *Current Biology, 28*, R232–R243.

Tajadura-Jiménez, A., Basia, M., Deroy, O., Fairhurst, M., Marquardt, N., & Bianchi-Berthouze, N. (2015). As light as your footsteps: Altering walking sounds to change perceived body weight, emotional state and gait. In *Proceedings of the 33rd Annual ACM Conference on Human Factors in Computing Systems* (pp. 2943–2952). New York, NY: ACM. Retrieved from https://www.youtube.com/watch?v=Dw_LC-YpjFw.

Tajadura-Jiménez, A., Väljamäe, A., Toshima, I., Kimura, T., Tsakiris, M., & Kitagawa, N. (2012). Action sounds recalibrate perceived tactile distance. *Current Biology, 22*, R516–R517.

Talis, V. L. (2015). New pages in the biography of Nikolai Alexandrovich Bernstein. In *Anticipation: Learning from the past* (pp. 313–327). Cham, Switzerland: Springer.

Tanenhaus, M. K., Spivey-Knowlton, M., Eberhard, K. & Sedivy, J. (1995). Integration of visual and linguistic information in spoken language comprehension. *Science, 268*, 1632–1634.

Taub, E., & Berman, A. J. (1968). Movement and learning in the absence of sensory feedback. In S. J. Freeman (Ed.), *The neuropsychology of spatially oriented behavior* (pp. 173–192). Homewood, IL: Dorsey.

Taub, E., & Uswatt, G. (2006). Constraint induced movement therapy: Answers and questions after two decades of research. *NeuroRehabilitation, 21*, 93–95.

Teufel, C., Kingdon, A., Ingram, J. N., Wolpert, D. M., & Fletcher, P. C. (2010). Deficits in sensory prediction are related to delusional ideation in healthy individuals. *Neuropsychologia, 48*, 4169–4172.

Thelen, E. (1995). Motor development: A new synthesis. *American Psychologist, 50*, 79–95.

Thelen, E., & Fisher, D. M. (1982). Newborn stepping: An explanation for a "disappearing" reflex. *Developmental Psychology, 18*, 760–775.

Thomas, L. E., & Lleras, A. (2009). Swinging into thought: Directed movement guides insight in problem solving. *Psychonomic Bulletin & Review, 16*, 719–723.

Thorndike, E. L. (1898). Animal intelligence: An experimental study of the associative processes in animals. *The Psychological Review: Monograph Supplements, 2*(4), 1–8.

Thorndike, E. L., & Woodworth, R. S. (1901). The influence of improvement in one mental function upon the efficiency of other functions. *Psychological Review, 8*, 247–261.

Tipper, S. P., Lortie, C., & Baylis, G. C. (1992). Selective reaching: Evidence for action-centered attention. *Journal of Experimental Psychology: Human Perception and Performance, 18*, 891–905.

Tolman, C. E. (1948). Cognitive maps in rats and man. *Psychological Review, 55*, 189–208.

Touvet, F., Daoud, N., Gazeau, J. P., Zeghloul, S., Maier, M. A., & Eskiizmirliler, S. (2012). A biomimetic reach and grasp approach for mechanical hands. *Robotics and Autonomous Systems, 60*, 473–486.

Treisman, A., & Gelade, G (1980). A feature integration theory of attention. *Cognitive Psychology, 12*, 97–136.

Tresilian, J. R. (1999). Visually timed action: Time-out for 'tau'? *Trends in Cognitive Sciences, 3*, 301–310.

Tresilian, J. R. (2012). *Sensorimotor control and learning: An introduction to the behavioral neuroscience of action.* London: Palgrave Macmillan.

Trommershäuser, J., Maloney, L. T., & Landy, M. S. (2008). Decision making, movement planning and statistical decision theory. *Trends in Cognitive Science, 12*, 291–297.

Turvey, M. T. (1990). Coordination. *American Psychologist, 45*, 938–953.

Tversky, A., & Kahneman, D. (1974). Judgment under uncertainty: Heuristics and biases. *Science, 185*, 1124–1131.

Tversky, B. (2019). *Mind in motion: How action shapes thought.* Boston: Hachette

Tweed, D., & Vilis, T. (1990). Geometric relations of eye position and velocity vectors during saccades. *Vision Research, 30*, 111–127.

van der Wel, R. P., Fleckenstein, R., Jax, S., & Rosenbaum, D. A. (2007). Hand path priming in manual obstacle avoidance: Evidence for abstract spatio-temporal forms in human motor control. *Journal of Experimental Psychology: Human Perception and Performance, 33*, 1117–1126.

van Swieten, L. M., van Bergen, E., Williams, J. H. G., Wilson, A. D., Plumb, M. S., Kent, S. W., & Mon-Williams, M. A. (2010). A test of motor (not executive) planning in developmental coordination disorder and autism. *Journal of Experimental Psychology, 36*, 493–499.

Vannasing, P., Florea, O., González-Frankenberger, B., Tremblay, J., Paquette, N., Safi, D., Wallois, F., Lepore, F., Béland, R., Lassonde, M., & Gallagher, A. (2016). Distinct hemispheric specializations for native and non-native languages in one-day-old newborns identified by fNIRS. *Neuropsychologia, 84*, 63–69.

Vangsness, L., & Young, M. E. (2020). Turtle, task ninja, or time waster? Who cares? Traditional task-completion strategies are overrated. *Psychological Science, 31*, 306–315.

Velasquez-Manoff, M. (2020, August 28). The brain implants that could change humanity. *New York Times Sunday Review*. Retrieved from https://www.nytimes.com/2020/08/28/opinion/sunday/brain-machine-artificial-intelligence.html.

Vogt, S., Buccino, G., Wohlschläger, A., Canessa, N., Shah, J., Zilles, K., Eickhoff, S., Freund, H. J., Rizolatti, G., & Fink, G. (2007). Prefrontal involvement in imitation learning of hand actions: Effects of practice and expertise. *Neuroimage, 37*, 1371–1383.

Volkmann, F. C., Riggs, L. A., & Moore, R. K. (1980). Eyeblinks and visual suppression. *Science, 207*, 900–902.

Volkmann, F. C., Schick, A. M., & Riggs, L. A. (1968). Time course of visual inhibition during voluntary saccades. *Journal of the Optical Society of America, 58*, 562–569.

VonderHaar, R. L., McBride, D. M., & Rosenbaum, D. A. (2019). Task order choices in cognitive and perceptual-motor tasks: The cognitive-load-reduction (CLEAR) hypothesis. *Attention, Perception, & Psychophysics, 81*, 2517–2525.

Wagman, J. B., & Blau, J. J. (Eds.). (2019). *Perception as information detection: Reflections on Gibson's ecological approach to visual perception.* Milton Park, Abingdon-on-Thames, Oxfordshire, England, UK: Routledge.

Wamsley, E. J., Tucker, M., Payne, J. D., Benavides, J., & Stickgold, R. (2010). Dreaming of a learning task is associated with enhanced sleep-dependent memory consolidation. *Current Biology, 20,* 850–855.

Warren, W. H. (1984). Perceiving affordances: Visual guidance of stair climbing. *Journal of Experimental Psychology: Human Perception and Performance, 10,* 683–703.

Warren, W. H., & Whang, S. (1987). Visual guidance of walking through apertures: Body-scaled information for affordances. *Journal of Experimental Psychology: Human Perception and Performance, 13,* 371–383.

Wasserman, E. A. (2018). Precrastination: The fierce urgency of now. *Learning & Behavior, 47,* 7–28.

Wasserman, E. A., & Zentall, T. R. (Eds.). (2006). *Comparative cognition: Experimental explorations of animal intelligence.* Oxford, UK: Oxford University Press.

Watson, J. B. (1930). *Behaviorism* (Revised edition). Chicago: University of Chicago Press.

Weir, A. A., Chappell, J., & Kacelnik, A. (2002). Shaping of hooks in new Caledonian crows. *Science, 297,* 981.

Weiskrantz, L. (1986). *Blindsight: A case study and implications.* Oxford, UK: Oxford University Press.

Welsh, T. N., & Elliott, D. (2004). Movement trajectories in the presence of a distracting stimulus: Evidence for a response activation model of selective reaching. *Quarterly Journal of Experimental Psychology, 57,* 1031–1057.

Willingham, D. B. (1998). A neuropsychological theory of motor skill learning. *Psychological Review, 105,* 558–584.

Wing, A. M. (1980). The long and short of timing in response sequences. In G. E. Stelmach & J. Requin (Eds.), *Tutorials in motor behavior* (pp. 469–486). Amsterdam, Netherlands: North-Holland.

Wing, A. M. (1993). The uncertain motor system: Perspectives on the variability of movement. In D. E. Meyer & S. Kornblum (Eds.) *Attention and performance XIV: Synergies in experimental psychology, artificial intelligence, and cognitive neuroscience* (pp. 709–744). Cambridge, MA: MIT Press.

Wing, A. M., & Kristofferson, A. B. (1973). The timing of inter-response intervals. *Perception & Psychophysics, 13,* 455–460.

Winters, J. M., & Kleweno D. G. (1993). Effect of initial upper-limb alignment on muscle contributions to isometric strength curves. *Journal of Biomechanics, 26,* 143–153.

Witt, J. K. (2011). Action's effect on perception. *Current Directions in Psychological Science, 20,* 201–206.

Witt, J. K., & Brockmole, J. R. (2012). Action alters object identification: Wielding a gun increases the bias to see guns. *Journal of Experimental Psychology: Human Perception and Performance, 38,* 1159–1168.

Witt, J. K., & Dorsch, T. E. (2009). Kicking to bigger uprights: Field goal kicking performance influences perceived size. *Perception, 38,* 1328–1340.

Witt, J. K., Linkenauger, S. A., Bakdash, J. Z., & Proffitt, D. R. (2008). Putting to a bigger hole: Golf performance relates to perceived size. *Psychonomic Bulletin & Review, 15*, 581–585.

Witt, J. K., Proffitt, D. R., & Epstein, W. (2005). Tool use affects perceived distance but only when you intend to use it. *Journal of Experimental Psychology: Human Perception and Performance, 31*, 880–888.

Wolbers, T., & Hegarty, M. (2010). What determines our navigational abilities? *Trends in Cognitive Sciences, 14*, 138–146.

Wolpert, D. M., & Flanagan, J. R. (2001). Motor prediction. *Current Biology, 11*, 729–732.

Woodworth, R. S. (1899). The accuracy of voluntary movement. *Psychological Review, 3*, 1–119.

Woodworth, R. S. (1938). *Experimental psychology*. New York, NY: Holt.

Woodworth, R. S., & Schlosberg, H. (1954). *Experimental psychology* (2nd ed.). New York, NY: Holt.

Wooster, M. (1923). Certain factors in the development of a new spatial co-ordination. *Psychological Monographs, 32, 1–96.*

Wright, C. E., Marino, V. F., Chubb, C., & Mann, D. (2019). A model of the uncertainty effects in choice reaction time that includes a major contribution from effector selection. *Psychological Review, 126*, 550.

Wulf, G., Landers, M., Lewthwaite, R., & Toöllner, T. (2016). External focus instructions reduce postural instability in individuals with Parkinson disease. *Physical Therapy, 89*, 162–168.

Wulf, G., & Lewthwaite, R. (2016). Optimizing performance through intrinsic motivation and attention for learning: The OPTIMAL theory of motor learning. *Psychonomic Bulletin and Review, 23*, 1382–1414.

Wunsch, K., Henning, A., Aschersleben, G., & Weigelt, M. (2013). A systematic review of the end-state comfort effect in normally developing children and in children with developmental disorders. *Journal of Motor Learning and Development, 1*, 59–76.

Yang, Y., & Wang, X. (2020). Odor modulates hand movements in a reach-to-grasp task. *Frontiers of Neuroscience, 14*, 560.

York, G. K., & Steinberg, D. A. (2011). Hughlings Jackson's neurological ideas. *Brain, 134*, 3106–3113.

Young, S., Pratt, J., & Chau, T. (2009). Target-directed movements at a comfortable pace: Movement duration and Fitts's law. *Journal of Motor Behavior, 41*, 339–346.

Zhang, Y., Kuhl, P. K., Imada, T., Iverson, P., Pruitt, J., Stevens, E. B., Kawakatsu, M., Tohkura, Y .I., & Nemoto, I. (2009). Neural signatures of phonetic learning in adulthood: A magnetoencephalography study. *Neuroimage, 46*, 226–240.

Index